STRENGTHENING THE HUMAN RIGHT TO SANITATION AS AN INSTRUMENT
FOR INCLUSIVE DEVELOPMENT

ACADEMISCH PROEFSCHRIFT

ter verkrijging van de graad van doctor

aan de Universiteit van Amsterdam

op gezag van de Rector Magnificus

Prof. dr. ir. K.I.J. Maex

ten overstaan van een door het College voor Promoties ingestelde commissie,

in het openbaar te verdedigen in de Agnietenkapel

op woensdag 16 mei 2018, om 12.00 uur

door

Pedi Chiemena Obani

geboren te Port Harcourt, Nigeria

Published by:
CRC Press/Balkema
Schipholweg 107C, 2316 XC, Leiden, the Netherlands
Pub.NL@taylorandfrancis.com
www.crcpress.com – www.taylorandfrancis.com
ISBN 978-1-138-61848-0

Promotiecommissie:

Promotor:	Prof. dr. J. Gupta	Universiteit van Amsterdam
Copromotor:	Prof. dr. O. Fagbohun	Lagos State University
Overige leden:	Prof. dr. I.S.A. Baud	Universiteit van Amsterdam
	Dr. C.M. Brölmann	Universiteit van Amsterdam
	Prof. dr. M.Z. Zwarteveen	Universiteit van Amsterdam
	Prof. dr. C.J.M. Arts	International Institute of Social Studies
	Dr. K.H. Schwartz	IHE Institute for Water Education

Faculteit: Faculteit der Maatschappij- en Gedragswetenschappen

Acknowledgements

When I first started my PhD research, my sole purpose was to make my own little contribution to the continuous development of legal research. I am thankful to NUFFIC for providing me the funding to enable me undertake my research, for the full duration, through the Netherlands Professional Fellowship grant. I am also thankful to the Federal Government of Nigeria, and the management of the University of Benin and the Faculty of Law, for all the support that I received in the course of my research. I started out knowing the PhD research was going to require a lot of hard work and dedication on my part, but I think at that point I did not realise the important role so many other people would play in the process. And, they indeed played their roles so well; I want to say 'thank you' for all your support! Words cannot sufficiently express my gratitude but I will do my best.

First, I am immensely grateful to my Promotor, Professor Joyeeta Gupta, and Copromotor, Professor Olanrewaju Fagbohun. I have learnt so much more from your supervision and mentorship than my PhD thesis can convey, and the goodwill I receive from colleagues, just for being your supervisee, has been tremendous. Joyeeta, your dedication to my PhD supervision, and the way you combined it so well with your other commitments will always be a driving force for me. You pushed me, cajoled me, and went beyond the call of duty so many times to support me every step of the way…even giving me very beautiful personal gifts that I will always treasure. I have added you to my list of superwomen because while my other PhD colleagues in the IHE Institute for Water Education were working with a Promotor, a Supervisor, and at least one Mentor, you closed the gap for me. 'Lanre, you have been an inspiration! You have redefined my perception of leadership, supervision and mentorship and taught me that the pursuit of excellence is indeed rewarding. Thank you for you immense contribution to my PhD supervision and professional development.

I am also very thankful to my family for being my friends, dependable support, and loudest cheerleaders. To my lovely mum, our nature girl, thank you for allowing me to pursue my goals in life and for your unending support even when I go against the norm. To my amazing dad, thank you for trusting my judgement and for sparing no resource to see me happy. Imma, and Osi, thanks for making me feel like an excellent scholar anyday…you really boosted my confidence when I needed it the most! Oria, and Chisa, thank you for all your encouragement. Ehi, thanks for always showing up in Delft whenever I needed you the most! Zindzi, you can't even imagine how you made life easy for me by doing just about anything I would ask for. Thank you for also letting me share your lovely parents with you; you let them take me as their own! Even as we celebrate daddy's legacy every day, mummy still provides a strong support system for me, Kora and Omole. To my best friend and husband Omole, I wish that I could gift my PhD degree to you because you have been my muse, firm support, research assistant, cook, baby sitter, cheer leader, and partner in every sphere of my life. Your mammoth support is the reason I have managed to balance my professional commitments with living a happy and fulfilling personal life and completing my PhD thesis… Meanwhile, you cost me just over the price of a Peugeot and no less!

My colleagues at IHE and UvA contributed immensely to making my PhD journey and my stay in the Netherlands smooth and pleasurable; for this, I am grateful. Raquel, Kirstin, Eva, Shakeel, Carol, Zarhra, and Mohan, thank you for making out time to provide detailed feedback and help with my work and much more, even when you had your own struggles and deadlines. You all, including Victor, helped me outside my PhD with just about anything I would care to ask for. Even when I did not ask, you would always offer your help! You are indeed special humans! I am also thankful to Professor Damir, Masoom, and Zaki, for the friendship and support especially at the beginning of my PhD, when I was trying to find my feet. To Jos, Jolanda, Selvi, and all the Secretaries in the Department of Integrated Water Systems Governance, the Socio-Cultural Office, the staff at the front desk, and the IT help desk, thank you for providing your valuable administrative services to ensure the smooth running of my research. Peter Heerings, I am glad that I could always count on your friendly hello and welcoming smile. Ed Gerritsen, thanks for keeping my belongings safe while I was away from the Netherlands. I am grateful to all the lecturers in IHE, both from within and outside my department, who would always encourage me and show interest in my research from the start to the end. Some were just curious about having a lawyer conducting PhD research in IHE; this gave me additional motivation to adopt an interdisciplinary research approach. I am especially grateful to Dr. Frank Jaspers for all the feedback on my work and for helping me translate my summary into Dutch. I also appreciate all my colleagues at the UvA, especially those in the Governance and Inclusive Development group, for all the support, inputs and friendship I enjoyed there. IHE, UvA, and the Netherlands indeed brought lovely humans like Thijs, Chris, Aline, Micah, Nirajan and Anita, Hauwa, Iman, Kaycee, Ayo, John Ogbodo, Charles, and Sylvester, into my life for good!

In the course of my PhD research, I found my 'home away from home' in the International Students Chaplaincy, Delft! Thank you our chaplains, Rev. Waltraut and Father Avin, for providing me with valuable spiritual guidance, and emotional support throughout my time in The Netherlands. To the choir members who have always shared my joys and celebrated my most important milestones in life, I can't thank you enough for also teaching me so many native songs from Spain, France, Indonesia, Germany, Brazil, Japan, Italy, South Africa, Kenya, and even Nigeria, to mention a few; songs that I may never have had the pleasure of singing without you! All the exciting activities we either planned together or engaged in like volunteering, the Global Meeting Point, dot painting, origami and kirigami, the Sister for Sisters meetings, and retreats truly enriched me. My beloved Peet Oom Henk, and Uncle Ruben thank you for all your love and care.

I am thankful to all my colleagues in the Faculty of Law, University of Benin, who supported me in one way or another during my research. I would especially like to appreciate the Deans of the Faculty, and Heads of Departments of Public Law, both past and present, including Prof. E. Chianu, Prof. L. Atsegbua, Prof. R. Idubor, Prof. N. Inegbedion, Prof. N. Aniekwu, and Dr. F. Osadolor, for all the support I enjoyed. I am also thankful to Dr. S. Ukuegbe for reading my drafts and making valuable contributions to my research; Prof. S. Ogbodo, Dr. E. Erhagbe and Dr. M. Ezekiel, for assisting me during my fieldwork; Prof. I. Omoruyi, Dr. G. Umoru, Dr. G. Arishe, and Dr. A. Enabulele, for always lending me a listening ear; and Prof. V. Onuoha, Prof. V. Aigbokhaevbo, Dr. E. Okojie, and Dr. F. Iyasere, for good counsel. To

Dr. Attah; Anwuli, Hadiza, Omuwa, Aisosa and Alero, soon to become Doctors too; and my visiting professor, Ngozi Finette, I cannot forget how you always gave me cause to smile and encouraged me to carry on with my work, even at times when I just did not feel like it. You have all played your parts well to define my past and future, and I cannot thank you enough for this!

Beyond the Faculty of Law, I deeply appreciate the invaluable friendship of inspiring academics like Prof. B. Agbonifo, Dr. J. Agbonifo, Dr. E. Ekhator, Dr. Ire, and Dr. S. Frank; the support of water and sanitation champions like Mrs. J. Wakaso; caring family friends like Dr. Inekigha, Uncle Mbadiwe, and Uncle Zagha, who encouraged me to dream big; and other special humans like Prof. J. Igene, Aminu Dahiru, Bem, Igwe, Osarobo, Austin, Elvis, and my colleagues from the UNIBEN Law Class of 2006, for making my data collection process run smoothly. Aunt Eby, thank you too for connecting me with Professor Fagbohun; you don't know how much good you did there! To the officers and members of Roger Bacon Chapter, Benin City, especially Oghale, Daddy Pat and Mummy Carol Onegbedan, you all inspire me! Thank you! Finally, to all my colleagues in the Earth System Governance Network, I cannot appreciate you enough for your intellectual companionship and support. I look forward to our continuing partnership.

Publications, presentations and trainings attended in relation to the thesis

Peer Reviewed Journal Articles:

- Obani, P. and Gupta, J. (2017). Inclusive development as an imperative for realizing the human right to water and sanitation. *Journal of Sustainable Development Law and Policy, 8*(2), 67-86. Doi:10.4314/jsdlp.v8i2.4.

- Obani, P. (2017). Inclusiveness in humanitarian actionaccess to water, sanitation & hygiene in focus. *Current Opinion in Environmental Sustainability, 24*, 24-29. (Impact Factor for 2016: 3.954)

- Obani, P. & Gupta, J. (2016a). Human right to sanitation in the legal and non-legal literature: the need for greater synergy. *Wiley Interdisciplinary Reviews: Water, 3*(5), 678-691.

- Obani, P. & Gupta, J. (2015a). The Evolution of the Right to Water and Sanitation: Differentiating the Implications. *Review of European, Comparative & International Environmental Law, 24*(1), 27-39. Doi:10.1111/reel.12095.

- Obani, P. & Gupta, J. (2014a). Legal pluralism in the area of human rights: water and sanitation. *Current Opinion in Environmental Sustainability, 11*, 63-70. (Impact Factor for: 3.954)

Peer Reviewed Book Chapters:

- Obani, P. & Gupta, J. (2016b). Human Security and Access to Water, Sanitation, and Hygiene: Exploring the Drivers and Nexus. In C. Pahl-Wostl, A. Bhaduri, & J. Gupta (Eds.), *Handbook on Water Security* (pp. 201-214). Massachusetts: Edward Elgar Publishing.

- Obani, P. & Gupta, J. (2014b). The Human Right to Water and Sanitation: Reflections on Making the System Effective. In A. Bhaduri, J. Bogardi, J. Leentvaar, & S. Marx (Eds.), *The Global Water System in the Anthropocene: Challenges for Science and Governance* (pp. 385-399). Cham: Springer Verlag.

Book Reviews:

- Obani, P. (2016). Water and Cities in Latin America: Challenges for Sustainable Development by Ismael Aguilar-Barajas, Jürgen Mahlknecht, Jonathan Kaledin, Marianne Kjellén & Abel Mejía-Betancourt (Ed.s). 2015, hardback, 282 pages. Routledge, Abingdon, UK ISBN: 978-0-415-73097-6. *Water Policy 18*(1), 251-254. Doi: 10.2166/wp.2015.13. (Impact Factor for 2016: 1.144)

- Obani, P. (2015). The human rights to water and sanitation in courts worldwide: a selection of national, regional, and international case law, WaterLex, 2014, ISBN: 978-2-

940526-00-0. *International Environmental Agreements: Politics, Law and Economics, 15*(2), 237-239. (Impact Factor for 2016: 1.651)

Policy Briefs:

- Obani, P. & Gupta, J. (2016b). Access to Sanitation, SDG Policy Brief #1. Amsterdam: Amsterdam Institute for Social Science Research (AISSR), University of Amsterdam.

- Gupta, J. *et al.* (2014). Sustainable Development Goals and Inclusive Development, Post 2015/UNU-IAS Policy Brief No. 5. Tokyo: United Nations University Institute for the Advanced Study of Sustainability.

Conference Presentations:

- Obani, P. and Gupta, J., 'Human Rights and Access in Earth System Governance: Case of Sanitation'. Presented at the Earth System Governance Conference 2014 held in Norwich on 1 - 3 July, 2014.

- Obani P., 'Legal Pluralism in the Area of Human Rights: Water and Sanitation'. Presented at the European Association of Development Research and Training Institutes 2014 held in Bonn on 23 - 26 June 2014.

- Obani P. et al., 'the Human Rights to Water and Sanitation: Willingness to Pay amongst Poor and Vulnerable Groups'. Presented at the Global Water System Project 2013 held in Bonn on 21 - 24 May, 2013.

Trainings Attended:

- United Nations Institute for Training and Research (UNITAR) and the University of Geneva e-learning course on International Water Law (2013)

- UNESCO-IHE Institute for Water Education, PILOT Water Communications Short course (2013)

- Research School for Socio-Economic and Natural Sciences of the Environment (SENSE) Writing Week (2012)

Table of contents

List of tables

List of figures

List of boxes

Acronyms and abbreviations

ACHPR	African Charter on Human and Peoples' Rights 1981
ACID	Africa Infrastructure Country Diagnostic
AfricaSan	African Conferences on Sanitation and Hygiene
AICHR	Association of Southeast Asian Nations Intergovernmental Commission on Human Rights 2009
AMCOW	African Ministerial Conference on Water
BATNEEC	Best Available Technology Not Entailing Excessive Cost
BITS	Bilateral Investment Treaties
CAP	Consolidated Appeal Process
CCCM	Camp Coordination, Camp Management
CEDAW	Convention on the Elimination of Discrimination against Women 1979
CERF	Central Emergency Relief Fund
CHC	Community Health Club
CLA	Cluster Lead Agency
CLTS	Community Led Total Sanitation
COE	Council of Europe
COM	Committee of Ministers
CRC	Convention on the Rights of the Child 1989
CRMW	Convention on the Rights of Migrant Workers & their Families 1990
CRPD	Convention on the Rights of Persons with Disabilities 2006
CtC	Child-to-child
CTCIDTP	Convention against Torture & other Cruel Inhuman or Degrading Treatment or Punishment 1984
CWIS	Core Welfare Indicator Survey
DFID	Department for International Development
Ecosan	Ecological Sanitation
ECOSOC	United Nations Economic & Social Council
ECRI	European Commission against Racism & Intolerance
EHS	Environmental, Health, & Safety
EIA	Environmental Impact Assessment

EMIS	Education Monitoring Information Systems
EPF	Emergency Programme Fund
ER	Early Recovery
ERC	Emergency Relief Coordinator
ES	Emergency Situation
ESD	Environmental Sanitation Day
EVD	Ebola Virus Disease
FMARD	Federal Ministry of Agriculture & Rural Development
FMEdu	Federal Ministry of Education
FMEnv	Federal Ministry of Environment
FMF	Federal Ministry of Finance
FMH	Federal Ministry of Health
FMWA	Federal Ministry of Women's Affairs, Youth Development, Information & Communication
FMWR	Federal Ministry of Water Resources
GDP	Gross Domestic Product
GIIP	Good International Industry Practice
GLAAS	UN-Water Global Analysis & Assessment of Sanitation & Drinking-Water
HC	Head of Cluster
HCT	Humanitarian Country Team
HR	Human Rights
HRC	Human Rights Council
HRS	Human Right to Sanitation
HRW	Human Right to Water
HRWS	Human Right to Water & Sanitation
IBNET	International Benchmarking Network for Water & Sanitation Utilities
IC	Industrialised Countries
ICCPR	International Covenant on Civil & Political Rights 1966
ICERD	International Convention on the Elimination of all Forms of Racial Discrimination 1965
ICESCR	International Covenant on Economic, Social & Cultural Rights 1966

ICPAPED	International Convention on the Protection of All Persons from Enforced Disappearances 2006
ICTY	International Criminal Tribunal for the Former Yugoslavia
ID	Inclusive Development
IDGEC	Institutional Dimensions of Global Environmental Change
IFC	International Finance Corporation
IFRCRCS	International Federation of Red Cross and Red Crescent Societies
IHL	International Humanitarian Law
JMP	World Health Organization & UNICEF Joint Monitoring Programme for Water Supply & Sanitation
MAPET	Manual Pit Emptying Technology
MDGs	Millennium Development Goals
MeHMKS	Menstrual Hygiene Management Knowledge Space
MENA	Middle East & North Africa region
MLG	Multi-level Governance
MPA	Methodology for Participatory Assessments
NAFDAC	National Agency for Food & Drug Administration & Control
NDHS	Nigeria Demographic & Health Survey
NEMA	National Emergency Management Agency
NESP	National Environmental Sanitation Policy 2005
NESREA	National Environmental Standards & Regulations Enforcement Agency
NEWSAN	Society for Water & Sanitation
NGOs	Non-Governmental Organisations
NOA	National Orientation Agency
NPC	National Planning Commission
NPWS	National Policy on Water & Sanitation 2000
NSWTG	National Sanitation Working Task Group
NWLR	Nigeria Weekly Law Report
NWRI	National Water Resources Institute
NWRMP	National Water Resources Management Policy 2003
NWSSP	National Water Supply & Sanitation Policy 2000

ODF	Open Defecation Free
OECD	Organisation for Economic Cooperation & Development
PHAST	Participatory Hygiene & Sanitation Transformation
POLR	Provider of Last Resort
PPP	Public Private Partnership
PPPHWS	PPP for Hand washing with Soap
PRA	Participatory Rural Appraisal
RBDAs	River Basin Development Authorities
RCS	Red Crescent Societies
RHRB	Regional Human Rights Bodies
RUWASSAs	Rural Water Supply & Sanitation Agencies
Sani-Centre	Sanitation Centre
SARAR	Self-esteem Associative Strength Resourcefulness Action Planning Responsibility
SCSL	Statute of the Special Court for Sierra Leone
SDGs	Sustainable Development Goals
SDI	Service Delivery Indicators
SEMAs	State Emergency Management Agencies
sEPA	State Environmental Protection Agency
SLTS	School Led Total Sanitation
sMoE	State Ministry of Environment
SOF	Strategic Operational Framework
SPT	Simplified Planning Tool
SSIPs	Small-scale Private Individual Providers or Operators
SWA	Sanitation & Water for All
TLA	Tribunal Latinoamericano del Agua
TrackFin	Tracking Financing to Drinking-water, Sanitation & Hygiene
TSSM	Total Sanitation & Sanitation Marketing
UDHR	Universal Declaration of Human Rights 1948
UN	United Nations
UNGA	United Nations General Assembly
UN-Habitat	United Nations Human Settlements Programme

UNHRC	United Nations Human Rights Council
UNICEF	United Nations Children's Fund
USD	US Dollars
USG	Under-Secretary General
VCLT	Vienna Convention on the Law of Treaties 1969
WASH	Water, Sanitation & Hygiene Services
WASHBAT	WASH Bottleneck Analysis Tool
WCC	WASH Cluster Coordinator
WES	Water & Environmental Sanitation
WESCOMs	Water & Environmental Sanitation Committees
WHO	World Health Organization
WHO	World Health Organization and UNICEF Joint Monitoring Programme for Water Supply & Sanitation (JMP)
WSMP	Water & Sanitation Monitoring Programme
WSSCC	Water Supply & Sanitation Collaborative Council
WTP	Willingness to Pay

Summary

In law, people can gain access to the basic necessities for human wellbeing by going to court to enforce their rights. However, despite the growing global consensus on the existence of a human right to sanitation in law, over a third of the current 7.3 billion people worldwide lack improved sanitation services (Baum, Luh & Bartram, 2013) caused by a variety of drivers that are not entirely connected to law. There are also spatial, group, and individual inequities in access even within countries that have presumably better records of improved access. The resulting social and relational inequities and environmental degradation make the realisation of the human right to sanitation (HRS) and inclusive development (ID – which includes social, ecological and relational inclusiveness) pressing concerns for all levels of governance, from the international to the local levels.

There are five main gaps in knowledge which justify my focus on the HRS, while making references to the human right to water (HRW) where necessary for analysis: (a) the combined scholarly analysis of the human rights to sanitation and water does not enhance the conceptualization of the HRS due to inherent differences between both rights that are inadvertently overlooked; (b) scholarly research on the HRS remains limited within the legal scholarship but other fields are more actively engaging in such research requiring more interdisciplinary analysis; (c) the legal definition of the HRS is limited at best and does not address environmental sustainability;   (d) the analysis of the HRS by legal scholars has minimal considerations of the impact of HRS instruments on the drivers of poor sanitation services and the options for adopting complementary non-human rights instruments to strengthen the governance framework; and (e) there is a paucity of measurable indicators for assessing the HRS. Further, the evidence in the literature suggests that while combining the rights to sanitation and water presupposes the use of water based sanitation services, this may not be feasible in some local contexts irrespective of water abundance, just as water-intensive hygiene practices may nonetheless occur in water scarce regions.

Hence, my research seeks to contribute to the debate on the viability of HRS for addressing the drivers of poor sanitation services and advancing ID through equitable access to environmentally sustainable sanitation services; this is valuable for policy makers, development partners, regulators, sanitation service providers and the general public who are the rightsholders. Its main research question is: *How can the human right to sanitation (HRS) be interpreted and implemented to promote inclusive development?* This question leads to five sub-research questions. Inspired by the institutional analysis methodology developed by the International Human Dimensions Programme's Institutional Dimensions of Global Environmental Change project, the sub-research questions focus on the HRS institution; instruments to help implement it; the effects of the instruments on key actors given the drivers of poor sanitation services; and based on the above, a redesign of the instruments where necessary to improve their performance. They include:

(i)   What are the drivers of poor sanitation services and how are these currently being addressed in sanitation governance frameworks?

(ii)   How has the human right to sanitation evolved across multiple levels of governance, from international to local; how do the human right to sanitation principles address the drivers?

(iii)   Which humanitarian law and any other non-human rights instruments, including principles and indicators, for sanitation governance promote the progressive realisation of the human right to sanitation, through addressing the drivers of poor sanitation services?

(iv)   How does legal pluralism operate in sanitation governance, with the implementation of the human right to sanitation, alongside non-HRS principles?

(v)   How can the human right to sanitation institution be redesigned to advance inclusive development outcomes across multiple levels of governance?

Chapter 2 expounds on the research methodology and theoretical framework of the research. I adopt a multi-disciplinary perspective (using both legal and social science research methods) because the rules, decision-making processes, and programs that define the acceptable sanitation standards, allot roles to key actors for achieving the standards, and steer interactions among the actors, emerge from law and many other disciplines. The drivers of poor sanitation services are also not confined to law, as shown in this thesis. Therefore, any effort to progressively develop the HRS inevitably implies leaving the comfort zone of a human rights lawyer to engage with other disciplines, policy approaches, and instruments that may also affect the HRS institution. My method consists of (1) a detailed literature review within the legal discipline as well as all articles on sanitation published in other scholarly journals; (2) content analysis on the subject; (3) adapting the Young *et al.* (2005) framework for multi-level institutional analysis for analysing HRS institutions, instruments, actors, drivers, and implications for ID; (4) applying my framework to study the operationalization of the HRS through a case study, with fieldwork in Nigeria; and (5) triangulating the research findings to inform my recommendations for redesign of the HRS normative architecture. My conceptual framework additionally incorporates multi-level governance type 1 theory and an adaptation of the heuristic typology for legal pluralism developed by Bavinck and Gupta (2014). My analysis and findings are contained in the remaining seven chapters of my thesis, summarised below.

In Chapter 3, I analyse the complex nature of the sanitation crisis, the meaning of sanitation; the economic character of sanitation goods and services; the economic, social and environmental drivers of poor sanitation services; and the main technologies for sanitation services. I reveal theoretical and practical complexities for inclusive sanitation governance that are poorly reflected in the scholarly literature on the HRS. First, there are different definitions of basic sanitation, environmental sanitation or improved sanitation that are value laden and incoherent. This makes monitoring and evaluating the level of access to sanitation and establishing a global uniform standard for sanitation best practices problematic in the absence of a common meaning. Overall, the contestations show that the definition of sanitation ought to address gender equality, accessibility for vulnerable users like children, hygiene and maintenance of the facility, affordability of tariffs, operating and maintenance costs, environmental sustainability, and social and relational equality between poor and rich users/households, to ensure inclusive outcomes. Second, my analysis indicates an ambivalence over the economic nature of sanitation components which may have the properties of a public good, common good, toll good or private good, in different contexts. The inherent complexity of classifying sanitation as an economic good mirrors the pluralistic foundations of sanitation governance and hints at the complex interactions between the different principles (like the human right to sanitation and neo-liberal principles) that converge in sanitation governance. This requires States to establish a strong framework for sustainable funding, equitable access for the marginalised and vulnerable, and environmental sustainability from the use of sanitation services and technologies. Additionally, Chapter 3 identifies seventeen direct and nine indirect drivers of poor sanitation services and analyses the predominantly technocratic response to the sanitation problem, within the context of sanitation technology and the sanitation ladder. It is because technological instruments may or may not promote ID or address context specific drives while delivering sanitation services that I propose a matrix for the selection of sanitation technology and a merger of the technology-based and function-based sanitation ladders, in order to maximise the public health and environmental benefits of sanitation services for users.

Chapter 4 mainly assesses the human rights (HR) governance framework at multiple levels, from international to local, as a background to analysing the HRS. It discusses the sources and meaning of HR, HR principles, and HR indicators. This highlights two key contestations which occur where marginalised rightsholders are denied access to HR or there are tensions between different HR (like health environment and development). The chapter further discusses five predominantly social HR principles that are mainly focused on improving transparency and the effectiveness of participatory processes while countering the power politics that leads to marginalisation or exacerbates inequalities in human development. The principles are mostly established in treaties and national laws. The limited application of HR principles to ecological inclusion points to the need to clarify how and under what circumstances the HR normative framework addresses environmental concerns. Nonetheless, their real impact depends on the way the principles are operationalised through instruments in practise, and social principles, like participation and access to justice, have played a significant role in the protection of the environment and the development of environmental rights. It is therefore important to formulate effective indicators for monitoring the performance of HR principles and HR compliance by the duty bearers. However, indicators by themselves do not guarantee the best outcomes for the most vulnerable rightsholders. This underscores the need for decentralisation of the processes for formulating indicators and the importance of opportunities for the weakest rightsholders to forge strategic political alliances that empower them to hold the duty bearers accountable for HR compliance.

Chapter 5 analyses the sources, meaning and principles of the HRS, HRS instruments and indicators, the impacts of the HRS on ID, and the manifestation of legal pluralism in HRS implementation. The HRS has evolved through implied recognition on the basis of the International Covenant on Economic, Social and Cultural Rights 1966 (ICESCR) provisions and has also emerged as a distinct right in various sources of international law and political declarations; this underscores the close links between sanitation and human dignity, the comparable importance of sanitation in relation to other expressly recognised rights, and evidence of both political and legal support for independent recognition. In the absence of express recognition of the HRS at the national and sub-national levels, HRS advocates can rely on the fundamental importance of HRS for the realisation of related economic, social and cultural rights. The HRS is also recognised in the decisions of courts at various levels of governance, and in the laws of around sixty-seven States around the world. The various sources generally define the HRS either narrowly (with a limited consideration of environmental impacts) or broadly (to ensure human wellbeing and environmental sustainability). At the international level, the definition of sanitation in HR terms as advanced by the former Special Rapporteur did not sufficiently address environmental sustainability concerns. There is a mixed attitude at the national level; some countries may further restrict the scope of services guaranteed under the HRS (like the United Kingdom's Statement on the human right to sanitation in 2012), strengthen the relational focus of the HRS (like South Africa's 1997 Water Services Act), or extend the ecological focus of the HRS beyond human excreta and wastewater (like the Guinea Bissau Water Code which includes industrial wastewater management within the aims of sanitation). The HRS principles are also mainly social, similar to the HR principles. The principles nonetheless address ten out of the seventeen direct drivers, and six out of the nine indirect drivers of poor sanitation services; mostly in a legalistic sense that requires other non-HR instruments like technology for effective practical implementation. This suggests a need to integrate complementary non-HRS instruments and principles in the sanitation governance framework. An ID assessment also shows that the HRS may not guarantee inclusiveness without specific measures to enhance social and relational inclusion (by protecting the needs of the poor and vulnerable rightsholders) and ecological inclusion (by integrating environmental sustainability). Hence, HRS requires measurable indicators for assessing its performance, and I derive two structural indicators, sixteen process indicators, and nineteen outcome indicators (n=37 indicators) from the literature. This offers a good starting point for developing HRS indicators in the policy process.

Chapter 6 examines the HRS in humanitarian situations. The HRS *stricto sensu* applies to humanitarian situations and may be used to complement the international humanitarian framework for the protection of vulnerable populations (like refugees, prisoners of war, internally displaced persons, and women and children) in humanitarian situations. Conversely, the humanitarian framework offers a broad approach to sanitation which includes water, sanitation and hygiene services (WASH), hygiene promotion, water supply, excreta disposal, vector control, solid waste management, and drainage, as well as principles, core standards, and minimum indicators that are capable of enriching the definition and implementation of the HRS, even outside humanitarian situations. The chapter presents ten social principles of humanitarian assistance; six of these are rooted in international humanitarian law which focuses mainly on the protection of vulnerable populations in situations of international and non-international armed conflict, but does not extend to humanitarian situations resulting from natural and climate change related hazards or power tussles that have not yet escalated into armed conflicts, for instance. This is a limitation because humanitarian situations that are not directly linked with armed conflicts are on the rise globally and capable of manifesting not only within the jurisdictions where they occur, but in host countries that receive the influx of the resulting refugee population. The combined humanitarian assistance and protection principles and instruments can address six direct and five indirect drivers of poor domestic sanitation services in practical ways (for instance, the humanitarian framework offers outcome and process indicators for monitoring progress on sanitation that extend beyond the indicators available under the human rights framework). This may be effective if sanitation is classified as a survival need in humanitarian settings. An ID assessment shows that although the humanitarian principles are predominantly social, the humanitarian framework contains indicators that address social, relational and ecological inclusiveness, while a legal pluralism analysis reveals that although the human rights and humanitarian frameworks prioritise human wellbeing, their interactions may be marred by strong forms of rule incoherence.

Chapter 7 analyses the non-HR framework for water and environmental management that are relevant for sanitation governance. It uncovers six principles that are spread across the social and environmental pillars of sustainable development. Although the social principles (like capacity building and subsidiarity) are still emerging in international law but have gained wide acceptance in international development and politics, the environmental principles (like polluter-pays and precaution) have largely gained formal legal recognition recognized in treaties, customary international law or soft law. Hence, the non-HR framework significantly offers environmental principles which may complement the predominantly social and relational HRS principles in practice. The non-HR principles however stem from different normative foundations with a potential for trade-offs in the absence of rules to address incoherence or contradictions in the implementation process. Overall, non-HR principles can address twelve direct and four indirect drivers of poor sanitation services, and this potential mainly stems from the wide array of environmental principles that is lacking in the other two frameworks already considered for sanitation governance. Nonetheless, the non-HR principles that are devoid of a relational component (with the exception of poverty eradication and equity) may exacerbate inequities in access to sanitation and therefore need to be complemented with the HRS to reduce contradiction. This makes it important to design sanitation governance instruments based on a good understanding of the drivers of poor sanitation services in each context, the interplay between the human and non-human rights principles, and the suitability of the principles for addressing each driver without undermining the HRS.

Chapter 8 presents the outcome of my fieldwork in Nigeria. Although sanitation governance and policy formulation in Nigeria trails behind water policy development, sanitation also remains strongly linked with water which is not expedient for the local context. There is a plurality of meanings of sanitation in the sanitation laws and policy documents, and among key stakeholders in Nigeria including: environmental sanitation, improved sanitation and sustainable total sanitation. However, it is the limited public health focus on improved sanitation facilities, championed by external partners (who play a prominent role in the sanitation sector reforms), that is largely adopted by the regulatory agencies in practise, although it neither reflects the local perceptions nor the formal definition of environmental sanitation in the National Environmental Sanitation Policy 2005. Further, Nigeria implicitly recognises the HRS in sanitation laws, policies and strategies, but the principles and instruments used do not fully support the realisation of HRS norm and are indeed sometimes competing. They are inadequate as they exclude informal settlements, have poor enforcement, lack non-judicial grievance mechanisms, and do not address the impact of non-sanitation policies. My fieldwork tries to fill some of the gaps in knowledge by assessing other HRS principles (like affordability, participation and accountability), and the performance of the HRS instruments, including principles, in formal settlements, and informal settlements and humanitarian situations to some extent.

Chapter 8 further proffers reflections for redesign of the normative framework of HRS governance in Nigeria, to better address the drivers of poor sanitation services and advance ID. Social inclusion requires the instruments to strengthen the targeted population through: (a) building their capacity to effectively participate in sanitation governance, (b) providing a structure for equity and public participation in sanitation governance at all levels, (c) including local knowledge in the design of sanitation infrastructure, and (d) protecting against all forms of overt and covert discriminatory practices. Relational inclusion requires instruments which: (a) protect vulnerable and marginalised populations from the negative consequences of securitization or privatization of the public or merit components of sanitation, (b) shield the poorest from the negative impacts of discourses like the economic goods discourse which may lead to the concentration of wealth in the hands of a few, (c) promote downward accountability of service providers and State agencies in charge of sanitation regulation and/or provision, and (d) ensure the equitable distribution of sanitation infrastructure and resources for sanitation governance. Ecological inclusion requires instruments which: (a) adopt a broad definition of sanitation components to ensure environment sustainability, (b) promote equitable access to sanitation within ecological limits, (c) distribute the benefits and risks of sanitation services equitably, and (d) strengthen human resilience and adaptive capacity for climate change and other bio-physical or human-induced humanitarian situations. The HRS could be clearly defined and recognised in relevant laws and policies to include a duty on the State to respect, protect and fulfil obligations with respect to universal access to sanitation as a public good, especially for the poor, vulnerable and marginalised rightsholders who cannot meet their sanitation needs by themselves. The current narrow set of indicators used for measuring access to sanitation at the national and sub-national levels also needs expansion and I further elaborate on possible structural, process, threshold and outcome indicators in Chapter 9.

Chapter 9 concludes by stating that the HRS potentially offers a strong normative framework and legal basis for improving public participation in sanitation governance processes, progressively realising access to sanitation for everyone, and seeking redress for violations. However, it fares poorly in addressing ecological inclusion and some of the key drivers of poor sanitation services. It therefore needs to be redesigned to improve the effect against the drivers. My conclusions on how to achieve this can be surmised in seven points. First, promoting ID would require clarifying conceptual issues relating to the meaning of the HRS and the economic classification of sanitation goods and services to ensure that at the minimum, sanitation is regulated as a predominantly public good and the services extend beyond excreta management only. Second, State intervention and budgeting is required, both through facilitating opportunities for universal access to sanitation and direct provision to poor, vulnerable or marginalised individuals or groups, who would otherwise be unable to access sanitation within a predominantly neo-liberal context. Third, shared sanitation facilities that are hygienically maintained can also address the sanitation needs of vulnerable populations in informal settlements and humanitarian situations, and could therefore be integrated in sanitation programming and funding as a viable short term solution at the minimum. Fourth, I propose a list of principles to complement the existing HRS principles which can address the multiple drivers of poor sanitation services. Fifth, I propose a list of indicators for monitoring these principles. The HRS indicators and monitoring process can also be adapted to existing mechanisms for the monitoring of relevant international development goals like the Sustainable Development Goals. Sixth, I use legal pluralism frameworks to argue that contradictions in policy frameworks need to be addressed. Seventh, the process of interpreting and implementing the HRS for ID across multiple levels of governance can be enriched through delinking the rights to sanitation and water, except where necessary for implementation given the local context, and strengthening the synergies between HRS and other disciplines and governance frameworks for sanitation.

In conducting this research, I adopted a multi-disciplinary perspective (using both legal and social science methods) because the rules, decision-making processes, and programs that define acceptable sanitation standards, allot roles to key actors for achieving the standards, and steer interactions among the actors stem from law and many other disciplines (including the social sciences, physical sciences and engineering). I also integrated both legal and non-legal publications to presents the current state of knowledge on HRS governance and reached beyond the current state of knowledge by combining quantitative and qualitative methods to evaluate the HRS framework against the drivers of poor sanitation services and the need for ID and proffering recommendations for redesigning HRS instruments, where necessary. The methodology may be improved upon in future research by engaging multiple researchers from other fields to validate the findings of this study, conducting ethnographic research to assess the official reports, conducting additional systematically designed case studies across different scales, and using official translations of legal documents. Other five potential areas for further research include: (a) investigating the political economy of sanitation at multiple levels of governance, from the international to the local, in order to demine an affordable rate for users in different social contexts and wealth quintiles; (b) linking the HRS to the food, water, and energy nexus discourse; (c) investigating structures for affordability and accountability of sanitation in humanitarian situations; (d) analysing the economic aspects of ID components in sanitation policy and programming; and (e) evaluating the import of power politics for HRS interpretation and implementation across multiple levels of governance.

Chapter 1. The Human Right to Sanitation and Inclusive Development under an Uncertain Future

1.1 INTRODUCTION

Over a third of the current 7.3 billion people worldwide lack improved sanitation services (Baum, Luh & Bartram, 2013) and there are spatial, group, and individual inequities in access even within countries that have presumably better records of improved access (United Nations Economic Commission for Europe & World Health Organization Regional Office for Europe, 2013). The resulting social and relational inequities and environmental degradation make the realisation of the human right to sanitation (HRS) and inclusive development (ID) pressing concerns for all levels of governance, from the international to the local levels (see 1.2).

This thesis focusing on the HRS is timely because it was only in 2010 that the UN General Assembly adopted Resolution 64/292 on 28 July 2010, with three paragraphs, that recognised "the right to safe and clean drinking water and sanitation as a human right that is essential for the full enjoyment of life and all human rights;" required "…States and international organizations to provide financial resources, capacity-building and technology transfer, through international assistance and cooperation, in particular to developing countries, in order to scale up efforts to provide safe, clean, accessible and affordable drinking water and sanitation for all;" and welcomed the work of the independent expert on human rights obligations related to access to safe drinking water and sanitation, respectively (see Annex A). Immediately thereafter, the UN Human Rights Council adopted Resolution 15/9 on 6 October 2010 (Annex B), that recalled the General Assembly Resolution 64/292 and further affirmed "that the human right to safe drinking water and sanitation is derived from the right to an adequate standard of living and inextricably related to the right to the highest attainable standard of physical and mental health, as well as the right to life and human dignity" (see Annex B).

The UN General Assembly and UN Human Rights Council resolutions corrected the situation that emerged following the adoption of the Millennium Development Goals (MDGs) in 2000/2002 which called on countries to halve the number of people without access to water and sanitation services, by stating that all people had the right to water and sanitation services. The Sustainable Development Goals (SDGs) adopted in 2015 emphasize the HRWS by stating the need to ensure availability and sustainable management of water and sanitation for all (Goal 6, see 3.2; 3.5.5). Further, in 2015, the UN General Assembly Resolution 70/169 of 17 December 2015 further affirmed "that the human rights to safe drinking water and sanitation as components of the right to an adequate standard of living are essential for the full enjoyment of the right to life and all human rights" (see Annex C). Remarkably, Resolution 70/169 marks the evolution of the human right to sanitation as a distinct right, that may or may not be linked to the right to water in practise due to the inherent similarities and differences between water and sanitation (see 1.3.1). In this thesis I only focus on the human right to sanitation.

There are three main justifications for my focus on the HRS, while making references to the human right to water (HRW) where necessary for analysis: (a) the combined scholarly analysis of the human rights to sanitation and water does not enhance the normative development of the HRS due to inherent differences between both rights that are inadvertently overlooked (Ellis & Feris, 2014); (b) scholarly research on the HRS remains limited and continues in parallel across various disciplines that address sanitation governance (Obani & Gupta, 2016); and (c) the analysis of the HRS by legal scholars has minimal considerations of the impact of HRS instruments on the drivers of poor sanitation services and the options for adopting complementary non-human rights instruments to strengthen the governance framework (Obani & Gupta, 2016).

Hence, my research seeks to contribute to the debate on the viability of the HRS for advancing ID through equitable access to sanitation services, and an understanding of how HRS instruments interact with the drivers of poor sanitation which is valuable for policy makers and development partners. It specifically explores the question: *How can the human right to sanitation (HRS) be interpreted and implemented to promote inclusive development*? As a background to the rest of the thesis, this chapter highlights the real life problem (see 1.2), and theoretical challenges in the conceptualization and articulation of HRS based on the gaps in scientific knowledge which I have identified from the literature (see 1.3), formulates the research questions for the thesis (see 1.4), defines the thesis focus and limits (see 1.5), and presents the thesis structure (see 1.6).

1.2   THE RISING COST OF POOR SANITATION SERVICES

Over and above the very obvious need to enhance human dignity (see 4.3) and security by enabling access to improved sanitation services (Obani & Gupta, 2016), this section aims to show that there is also an economic argument for providing such services. On average, every adult human being generates 250 grams of faeces and 1 litre of urine daily (Freitas, 1999). With the world population estimated to reach 8.5 billion by 2030 and 9.7 billion in 2050 (United Nations Department of Economic and Social Affairs [UNDESA], Population Division, 2015), equitable access to sustainable sanitation simultaneously poses complex challenges and opportunities for the international development agenda.

There are widespread harmful effects on humans and the environment due to poor sanitation services. For instance, poor sewage management is commonly the largest source of environmental pollution (Evans & Bartram, 2004) and water contamination which could also result in the death of plant and animal life in rivers due to lower oxygen levels (Abraham, 2011; Hamner et al., 2006). Micro-pollutants from untreated waste also contaminate food chains, leading to public health risks (Joss et al., 2006). As a result, sanitation and water-related diseases are leading causes of mortality and morbidity (Prüss-Ustün et al., 2014), and could impair children's health, development and education (Dangour et al., 2013; Spears et al., 2013). Poor sanitation also affects human dignity (Joshi, Fawcett & Mannan, 2011), and fosters gender violence and inequities (Amnesty International, 2010; Srinivasan, 2015) and social unrest (Robins, 2014).

Further, the available data from official sources and the literature strongly suggest that the cost of poor sanitation is rising. At the international level, the economic cost of lack of access to sanitation services was USD 222.9 billion in 2015, 22% higher than the cost in 2010 (USD 182.5 billion) (Lixil, WaterAid Japan & Oxford Economics, 2016). India (USD 106.7 billion) accounts for almost half of the international loss (Lixil et al., 2016). While over half (55%) of the costs of poor sanitation result from premature deaths, an additional quarter is due to the treatment of sanitation-related diseases, and other costs are from reduced labour productivity as a result of sickness (Lixil et al., 2016). At the regional level, the Asia Pacific bears the most economic losses from poor sanitation which stands at USD 172.3 billion of the total international loss, Latin America and the Caribbean, and Africa account for about 10% of the total international loss, with Africa also having as high as 75% of its loss resulting from premature deaths (Lixil et al., 2016). Poor sanitation also leads to economic losses at the national level; from 1999 to 2008, the cost of environmental degradation in the Middle East and North Africa region (MENA) for instance, stood at an average of 3.6% of GDP with water and air being the main degradation categories (Doumani, 2014). When weighed against the evidence that the MDGs sanitation and water targets could have been met with an annual investment of around USD 60 billion, with sanitation accounting for USD 54 billion (Trémolet & Rama, 2012), this cost differential justifies increased investment in improving sanitation services.

1.3   GAPS IN SCIENTIFIC KNOWLEDGE ABOUT THE HUMAN RIGHT TO SANITATION

I identified five gaps in the scholarly knowledge, following a systematic literature review on sanitation from a human rights perspective (see Chapter 5): (1) limited scholarly literature on the HRS; (2) contestations over the meaning of the HRS; (3) inchoate analysis of the drivers of poor sanitation and their interaction with the HRS; (4) paucity of indicators for measuring assessing the HRS; and (5) incoherence between the legal and non-legal literature.

1.3.1 Limited scholarly literature on the human right to sanitation, compared to water

There is limited scholarly literature on the HRS, compared to water; the former is often subsumed under the HRW. Both rights share some similarities, to the extent that they are essential to the realisation of other rights and human security, require capital investments in infrastructure for their realisation, and are public goods. Nonetheless, the differences illustrated in Table 1.1 support calls by scholars for delinking the understanding and analysis of the HRS from the HRW (Ellis & Feris 2014; Obani & Gupta 2016), and justify my distinct consideration of the HRS in this thesis. In practical terms, water may not be crucial for sanitation but water-intensive hygiene practices nonetheless occur in water scarce regions (Rusca, Alda-Vidal, Hordijk & Kral, 2017).

Table 1.1 Differences between sanitation and water

Sanitation	Water
Taboo subject	Openly discussed
Adequate sanitation is contextual	Safe drinking water is non-negotiable
Facilities are stationary	Facilities are stationary but water is also potable
High infrastructure & maintenance cost	Lower infrastructure & maintenance costs
Essential for water quality	Non-essential for dry sanitation systems

The relationship between sanitation and water dates back to early civilizations like the Minoan, Mycenaean and Roman, where water played a significant role with respect to flush systems and storm drainage systems (Juuti, 2007); this predates the modern day formulation of the human rights to sanitation and water. The close link between sanitation and water further deepened with the sanitation revolution in mid-nineteenth century Britain, significantly with the publication of the *Report on the Sanitary Condition of the Labouring Population of Great Britain*. The report, published by Sir Chadwick, mainly presented disease carrying vapours caused by "insufficient sewers and drains, uncovered or stagnant drains, open stagnant pools, undrained marshes . . . and exhalations from cesspools" (Hamlin 1998, p.108) as the main cause of diseases and social ills including general immorality; the solution was to build universal water based systems that would wash away all the filth from the environment to be drained through sewers.

This linking of water and sanitation in mid-nineteenth century Britain gradually eliminated the disease causing vapours and reduced the cost of evacuating waste from the living environment (Hamlin, 1998). While both benefits appear to be important for public health and human wellbeing, the reforms introduced by the sanitation revolution were mainly geared towards preventing a civil revolution, maintaining the status quo of class relations and reducing the lost profits as a result of ailing workers (Hamlin, 1998; Morley, 2007). Only a few critics attributed the unsanitary conditions to a systemic violation of the rights of all British citizens rather than just the working class (Hamlin, 1998; Susser, 1993); which is similar to the essence of the HRS and the theoretical basis for linking HRS to ID in this thesis.

Nonetheless, the use of technologies for water-borne sanitation quickly extended beyond Britain, reducing the public health risks from poor waste management, and waterborne diseases are still widely perceived as the standard for adequate sanitation in many parts of the Westernised world (Kramek & Loh, 2007; Robins, 2014; Schuster, 2005). Furthermore, with population growth and increasing demand for efficient sanitation technologies in the twentieth century, the linking of water and sanitation has continued in various responses to the global sanitation crisis, including the human rights discourse (see 5.2.1) and the SDGs (see 1.1, 3.2, 3.5.5, 7.3.4). But the twenty first century has witnessed scholarly debates for and against the continued linking of water and sanitation. The alternative views uncovered from experts are that water issues may be addressed before sanitation issues due to the high cost of sewage management;[1] both can be separated to maximise funding and coverage for sanitation that would otherwise be subsumed under access to water only;[2] and that while it

---

[1]   Interviewee 28.
[2]   Interviewees 44 and 45.

would be better to address both together the appropriate response would depend on the local context[3] and the expediency of combined regulation of water and sanitation infrastructure.[4]

Based on the foregoing, there are three main reasons canvassed for combining the rights to sanitation and water. First, international development organisations prioritise water over sanitation projects and this makes the practise of piggybacking sanitation onto water expedient for the purpose of attracting funding and a higher level of focus (Chaplin, 2011). However, there is little evidence to support this rationale given that at the international level, the MDGs sanitation target was not achieved yet the related MDGs water target was achieved in advance of the schedule (United Nations [UN], 2017). There is also no evidence that combining the HRWS results in better implementation within national jurisdictions (Ellis & Feris, 2014). Second, sanitation is highly capital intensive but also has high indirect economic returns on investments which mainly accrue from healthcare savings (see 1.2) (Hutton, 2012; 2013; Salter, 2008; Toubkiss, 2006). However, a purely narrow economic analysis suggests that sanitation investments may not give sufficient returns to the investor and could discourage investment; HRS would therefore suffer from standing independently from the HRW which is relatively cheaper to implement (Ellis & Feris, 2014). Third, the rights discourse enriches sanitation programmes with normative content that is critical for prioritising human wellbeing and which would otherwise be missing from non-human rights based approaches.

Conversely, there are also three arguments against continued combination of sanitation and water as human rights. First, the combination makes it difficult to develop the distinct normative content of the right. For instance, the quantity of water required to meet the right to sanitation is context specific; the quality of water required for flushing toilets (and whether flush toilets are needed) is not necessarily the same as that required for drinking; and the combination does not account for the specific amount of water necessary for sanitation and hygiene distinct from drinking water needs. Second, combining both rights virtually implies the right to water for water-based sanitation services and thereby potentially undermines the relevance for people who rely on dry sanitation systems whereas the normative contents of the HRS apply to all forms of sanitation and hygiene systems (see Chapter 3) (Ellis & Feris, 2014). Third, sanitation and water differ in terms of perception (sanitation is often a taboo issue) (Black & Fawcett, 2008; IRIN, 2012); potability (while individuals can take drinking

---

[3]   Interviewees 30, 33, 38, 40, 41, and 42.
[4]   Interviewee 36.

water with them wherever they go, sanitation facilities are not equally potable and poor sanitation affects both non-users and the wider community) (Obani & Gupta, 2015); source and responsibility for service provision (water services are commonly regarded as the responsibility of the government while sanitation is regarded as a private matter) (Pories, 2016); the cost of infrastructure and the rates of return on investments for both differ (relatively higher for sanitation) (Hutton, 2013); and while sanitation is essential for water quality, some sanitation systems do not require water for their operation (see Table 1.1).

Hence, although the HRS is still often combined with water, the purpose of the HRS extends beyond ensuring water quality to ensuring human wellbeing through health for instance. And the development of the HRS norm has continued to suffer and billions of people around the world lack access to sanitation with the current combination, yet the right to water norms have been extensively developed and water targets, including the global MDGs target, have been achieved although this is subject to debate (Bain et al., 2014). It is therefore necessary for the HRS to be analysed as an independent right to strengthen the normative content, albeit related to the right to water to the extent that this is expedient for successful implementation of HRS. The rest of this thesis therefore focuses mainly on the HRS, with reference to the HRW where necessary for analysis and illustration.

1.3.2 Contested Meaning

Sanitation services evolved as engineering or technological solutions to the public health risks resulting from excreta or waste contamination (Eyler, 2001; Juuti, 2007; Ramsey, 1994). The scholarly literature and development agenda are therefore awash with technocratic definitions of sanitation, and technical solutions and innovations. Nonetheless, access to improved sanitation facilities alone has failed to guarantee safe excreta containment (Baum et al., 2013; Irish et al., 2013), just as measuring access to sanitation solely through the use of improved facilities may not capture important environmental, economic, and social concerns that are critical for sustainable outcomes (Kvarnström, McConville, Bracken, Johansson & Fogde, 2011; Obani & Gupta, 2016). It is therefore necessary to explore a better understanding of what sanitation means and how technology can be adopted within national jurisdictions to effectively promote the HRS without compromising on ID. More so because, despite a clear set of HRS norms established at the international level least, the HRS continues to be interpreted and implemented locally using instruments that may compromise on social, relational, and ecological inclusion. The meaning of sanitation is one of the determinants of the scope of obligations arising from the HRS, and therefore requires further investigation.

There now exist many definitions of sanitation, including basic sanitation, adequate sanitation, improved sanitation and environmental sanitation (see 2.2). Some definitions are limited to just access to toilet facilities, while others include emptying, transport, treatment and disposal of excreta. Other discourses enmesh sanitation in individual aspirations for a clean and healthy physical environment, dignity, and privacy (Joshi et al., 2011); complex interactions of culture, politics and institutions at various levels of governance (Akpabio, 2012; Akpabio & Subramanian, 2012; Chaplin, 1999); climate change (Geels, 2013; Lopes, Fam & Williams, 2012); the emerging story of Peak Phosphorus (Cordell et al. 2009); and concerns for environmental sustainability (Feris, 2015; Kvarnström et al., 2011). Therefore, the meaning of the human right to sanitation requires further investigation.

1.3.3 Inchoate Consideration of the Drivers of Poor Sanitation

In addition to clarifying the meaning of the HRS and adopting HRS principles in the domestic legal and policy framework, it also important to ensure that the principles and instruments through which they are operationalized effectively address the drivers of poor sanitation services. This is because the challenges of human rights realisation and human development are not confined to the legal framework (Arts, 2014; McGranahan, 2013; Obani & Gupta, 2014b; Sosa & Zwarteveen, 2016). However, there is no comprehensive scholarly analysis of whether and how the HRS addresses the drivers of poor sanitation (see 3.4). Rather, scholars often analyse the drivers that fall within their domains without reflecting broadly on how these interact with other drivers and instruments identified by other scholarly domains.

The political ecology literature from specific case studies emphasize the political, economic, and social drivers of poor sanitation and largely advocate tailoring sanitation interventions to reflect the local context and practices, rather than standardized technological responses (Aguilar-Barajas et al., 2015; Akpabio & Subramanian, 2012; Jewitt, 2011; Joshi et al., 2011; McFarlane, 2008). Nonetheless, there is still a lot of investment in technocratic engineering sanitation solutions and a significant body of literature on promoting demand for predetermined sanitation goods and services (Barrington et al., 2016, 2017; Evans et al., 2014; Jenkins & Scott, 2007). The environmental sciences and public health literature analyse the technological drivers of poor sanitation, such as sewerage connections without treatment

(Baum et al., 2013) and the limits of monitoring progress towards sanitation targets based on a hierarchy of predefined technologies (Kvarnström et al., 2011). The engineering literature emphasizes the need to eliminate challenges in the bio-physical environment (environmental drivers), like weather variability, water scarcity and the topography of the service area, which affect the operations of sanitation infrastructure (Cairncross & Valdmanis, 2006; Dagdeviren & Robertson, 2009; Ensink, Bastable & Cairncross, 2015; Schouten & Mathenge, 2010). From the demographic perspective, growing human population and changes in the migratory patterns make it important to also consider social drivers affecting sanitation. In the governance literature, drivers have been analysed in the context of internalising the HRS by local actors, including system operators, utilities, and management boards who are directly involved in service provision but least connected to the international human rights debate (Meier et al., 2014).

The legal literature identifies the following drivers of poor sanitation: poverty, latent demand and the low prioritisation of funding for the poor (economic); unclear roles of actors (socio-political), and unsustainable solutions (technological) (Centre on Housing Rights and Evictions [COHRE], UN-HABITAT, WaterAid & Swiss Agency for Development and Cooperation [SDC], 2008). Additionally, the legal literature differentiates between the direct drivers like poverty (economic); poor maintenance, taboo and non-prioritisation (social); and weather variability and natural disasters (environmental), from indirect drivers like risk aversion, privatization, foreign debt, and sanctions (economic); population growth, weak institutions, demographic trends, and conflicts over transboundary resources (social); and climate change (environmental) (Obani & Gupta, 2016b). The analysis is however often limited to considering drivers as a justification for the human rights approach to sanitation without further assessment of the performance of HRS instruments in addressing the drivers.

Further, the drivers are not delineated according to various social contexts. For instance, technocrats and policy makers identify (un)willingness to pay as a driver of poor sanitation services (Seraj, 2008; Van Minh, Nguyen-Viet, Thanh & Yang, 2013) because it reduces the funds available for sanitation services, and this is mostly supported by case studies in formal settlements (Arimah, 1996; Rahji & Oloruntoba, 2009), informal settlements (Goldblatt, 1999), and rural areas (Gross & Günther, 2014). However, the increasing number of humanitarian situations raises the question of whether or not refugees and internally displaced persons can be made to pay for their sanitation needs, and how, but this is yet to be addressed in the literature. This is a significant gap in scholarly knowledge, especially in the light of recent news reports that the authorities in Germany and Denmark relieve refugees of their valuables as payment for their accommodation (Dearden, 2016). This research goes further to evaluate the impact of HRS principles and instruments by positioning ID as a directional objective for the HRS. My reasons for choosing ID are further explained in my methodology chapter (see 2.4.3).

1.3.4 Paucity of Measurable Indicators for the HRS

Indicators can be described as "a question or series of questions, which assist in determining the extent to which the target has been met" (Roaf, Khalfan & Langford, 2005, p.9). I use indicators to evaluate how instruments, including principles (see 2.5.1) advance the HRS; this facilitates measuring and monitoring progress on the HRS and ensures the accountability of actors (Jensen, Villumsen & Petersen, 2014; Roaf et al., 2005). Developing suitable HRS indicators that are both contextually relevant and allow for comparison remains a policy challenge. The legal literature mainly proffers qualitative indicators but scarcely discusses quantitative indicators for monitoring the HRS.

The non-legal literature further offers a variety of qualitative and quantitative indicators for excreta containment, access, grey water management, pathogen reduction, nutrient reuse, eutrophication risk reduction, and integrated resource management (Kvarnström et al., 2011); proxy indicators for monitoring international and national sanitation targets, including the safety of sanitation technologies (Baum et al., 2013) and wastewater reuse options (Ensink et al., 2015; Ensink, Blumenthal & Brooker, 2008; Irish et al., 2013); and sanitary inspections for complementary safe assessments and identify corrective actions for water safety (Bain et al., 2014).

The indicators in the literature are often either proposed or operationalized independently, without strong links and feedback between the legal and non-legal fields to ensure progressive realisation of the HRS. It is therefore important to synergise the indicators for measuring the progressive realisation of the HRS (see Chapter 9).

1.3.5 Incoherence between Legal and Non-Legal Research on the Human Right to Sanitation

As can be seen from the previous sections, there is a growing body of scholarly literature on the HRS, by both lawyers and non-lawyers. However, the existence of knowledge does not unequivocally mean effective application to solving real life problems because they often run in parallel, and are poorly integrated in the institutional structure (Baud, 2016). Second, the developments in the legal and non-legal literature are poorly linked in addressing sanitation governance problems in real life and therefore do not optimally influence key actors in sanitation governance (Obani & Gupta, 2016). Third, there are different discourses and types of knowledge on sanitation that are affected by legitimacy issues (Karpouzoglou & Zimmer, 2016); the outcomes may either support or contradict the HRS and therefore need to be clarified through a legal pluralism diagnostics at multiple levels of governance (Obani & Gupta, 2014b).

The non-legal literature critiques the narrow formulation of the HRW which affects the realisation of HRS norms especially for people relying on various wet sanitation systems (Joshi et al., 2011; Hall, van Koppen & van Houweling, 2014); analyses the central role of knowledge in the re-production of the urban waterscape, including wastewater (for instance, Bakker, 2003; Castro, 2004; Kaika, 2003; Karpouzoglou & Zimmer, 2016; Swyngedouw, Kaïka & Castro, 2002); assesses sanitation governance processes and the motivations and accountability of key actors (Abeysuriya, Mitchell & White, 2007; Giles, 2012; Meier et al., 2014), even using human rights standards (Galvin, 2015); and proffers quantitative and qualitative indicators for measuring access levels (Bain et al., 2014; Baum et al., 2013; Ensink, et al., 2008; Ensink et al., 2015; Irish et al., 2013; Kvarnström et al., 2011). The non-legal literature however does not sufficiently address the HRS norms beyond a cursory consideration at best but has instead focused largely on the technological and economic aspects of sanitation services (Obani & Gupta, 2016 analyses the coverage of the HRS in the legal and non-legal literature in detail).

Likewise, the legal literature expounds on the meaning, normative content and limitations of the current HRS construct (Feris, 2015; Obani & Gupta, 2015), and prospects for applying human rights norms to other aspects of resource governance like the management of transboundary aquifers (Gavouneli, 2011). The legal literature also analyses the legal framework for the implementation of the right at various levels of governance, from the international (McIntyre, 2012; Misiedjan & Gupta, 2014; Obani & Gupta, 2015; Salman, 2014) to the national and local levels (Baer, 2015; Bhullar, 2013; Cullet, 2013; Martin, 2011; Tignino, 2011; Wekesa, 2013); and have mainly developed qualitative indicators either on the basis of existing human rights indicators or new proposals specifically tailored to the HRS (de Albuquerque, 2012). Despite the similar themes covered, the linkages between legal and non-legal research on HRS remain feeble at best and the HRS is even sometimes seen as an extra impediment to achieving (economic) development by some professionals outside the legal field.

Nonetheless, particularly following the momentum generated by the International Year of Sanitation (2008), the International Decade for Action 'Water for Life' (2005 – 2015), and the various UNGA and HRC resolutions on the human rights to water and sanitation since 2010 (see 1.1, 5.2.1), amongst others, there has been increasing emphasis on mainstreaming HRS norms in sanitation programming and interventions. The sanitation target under the 2030-bound Sustainable Development Goals largely imbibes the HRS norms into the international development agenda and is reminiscent of the increasing mainstreaming of HRS norms in sanitation programming and interventions, much unlike the 2015-bound Millennium Development Goals sanitation target that was mainly focused on 'improved sanitation facilities' and poorly reflected the social and ecological aspects of the sanitation challenge (see 3.2).

At the national level, even countries who previously abstained from voting on the UNGA resolution (see 5.2.2) have subsequently clarified their support for the HRS, at least within their national borders (like Canada and the United Kingdom) (see Chapter 5). Meta-analysis sometimes shows marked differences in the conception of the HRS norms like participation and accountability, between lawyers and non-lawyers for instance (Klasing, Moses & Satterthwaite, 2011). Nonetheless, there are positive prospects for collaboration between legal and non-legal researchers working on HRS issues (Obani & Gupta, 2016) and the potential conflicts can be approached through integrating a common understanding of inclusive development as a directional objective for the HRS and resolving rules incoherence, as demonstrated in this thesis.

1.4 RESEARCH QUESTIONS

This thesis conducts an institutional analysis of the HRS, focusing on the effect of HRS and the instruments used to operationalize it for ID. In other words, the thesis analyses how HRS and complementary instruments to operationalize it contribute to ID. The main research question is therefore:

> *How can the human right to sanitation be interpreted and implemented to promote inclusive development?*

The main research question is further broken into five sub-research questions about the HRS institution, instruments to help implement it and their effects on key actors given the drivers of poor sanitation services:

(i)   What are the drivers of poor sanitation services and how are these currently being addressed in sanitation governance frameworks?

This sub-question contextualises the sanitation problem and establishes the background for the thesis and the linking of the legal and non-legal literature by: (a) highlighting the drivers of poor sanitation services, (b) analysing the predominantly technocratic approach to addressing the sanitation problem, and (c) highlighting existing contestations over the meaning and economic characteristics which make sanitation governance all the more complex. I address this question in Chapter 3.

(ii)   How has the human right to sanitation evolved across different levels of governance, from international to local; how do the human right to sanitation principles address the drivers?

This sub-question focuses on the evolution of the HRS institution as a normative instrument for addressing the sanitation crisis at different levels of governance, and the interactions between the HRS and humanitarian law and non-human rights governance frameworks for sanitation as an indication of legal pluralism. In addressing this question, I conduct a literature review on the HRS and identify both synergies and contestations in the current state of knowledge on the HRS and the main principles for HRS and sanitation governance generally (from both the legal literature and literature from other relevant fields). I thereby also address the knowledge gap relating to limited scholarly focus on HRS, and the poor integration of knowledge from other fields in the existing legal literature on HRS in Chapter 5, having laid the foundation with my analysis of the human rights principles in Chapter 4.

This sub-question also links the drivers to the relevant HRS instruments and principles in Chapter 5. In my analysis, I go beyond the existing literature by distinguishing between the direct and indirect drivers of poor sanitation services (see Chapter 3). Such an in-depth understanding of the drivers is critical to formulating inclusive HRS instruments.

(iii) Which humanitarian law and any other non-human rights instruments, including principles and indicators, for sanitation governance promote the progressive realisation of the human right to sanitation, through addressing the drivers of poor sanitation services?

This sub-question investigates other (non-HRS) instruments, including principles and indicators that are applied in sanitation governance framework, focusing on humanitarian law and humanitarian situations in Chapter 6, and other frameworks that are widely applied for water and environmental management and their performance against the prevailing drivers of poor sanitation services in Chapter 7.

(iv) How does legal pluralism operate in sanitation governance, with the implementation of the human right to sanitation, alongside non-human rights instruments and principles?

This sub-question analyses the interactions between the HRS principles and non-HRS principles at different levels of governance, through the theory of legal pluralism. It highlights the quality and intensity of the interactions to show whether there is indifference, competition, accommodation or mutual support (see 5.6.3, 6.6.3, 7.4.3 and 8.6.3) between the HRS principles and non-HRS principles, at any given level of governance and the consequences for realizing the HRS and advancing ID.

(v) How can the human right to sanitation institution be redesigned to advance inclusive development outcomes across multiple levels of governance?

This sub-question builds on the understanding of the operation of the   HRS instruments/principles, in relation to drivers and ID, already established by the preceding sub-research questions to: (a) evaluate the performance of the HRS institution against the social, relational and ecological component of ID, (b) identify HRS principles/instruments that advance ID in my case study in Chapter 8; and (c) proposes a redesign of the less effective HRS instruments through integrating either HRS principles or complementary principles form humanitarian law and other sanitation governance frameworks in my Chapters 9 and 10.

1.5 FOCUS AND LIMITS

This thesis on the HRS is written by a legal scholar while recognizing that the implementation of this right involves many other disciplines and policy actors and any effort to progressively develop this right will inevitably imply moving out of the comfort zone of a human rights lawyer to engage with knowledge and policy approaches and instruments that are complementary to the HRS and help to implement it. This has informed alternating between the internal and external points of view in the research process. I adopt a panoramic view in my analysis of the HRS institution as it has evolved at different levels of governance, because although the HRS has emerged strongly from international law, its origin and development is deeply rooted in local environmental justice and water and sanitation rights movements and its application is nuanced in various jurisdictions and multiple levels of governance. My approach is therefore unique and affords me a holistic consideration of the evolution of the HRS, its current legal status, and future trajectory. I holistically investigate the HRS in relation to poor, marginalised and vulnerable groups generally, with specifics of how the drivers of poor sanitation services affect women, children, refugees, migrants, internally displaced persons, detainees, and residents of informal settlements, for instance (in Chapters 3, 5, 6 and 8). Although, I do not focus on the rights of nature in my analysis, because the HRS is mainly anthropocentric, I highlight the impact of the HRS on ecological inclusion and the need to meet human sanitation needs within ecological limits in order to ensure environmental sustainability (see 9.3.2 and 9.5).

Further, I use a case study for an in-depth analysis of the operation of the HRS and from this I obtain results that I extrapolate into my final recommendations for interpreting and implementing the HRS to advance ID (see Chapter 9). A common criticism of case studies is that the context specific character hampers generalisation. Though no set of cases can be representative enough to avoid this criticism, through the application of replication logic, elements of sanitation governance under peaceful or relatively stable conditions (like formal and informal settlements) and humanitarian situations (resulting from armed conflicts or emergencies, for instance) are considered within Nigeria, to obtain evidence of convergence, or differences in the outcomes (Yin 2009). As a result, some aspects of my results from the case study can be generally applied to advancing ID through the human rights construct in other aspects of human development beyond sanitation.

There are three main data limitations which I encountered in this thesis: (a) my use of some unofficial translations of laws and policy documents for my content analysis; and (b) my sample size which though sufficient for the purpose of this thesis may not be sufficient to draw universal conclusions on the impact of HRS institutions on ID given that there are unique local circumstances in every domestic jurisdiction across the world; (c) the covert discrimination against informal settlements in my case study which limits the generalizability to jurisdictions with overt discriminatory policies against informal settlements. I elaborate on these limitations below.

First, some of the national laws and policies used for the content analysis were in Arabic, English, French, and Spanish. I had to rely on unofficial translations in some instances but compared the results across legal databases to improve reliability. At the international level, mainly hard sources of law were assessed, while soft sources of law were also assessed where they formed evidence of customary international law. At the national level, national constitutions and laws were mostly assessed because they provide the legal basis for policy direction in domestic jurisdictions.

Second, the sample size and stakeholders for the case study were constrained by limited time, financial resources and bureaucratic bottlenecks. Though the sample size achieved for the household survey was sufficient for qualitative and statistical analysis for this thesis, the research findings would have been further reinforced by a larger representative sample size. It was particularly difficult to interview more households in humanitarian situations either because of their physical or psychological vulnerability or unwillingness to participate in the study. Otherwise, questions about affordability and accountability could be interesting issues for further research on the application of HRS in humanitarian situations. It was also impossible to set up interviews with informal service providers at the local level because they were sceptical of the research objectives and feared being identified and prosecuted by the authorities. Further, some important regulatory agencies with mandates related to sanitation (for instance, school sanitation) declined to be interviewed stating that they were not directly responsible for household sanitation services. This underscores the fragmentation of sanitation governance and interviewing such agencies may have contributed further insights into sanitation governance which could be explored in future research.

Third, Nigeria does not have a national policy covering public services in informal settlements. As such, the question of whether or not to provide/facilitate public sanitation services in informal settlements is often political and left to the discretion of the relevant agencies. This is unlike the case in some other jurisdictions like India, Brazil and South Africa which have clear national policies on the inclusion or exclusion of informal settlements from public service provision. As such, the results of my case study of HRS may not be generalizable to jurisdictions where informal settlements are expressly excluded from or included in public services. However, with the increasing recognition of HRS across various levels of governance and the far-reaching SDGs sanitation target (see 1.3, 3.2 and 3.5.5), it is expected that domestic laws and policies entrenching various forms of overt or covert discrimination against informal settlements and other vulnerable/marginalised groups may soon give way to mechanisms for universal access. Further, I augment my case study results with findings from a literature review on the performance of HRS instruments in informal settlements in other jurisdictions.

1.6 THESIS STRUCTURE

This chapter sets the tone for the thesis having established the real life sanitation crisis, gaps in scientific knowledge about the HRS, the research questions based on the gaps, and the focus and limits of my research. In Chapter 2, I elaborate on the research methodology and theoretical framework which result in the conceptual framework of my thesis. Chapter 3 contextualises the sanitation problem, focusing on the meaning of sanitation, its economic characteristics, and drivers of poor sanitation services as well as the predominantly technocratic instruments for addressing the sanitation problem. Chapter 4 generally analyses human rights principles for tackling human development challenges, as a precursor to analysing the HRS principles and instruments as they apply to formal and informal settlements, in Chapter 5. Further, Chapter 6 analyses humanitarian law principles for sanitation services in humanitarian situations and Chapter 7 analyses other non-human rights principles/instruments for sanitation governance at multiple levels of governance. In Chapter 8 I present my case study findings, and recommendations for redesign of the instruments encountered in the case study. Chapter 9 combines my key research findings, presents my conclusions and recommendations on interpreting and implementing the HRS to advance ID. The logical framework for the structure of my thesis is further outlined in Annex D. There are some substantive overlaps between the content of the chapters and papers/book chapters that

I have published in the course of my PhD research, to the extent that the papers/book chapters individually address some parts of the research questions (see p. v).

Chapter 2.   Research Methodology and Theoretical Framework

2.1   INTRODUCTION

This chapter presents the research methodology and theoretical framework of this thesis. First, I expound on the ontological and epistemological basis of my research (see 2.2). On the basis of the systematic literature review I conducted during the proposal writing phase of my research, I formulated a research design that involves a combination of methods from law and the social sciences. For my data collection, I relied on systematic literature review (see 2.3.1), case study, including interviews and household surveys (see 2.3.2). Further, I analysed the data through content analysis (see 2.3.3), and legal reasoning and argumentation (see 2.3.4). Also based on the outcomes of my preliminary systematic literature review while writing my research proposal, I draw from multi-level governance (see 2.4.1) and legal pluralism (see 2.4.2) to specifically evaluate the performance of the HRS, with inclusive development (ID) (see 2.4.3) as the guiding norm. In Section 2.5 I elaborate on my conceptual framework and how I synthesize my research methods and main theories to address my research questions. The chapter ends with an explanation of the ethical considerations (2.6) guiding my conduct of the research.

2.2   ONTOLOGY AND EPISTEMOLOGY

The problem of poor sanitation services is one that affects multiple facets of human existence and has been researched by scholars from different disciplines who often adopt either a disciplinary or multidisciplinary approach with minimal integration of methods or knowledge from other relevant disciplines (see 1.3). Given the myriad drivers, actors, and instruments from law, social sciences, and natural sciences that converge in sanitation governance, this thesis adopts an interdisciplinary approach to investigate the drivers of poor sanitation services *vis-à-vis* a variety of human rights, humanitarian law and non-human rights principles, instruments and indicators for sanitation governance that co-exist both *de facto* and *de jure* to affect access to and allocation of sanitation services in the real world. To achieve this, I draw on my interdisciplinary academic background to identify multiple methods of knowing that are suited to my research questions (see 2.3). I also rely on institutional analysis to identify aspects of the multiple disciplines that are relevant to addressing my research question (see 2.4).

I operationalize my interdisciplinary perspective through: (a) designing my research questions to reflect the multiple disciplinary perspectives integrated in my research; (b) combining theoretical underpinnings and qualitative and quantitative research methods or mixed methods that are drawn from otherwise distinct disciplines, including law, international relations/international development studies, natural sciences, and economics in my conceptual framework (see 2.3 and 2.4); and (c) alternating between internal and external points of views to better understand the complex interactions between the natural and social sciences (McKay 2014 distinguishes between the use of the internal and external points of view by positivists and empirical legal theorists). In my research, I found both the internal and external points of view to be complementary and applying the two enabled me to better understand the interactions between human rights, humanitarian law and other non-HR principles, instruments, and indicators that are relevant for sanitation governance, thereby enriching my institutional analysis.

2.3   DATA COLLECTION AND ANALYSIS

This section explains the methods I used for my data collection and analysis: (a) systematic literature review (see 2.3.1), (b) case study (see 2.3.2), (c) content analysis (see 2.3.3), and (d) legal reasoning and argumentation (see 2.3.4).

2.3.1 Systematic Literature Review

I conducted a systematic literature review to ascertain the current state of knowledge about the research focus. This informed the research problems (see 1.2), gaps in scientific knowledge (see 1.3), research questions (see 1.4), focus and limits of this research (see 1.5), the thesis structure (see 1.6), and the research methodology and theoretical framework presented in this Chapter. It is also through the literature review that I uncover the drivers of poor sanitation services and various contestations over the meaning and economic characteristics of sanitation (see Chapter 3), and principles and instruments (see 2.5.1) and indicators (see 1.3.3) for sanitation governance from different disciplines (see Chapters 3, 4, 5, 6 and 7). My findings from the literature review substantiate my content analysis (see Chapters 4 and 5), case study results (see Chapter 8), and recommendations for redesign (see Chapter 9).

The publications I reviewed include peer-reviewed journal articles, books, and grey literature including official reports and communications (containing important empirical data on access to sanitation at multiple levels of governance and the principles applied by key actors for sanitation governance) from especially the legal literature but also complemented by the non-legal literature which I sourced through scholarly databases.[5] I used different combinations of keywords/concepts focused on the human right to sanitation principles, instruments, and indicators; the Sustainable Development Goals on sanitation and relevant indicators; and the human right to sanitation in humanitarian situations (see Annex E) to obtain an extensive data set to support my evaluation of HRS instruments and non-human rights instruments for sanitation in the case study and in relation to the global development agenda (see Chapters 4, 5, 6, 7, 8 and 9). The literature collected was then short-listed to ensure relevance for addressing my research questions.

2.3.2 Case Study

Legal research traditionally consists of analysing the conventional sources of law recognised by positivist theory, and relying on secondary sources like the academic literature where necessary for an elucidation of primary sources of law like treaties and national laws. It mainly involves desk study and emphasises the content of the law as it is. This does not offer an opportunity to explore the performance of the HRS and the drivers (see 3.4) which affect access to sanitation on the ground. As a result, I had to adapt some social science methods for empirical legal research, including a case study with interviews and field visits (Epstein & Martin, 2014). I use a layered single case study design to gather in-depth information on the performance of HRS instruments at the national and sub-national levels of governance. The case study enables me to localise the HRS and demonstrates that the instruments required to advance the HRS differs from one setting (like formal settlements), to another (like informal settlements or humanitarian situations) because of the operation of a different set of drivers in each setting (see 3.4). Despite the criticism that a single case study is not generalizable and lacks external validity (Easton, 2010), the findings from a single case study can be generalizable and contribute to theory formation (Woodside, 2010; Yin, 2014), and is appropriate for researching complex subjects (Creswell, 2012).

---

[5]   This includes Google Scholar, Science Direct, Scopus, Wiley Online, and the online library search functions of the University of Amsterdam and IHE Institute for Water Education.

Selection of the case

The problem of poor access to infrastructure like sanitation services that often characterise developing countries makes it important to investigate the operationalization of the HRS in such settings. Urban settlements in many developing countries also embody spatial, individual, and group inequities in access to sanitation, despite interventions by states and non-state actors (World Health Organization [WHO] & UNICEF, 2015). I therefore selected Nigeria, a developing country, for my case study, to illustrate the evolution and performance of the HRS. Within Nigeria, there also exist a variety of principles and instruments for sanitation governance which offers a good case for analysing the effects of pluralism (see 2.4.2) in sanitation governance. Furthermore, there is an increasing tendency towards liberalisation in the current sanitation reforms in various states across Nigeria which is actively shaping the trajectory of the sanitation sector and influencing the policy arena for the implementation of the HRS as a predominantly social and relational norm. A second element of my case study was the household surveys I conducted within Benin City, Nigeria. I chose Benin City partly because its sanitation governance institution is representative of the sanitation governance architecture in many cities across Nigeria. Additional factors which I considered in selecting my case study include my understanding of the official and local languages, personal affiliations, and prospects of accessing the necessary information for my research.

Steps followed in conducting the case study

I followed three preliminary steps to design my case study. First, I assessed whether a case study approach was appropriate. Next, I selected an appropriate case, Nigeria. I then developed my research design based on a desk study, before proceeding to data collection from the field. In selecting my case study, I was careful to choose a jurisdiction that offered a wide range of drivers of poor sanitation services to assess the performance of the HRS. Although Nigeria is only one among many countries with poor access to sanitation services, I selected it for my case study for three main reasons, namely: (a) Nigeria ranks low in the human development index and therefore offers a good case to investigate many of the challenges which low and middle income countries may face in implementing the HRS; (b) is experiencing rapid urbanisation and humanitarian crises (resulting from insurgency, flooding, drought, and ethnic conflicts); and (c) failed to meet the MDGs target for sanitation, even though it met the target for water (see 8.2.3).

Having selected my case study, I proceeded to conduct a desk study to ascertain the legal framework for the HRS in Nigeria. Many of the national and local laws on sanitation in Nigeria are not digitally available and I had to travel to various parts of Nigeria in order to access hard copy materials for my analysis. I also analysed current state of knowledge on the HRS in Nigeria through desk study. Most of the available literature was grey-literature on sanitation governance instruments published by NGOs like WaterAid, international financial institutions like the World Bank, and project reports by regulatory agencies like the Ministry of Water Resources. I reviewed the academic literature on the HRS written by lawyers and other disciplines (see 1.3). The available peer-reviewed academic literature covering Nigeria provided useful insights on waste management practices and challenges but did not address the implementation/performance of the HRS in Nigeria and the effects on ID. It was therefore also important to embark on a field study to be able to address the research questions (see 1.4) in the context of the case study.

It was necessary for me to conduct expert and stakeholder and expert interviews because the information on the HRS in Nigeria is not readily available in the literature. Information was collected through a number of interviews with government officials, political leaders, NGOs, communities, and other relevant parties. I conducted 47 semi-structured interviews with water, sanitation and hygiene (WASH) and governance experts, government officials, NGOs, and the media, at different levels of governance. Further details of the interviewees (excluding names), including their professional backgrounds and the level of governance country of residence are coded in Annex G. I identified my interviewees using a combination of purposive sampling and snowball sampling to reduce the potential for bias in purposive sampling and the risk of omitting key interviewees by solely relying on snowball sampling. The majority of my interviewees (n=41) were selected through purposive sampling, and I used work experience in the sanitation sector, expertise in sanitation governance and policy processes for instance based on academic publications, and gender, disciplinary, and geographic diversity as my selection criteria. The additional interviewees (n=6) were selected based on referrals from other interviewees. I conducted the interviews in three phases from 2013 until 2014. The interviews lasted between 1 and 2 hours each, covering questions pertaining to: a) the impacts of HRS norms and instruments, b) the drivers of lack of universal access, and c) human rights and non-human rights instruments that can potentially address the identified drivers.

At the local level, I selected 450 households in Benin City through random area sampling. The selected households were administered questionnaires containing open and close ended questions about the definition and status of access to sanitation, knowledge of good sanitation and hygiene practices, the quality of services based on HRS standards, the instruments for sanitation governance, and the drivers of poor sanitation services (see 3.4). Of all the questionnaires administered, a total of 254 questionnaires were retrieved and utilised in this research. The background information on the respondents is presented in Annex H. I also validated the data from the household survey through my personal observation of sanitation governance in Benin City and three other cities across Nigeria: Port Harcourt, Abuja, and Lagos. It was not possible to administer a significant number of questionnaires to households in humanitarian situations and informal settlements because most declined to participate in the survey due to physical safety, psychological, and legal concerns (especially in the informal settlements), as well as other personal reasons that they may not have disclosed. Nonetheless, I interviewed humanitarian actors, local leaders, experts and other stakeholders for informal settlements. I also visited 2 camps for internally displaced persons fleeing from the Boko Haram insurgency in the North Eastern part of Nigeria (see 8.2.2). Data from the expert and stakeholder interviews and household survey were analysed within my conceptual framework and triangulated with the result of the literature review and content analysis. This provided some insight into gaps in knowledge about HRS performance in Nigeria and also some generalizable results that were useful in developing my final recommendations and conclusions.

2.3.3 Content Analysis

I analysed the content of laws and policy documents to identify: (a) contestations over the meaning and economic character of sanitation across multiple levels of governance, (b) the drivers of poor sanitation services, (c) the principles of the HRS and how they are operationalised within the HRS institution through instruments, (d) other non-human rights principles and instruments for sanitation governance, (e) the operation of legal pluralism, and (f) the components of inclusive development.

Content analysis comprises the set of procedures used for collecting and organising information obtained from any sign vehicle (any form of communication that carries a meaning) in a standardised format (Krippendorff, 2013). This allowed me to make inferences about the characteristics and meaning of the information contained in any sign-vehicle including recorded materials, and supports replication (Stemler, 2001; United States General Accounting Office [US GAO], Program Evaluation and Methodology Division, 1996). Content analysis is also suitable for determining the major trends in a society as well as to observe any changes in the trend or in the meaning of concepts as they are used in different texts (Aburdene, 2007; Carley, 1993; McLellan & Porter, 2007; Naisbett & Aburdene, 1990). In my thesis, it showed how sanitation is defined and how sanitation governance principles have emerged from various levels of governance and are operationalised through instruments, the underlying social factors which have shaped the evolution process, and trends for future evolution.

There are two main methodological limitations of content analysis; the process of coding and counting is not appropriate for: (a) identifying latent contents, (Altheide & Schneider, 2013) and (b) distinguishing between legal opinions, sources, or other variables that carry unequal weight (Babbie, 2012). The first limitation does not pose a challenge because my research is primarily concerned with the main principles of sanitation governance and not latent content. I avoid the second limitation by expressly applying the traditional methods/tactics of legal reasoning and argumentation (see 2.3.4) during codification.

I followed five main steps to conduct my content analysis. First, I defined the recording units for my analysis based on my literature review. This resulted in: (a) three categories of drivers (economic, social, and environmental); (b) three categories of principles based on the three main components of ID (economic/relational, social, and ecological); and five categories of instruments (regulatory, economic, management, and suasive, including technologies) that operationalize the principles. The drivers, principles, and instruments contained in the literature were then coded accordingly. Next, I identified and coded additional principles and instruments contained in the laws, policy documents and governance texts at multiple levels of governance, from international to local, also taking into consideration the legal status (binding, persuasive, or non-binding) of each of the documents. This resulted in an additional category of instruments: administrative instruments. I then updated my coding scheme accordingly and conducted additional rounds of review to ensure that all the drivers, principles, and instruments covered in the literature and/or laws, policy documents and governance texts had been captured and coded. The information was then ploughed back into my conceptual framework for further analysis.

2.3.4 Legal Reasoning and Argumentation

I further analysed my laws, policy documents and governance texts through the application of legal reasoning and argumentation methods (see Table 2.1). The legal reasoning process involved reasoning by analogies, applying the doctrine of precedents, and inductive and deductive reasoning (Cook 2001; Farrar and Dugdale 1990). It enabled me to identify and compare the legal status of each of the laws, policy documents and governance texts that I analysed. I then applied an intuitive legal argumentation process, based on my training as a lawyer to methodically deploy arguments built on legal rules in order to resolve the legal angles to the theoretical problems (Hanson 2003).

For instance, in order to analyse the legal basis for the HRS at various levels of governance, I first relied on the literal meaning of the words contained in the document I was analysing. Where this led to absurdity, I proceeded to interpret the provisions in such a way as would promote the intentions of the legislature (golden rule of statutory interpretation)[6] by relying on both internal aids of statutory interpretation like the title of a statute,[7] preamble,[8] headings,[9] marginal notes,[10] interpretation sections,[11] and provisos;[12] and external aids like the Interpretation Act or other equivalent law.[13] I also crosschecked with the relevant literature to ensure reliability.

---

[6]   Becke v Smith (1836) 150 ER 724 at 726; Coutts & Co. v IRC (1953) 2 WLR 364 at 368; Awolowo v Federal Minister of Internal Affairs (1962) LLR 177.
[7]   Okeke v A.G. Anambra State (1992) 1 NWLR Pt. 215, p. 164; Agbara v Amara (1995) 7 NWLR Pt. 410, p. 712.
[8]   I.R.C. v Pemsel (1891) AC 531 at 542 per Lord Halsbury; Opeola v Opadiran (1994) 5 NWLR Pt. 344, p.368 SC; Okeke v A.G. Anambra State (1992) 1 NWLR Pt. 215, p. 164; Osawe v Registrar of Trade Unions (1985) 1 NWLR Pt. 4, p. 755 SC.
[9]   Lawal-Osula v Lawal-Osula (1993) 2 NWLR Pt. 274, p. 158 CA; Ondo State v Folayan (1994) 7 NWLR Pt. 354, p.1 SC.
[10] Yabugbe v COP (1992) 4 NWLR Pt. 234, p.152 at 171; Oloyo v Alegbe (1983) 2 SCNLR 35 at 37; UTC Nig Ltd v Pamotei (1989) 2 NWLR Pt. 103, p.244 SC.
[11] Board of Custom & Excise v Barau (1982) 10 SC 48; Utih v Onoyuvwe (1991) 1 NWLR Pt. 166 SC; Ejoh v IGP (1963) All NLR 248.
[12] Abasi v State (1992) 8 NWLR Pt 260, p. 383; Nabham v Nabham (1967) All NLR 51; NIPC Ltd v Bank of West Africa (1962) 1 All NLR 551.
[13] Ibrahim v JSC (1998) 14 NWLR Pt. 584, p. 1SC; Osadebay v A.G. Bendel State (1991) 1 NWLR Pt 169, p. 525 SC.

Table 2.1 Steps and some data sources for legal reasoning and argumentation

Steps	**Some data sources**	**Legal reasoning tactic**	**Contribution to the research**
1. Analyse key international law sources	International human rights treaties with provisions on sanitation; General Comments & Resolutions of UN bodies	Apply the Rome Statute of the International Criminal Court 1998 (to analyse sources of international law) and Vienna Convention on the Law of Treaties 1969 (for treaty interpretation)	Determine the legal status and normative content of the HRS in international law
2. Analyse national human rights institutions, laws and policy provisions	WaterLex publication on National Human Rights Institutions; law databases	Various rules of statutory interpretation; external & internal interpretation aids	Understand the evolution of HRS in national laws and policies
3. Analyse constitutional provisions	National Constitutions	Investigate constitutional convergence	Understand the evolution of HRS in national constitutions
4. Analyse court decisions	WaterLex publication on court cases worldwide involving the human right to water and sanitation	Judicial precedents; analogies; inductive and deductive reasoning	Understand the evolution of the HRS principles in court cases

Applying the literal rule allowed me identify explicit sources of the HRS for further analysis. In the absence of express recognition of HRS, applying the golden rule also highlighted implicit recognition of the HRS in various sources of law, and the principles and instruments for sanitation governance contained in laws, policy documents, and other governance texts. Other rules that I applied for my analysis include the *ejusdem generis* rule, that where particular words are followed by general words, the general words are to be limited to the same kind or class as the particular words;[14] *noscitur a sociis* to the effect that where the meaning of a word is doubtful it was interpreted with reference to the context in which it appeared.[15] I also had to read the whole provisions as a further check to ensure that my interpretation was consistent with the objective(s) of the document when considered as a whole.[16]

---

[14] Tillmans & Co. v S.S. Knustford (1908) 2 QB 385; Nasr v Bouari (1969) 1 All LNR 35; A.G. Cross River State v Esin (1991) 6 NWLR Pt. 197, p. 365 CA.

[15] F.C.S.C. v Laoye (1989) 2 NWLR Pt 106, p. 652 SC; Okeke v AG Anambra State (1992) 1 NWLR Pt. 215, p. 164.

[16] A.G. Bendel State v A. G. Federation & 22 Others (1982) 3 NCLR 1 SC; Ishola v Ajiboye (1994) 6 NWLR Pt. 352, p. 506 SC; Fasakin v Fasakin (1994) 4 NWLR Pt. 304, p. 597 SC.

2.4 MULTI-LEVEL INSTITUTIONAL ANALYSIS, LEGAL PLURALISM AND INCLUSIVE DEVELOPMENT

This section discusses the theoretical basis for my analysis, including: (a) multi-level governance theory (see 2.4.1), legal pluralism (see 2.4.2), and inclusive development (see 2.4.3).

2.4.1 Multi-Level Governance of the Human Right to Sanitation

The HRS institution and indeed sanitation governance generally is composed of norms, principles, and instruments that operate at multiple levels of governance. In order to understand the performance of the HRS at different levels of governance, and interactions with non-HRS norms/principles for sanitation governance given pluralism, I have drawn on Multi-Level Governance (MLG) theory. MLG essentially refers to an arrangement for making decisions which engages multiple interdependent but politically independent actors, both public and private, "at different levels of territorial aggregation in more-or-less continuous negotiation/deliberation/ implementation, and that does not assign exclusive policy competence or assert a stable hierarchy of political authority to any of these levels" (Schmitter, 2004, p.49).

The theory was first developed within the European Union (Marks, 1992), and is useful when analysing government processes and the changes in decision-making processes in cases of declining state sovereignty (Piattoni, 2010). Nonetheless, as State sovereignty and centralisation of government become increasingly outmoded globally, especially in the context of supra-nationalism, due to new challenges that traverse administrative boundaries, and because of the new emerging concepts of legitimacy, there are also new forms of solidarity emerging to replace traditional models such as state-society and state-solidarity, as well as new connections emerging between the different actors at different levels (Piattoni, 2010). As a result, multi-level governance has become increasingly prevalent, also outside the European Union where it first emerged, as a concept for analysis of government processes and reflection on the changes in decision-making processes in many cases where state sovereignty is diminishing (Piattoni, 2010). This is particularly relevant for my thesis because state sovereignty and centralisation of government are increasingly becoming outmoded in addressing development challenges like poor sanitation; the result is an emergence of new forms of solidarity and new connections between actors at different levels (Piattoni, 2010).

The literature distinguishes MLG Type I (focused on direct links between the different levels of governance) from MLG Type II (focused on multiple links between the levels (Piattoni, 2010). For my research question, it is more relevant to understand the direct links between MLG in operationalizing the international HRS and its performance in addressing the sanitation challenge across multiple levels of governance. I therefore focus on the direct links between multiple levels of governance: international, regional, national, and sub-national; that is, Type 1 MLG theory. The structure of international human rights has similar weaknesses as MLG, particularly: lack of coordination between different policies and sub-systems, and the challenges of legitimating and ensuring compliance (Enderlein, Wälti & Zürn, 2010). The application of MLG theory therefore enables analyses of the HRS governance framework, contributing to input legitimacy (political equality) and output legitimacy (policy implementation). Although the analyses conducted were primarily qualitative, this was complemented with quantitative data from official reports and the case study. The outcome is integrated in the recommendations for redesign of HRS instruments where necessary, to avoid the weaknesses of global multi-level governance in the policy implementation process (see Chapter 9).

2.4.2 Legal Pluralism in Human Right to Sanitation Governance

Given that the HRS operates alongside other HR, as well as non-HR norms for sanitation governance at different levels, I also draw on legal pluralism theory for my analysis. A legal rule is an oral or written statement that guides human conduct, and the infringement of which may result in compulsory or discretionary action being taken to enforce its observance (Hanson, 2003). Rules often contain value statements and may be prescriptive, normative, or facilitative where they stop, guide, or allow certain actions, respectively (Hanson, 2003). Generally, legal rules are issued through formal State authorised procedures and enforced through the appropriate formal institutions in specific contexts. Nonetheless, there are situations in which multiple sets of rules become applicable to the same actors or in the same jurisdiction, either as a result of historical evolution (Farran, 2006), or a shift from government to governance (Gupta & Bavinck, 2014) leading to legal pluralism.

Legal pluralism is defined as "the co-existence *de jure* or *de facto* of different normative legal orders within the same geographical and temporal space" (Quane, 2013, p.676). Legal pluralism could be *de jure*, where co-existing multiple legal orders and their linkages are formally recognised by the State, in order to enhance the efficiency and effectiveness of the formal justice system; (Quane, 2013) or *de facto*, where the State does not recognize non-state or informal legal orders, but may implicitly allow their operation (Quane, 2013; McGoldrick, 2009). The literature investigates legal pluralism as the outcome of: (a) the historical evolution of legal systems (Farran, 2006); (b) interactions between traditional systems and other legal systems (Benda-Beckmann, 2001) within any given state (International Council on Human Rights Policy [ICHRP], 2009a), or jurisdiction (Tamanaha, 2011; Twining, 2009); (c) fragmentation of the international legal order (Koskenniemi & Leino, 2002); sometimes distinguishing between legal pluralism occurring at the horizontal level (multiple rules operating within the same level of governance) and the vertical level (multiple rules operating across different levels of governance (Conti & Gupta, 2014; Gupta & Bavinck, 2014; Obani & Gupta, 2014b).

There is an increasing scholarly focus on legal pluralism in surface (Prakash & Ballabh, 2005), groundwater governance (Conti & Gupta, 2014), and the human right to water and sanitation (Obani & Gupta 2014b). In the area of human rights especially, the literature links legal pluralism to principles of equity and justice in the operation of formal and informal legal orders in the sense that legal pluralism "may reflect a pragmatic response to resource or other constraints that are perceived to impede a population's right of access to justice" (Quane, 2013, p.677). The legal literature further analyses interactions between international human rights standards and other formal and informal legal orders (ICHRP, 2009a; Nelson, 2010); the role of States in managing the outcomes of such interactions and the importance of protecting individual rights in the process (Quane, 2013; Nelson, 2010), and specific tools for addressing the challenges of pluralism in implementing the human right to water and sanitation (Obani & Gupta, 2014b).

In my thesis, I apply legal pluralism to analyse: (a) the different definitions of HRS in laws and policies, (b) the different principles for sanitation governance within the HRS and other non-HR governance frameworks, and (c) the quality of the relationships and the intensity of the interactions between the HRS and other existing legal rules, both across multiple levels of governance (vertical pluralism) and within my case study (horizontal pluralism) (see 6.6.2, 7.4.2 and 8.6.2). I do this using four heuristic types (competition, indifference, accommodation, and mutual support), as developed by Bavinck and Gupta (2014) and applied by Obani and Gupta (2014b). Competition occurs where there are strong tensions between the HRS and other rules for sanitation governance in a given context. Indifference exists where the various rules, including the HRS, are technically applicable within a given context but there is no overlap in their operation in practice. Accommodation highlights the recognition of the HRS alongside other rules for sanitation governance and some degree of coadaptation between the various rules, but minimal institutional integration. Mutual support reiterates a strong positive relationship between the HRS and other rules for sanitation governance in any given context.

2.4.3 Inclusive Development as the Overarching Norm for Evaluating the Human Right to Sanitation

A fundamental reason behind the HRS is to provide sanitation services to the most vulnerable and poor in order to enhance their ability to live their lives in dignity, and to participate in development processes. Scholars further propose that the HRS can address environmental sustainability simultaneously (for instance, Feris, 2015) and that addressing the challenges of the poor inevitably requires renegotiating the political context (for instance, McGranahan, 2013). Since the concept of inclusive development (ID) calls for social, ecological and relational inclusiveness (see Gupta, Pouw & Ros-Tonen, 2015), I adopt ID as an overarching or guiding norm for the HRS in my thesis. For the purpose of clarity, I use norms in my thesis to include principles, values or ideals that dictate the appropriate standards of behaviour within an institution or any governance structure (Finnemore & Sikkink, 1998).

The evolution of the concept of development is traceable to the 1940s when development economists were preoccupied with post-war recovery and State intervention aimed to balance the goal of industrialization and full employment without compromising on economic productivity, in other words balancing competing social and economic goals. In the 1960's the need to ensure that all citizens could enjoy some basic aspects in democratic societies led to the adoption of economic, cultural, and social rights and political and civic rights, respectively, in the International Covenant on Economic, Social and Cultural Rights 1966 and the International Covenant on Civil and Political Rights, 1966.

By the 1970s, environmental issues also gained prominence in development discourse which led to the emergence of the concept of sustainable development, focused on reconciling social and economic development and environmental protection, for the benefit of both the present and future generations (Schrijver, 2008).[17] Subsequent discourses however tend to focus on a

---

[17] Gabčíkovo-Nagymaros Project (Hungary v. Slovakia), Judgment, ICJ Reports 1997, pp. 7–84.

combination of the social, economic, and environmental pillars of sustainable development rather than integrating all three.

On the one hand, discourses and theories on social movements (Escobar, 2012), participatory development (Cornwall & Scoones, 2011), and the entitlements, capabilities and freedom (Sen, 2011) mainly focus on social and economic equity. On the other hand, other scholars focus on planetary boundaries (Rockström et al., 2009; Rockström et al., 2009; Steffen et al., 2015), investigations of safe and just ways of achieving human development within the planetary boundaries (Raworth, 2012) focus on environmental issues. These have been merged in the 2015 adoption of the universal Sustainable Development Goals, including those reflected in the United Nations Conference on Environment and Development Agenda 21 (United Nations, 1992), and aim to address both social floors and planetary boundaries.

In particular, since 2010, there has also been a return of the growth and employment discourse through theories like green economy and inclusive growth. Although inclusive growth is easily misunderstood as entailing inclusive development, the former however mainly focuses on economic growth and equal opportunities for the poor (Abosede & Onakoya, 2013; Commission on Growth and Development, 2008; Rauniyar & Kanbur 2010a), but does not fully integrate environmental sustainability which is a key aspect of this thesis. Although ID is also similar to sustainable development, the operationalization of the latter often prioritises economic growth and compromises on social and ecological sustainability and often does not address the politics of inequality or ecological exclusion (Gupta et al., 2015; Gupta & Vegelin, 2016; cf. Voigt, 2009). The Sustainable Development Goals (SDGs) has an inclusive development focus that is inspired by human rights, but it also fails to fully integrate the social, economic and ecological dimensions of development (Arts, 2017). Hence, this thesis employs the inclusive development approach instead but also draws on sustainable development. ID is however less covered in the literature; exceptions to this include Arts, 2017; Bos & Gupta, 2016; Gupta et al., 2015; Gupta & Vegelin, 2016; Rauniyar & Kanbur 2010a, 2010b.

Definition and components

ID was first introduced as a strategy for promoting human equity and empowerment and it has been elaborated to encompass social, economic, and environmental equity and empowerment (Gupta et al., 2015; Rauniyar & Kanbur, 2010a). It is "development that includes marginalized people, sectors and countries in social, political and economic processes for increased human well-being, social and environmental sustainability, and empowerment" (Gupta et al., 2015, p. 546). It has three main components, namely: inclusiveness *per sé* or social inclusiveness, inclusiveness as a relational concept or relational inclusiveness, and ecological inclusiveness (Gupta et al., 2015; Gupta & Vegelin, 2016). Social inclusiveness promotes overall well-being, particularly for the poorest, most marginalised, and most vulnerable people, through capacity building (Fritz, Miller, Gude, Pruisken & Rischewski., 2009), and institutionalising equitable principles, participatory approaches including the use of local knowledge, and non-discrimination (Lawson, 2010). Inclusiveness as a relational concept tackles the direct and underlying drivers of inequitable development (Laven, 2010), including through ensuring the affordability of basic services (Arthurson, 2002). Finally, environmental inclusiveness promotes development within the earth's carrying capacity and improved human resilience (Crutzen & Brauch, 2016). Though the neo-classical concept of efficiency suggests that the efficient allocation of natural resources would result in sustainable development, the realities show that the uncertainties of time preference and the strong bias in favour of economic growth and the substitution of natural capital, using technology and innovation for instance, results in exclusion and inequities in development (Basu & Shankar, 2015). I therefore adapt the three components of ID in my analytical framework to minimise trade-offs between them.

Inclusive development assessment framework

To operationalize ID as the overarching norm for my thesis, I develop a simple framework as an assessment tool for evaluating the performance of the HRS instruments/principles against social and relational, and ecological inclusion (see Figure 2.1). A variety of assessment frameworks exist for analysing urban water and wastewater systems for sustainable development, including non-integrated analysis (Hanley & Spash, 1993; Jeppsson & Hellström, 2002; Rebitzer, Hunkeler & Jolliet, 2003; Tillman, Svingby & Lundström, 1998), and integrated multi-criteria decision making and participatory approaches (Motevallian & Tabesh, 2011), and statistical frameworks (Sahely, Kennedy & Adams, 2005) that are aimed at supporting decision making and policy development. More recently, diverse sustainability assessment approaches have been synthesised based on critical decision-making elements as determined by domain experts, with clear indications of how to move from integrated assessment to sustainability assessment and the practical application for decision making in various contexts (Sala, Ciuffo & Nijkamp, 2015). The foregoing sustainability assessments are generally suitable for assessing the entire system for water and wastewater management but inappropriate for evaluating specific instruments in isolation of others; this does not fit in with the theme of my thesis. As a result of this, I had to develop a simple ideal typical framework to enable me to evaluate the impacts of HRS instruments at various levels of governance through four quadrants that are based on the social, relational, and ecological components of ID (see Quadrants 1-4). The x axis ranges from ecological exclusiveness (that hampers environmental sustainability) to ecological inclusiveness (that promotes environmental sustainability), and the y axis ranges from social and relational inclusiveness to social and relational exclusiveness. The fundamental argument is that the implementation of HRS through different instruments will take place differently in different contexts leading to different outcomes.

Instruments, including principles, falling into Quadrant 1 (Q1) potentially aggravate social, relational and ecological exclusion; Quadrant 2 (Q2) hamper social and relational inclusion but promote ecological inclusion; Quadrant 3 (Q3) only promote social and relational inclusion without addressing environmental sustainability concerns; those falling into Quadrant 4 (Q4) offer the ideal integration of social, relational and ecological inclusiveness for ID (see Figure 2.1).

The proposed framework is easily adaptable to any choice of sanitation governance instruments, including human rights instruments (see 4.5), humanitarian instruments (see 5.5) and other non-human rights instruments (see 7.5), and across multiple levels of governance, from the international to sub-national levels (see 8.5.6). The results from the framework can be integrated into existing sustainability assessments that address water and wastewater systems. It further has the potential to enrich sustainability assessments by balancing competing social, relational and ecological inclusion in the evaluation of instruments' performance and can be adapted to other development issues beyond water and sanitation. The outcome of the analysis, however, largely depends on how each instrument is formulated and adapted to the prevailing drivers in any given context, and may therefore vary for the same instrument depending on the context and drivers.

Application

I rely on the three components of ID (social, relational and ecological inclusion) to test the performance of HRS instruments at multiple levels of governance (MLG) in Chapters 5, 6, 7 and 8 of my thesis. To operationalize this, I merge social and relational inclusion along the y-axis of my ID assessment framework with two indicators for both components, namely: (i) public participation, defined as improvements in mechanisms for the effective participation for marginalised individuals and groups such as residents of informal settlements and (ii) equitable pricing of services, defined as pricing mechanisms that create a positive incentive for the adoption of sustainable sanitation and hygiene practices while also protecting access to adequate sanitation for the poor and people living in humanitarian situations. Further, I position environmental inclusion along the x-axis of my ID assessment framework with one indicator: (i) sustainable sanitation system, defined as a combination of sanitation infrastructure and services which ensures the safe management of all waste streams with minimal negative impacts on the ecosystem and vulnerable groups.

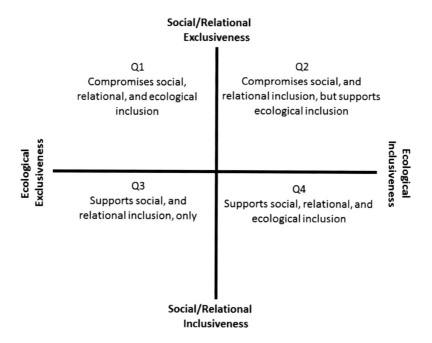

Figure 2.1 Inclusive development assessment framework (Adapted from Obani & Gupta, 2017)

2.5   CONCEPTUAL FRAMEWORK AND INTEGRATED ANALYSIS

This section explains my conceptual framework (see 2.5.1), and how I conducted my integrated analysis (see 2.5.2).

2.5.1 Conceptual Framework

The terms conceptual framework and theoretical framework have been used interchangeably in the literature, while some authors rely on one of both terms (Green, 2014; Maxwell, 2012). One perspective is that theoretical framework is the appropriate term when research is based on a single theory, while conceptual framework applies when research is guided by different theories or findings (Parahoo, 2006). A second slightly different perspective is that a conceptual framework consists of different concepts built into a theory, and where the framework is based on concepts it is a conceptual framework; if it is based on theories, it is a theoretical framework (Fain, 2017). Despite the confusion over the meaning of conceptual framework *vis-à-vis* theoretical framework, a conceptual framework is a useful guide that identifies the researcher's world view of the research topic and defines their assumptions and preconceptions (Lacey, 2010). Having already identified the theories I apply in my research (see 2.4), I now illustrate the conceptual framework for my research in order to explain the key concepts I use and how they relate to each other.

As mentioned earlier, I am adopting an institutional approach to evaluating the HRS. Institutions are "sets of rules, decision-making procedures, and programs that define social practices, assign roles to the participants in these practices, and guide interactions among the occupants of individual roles" (Young 2002, p. 5). To achieve my institutional analysis of the HRS, I adapt the framework for the institutional analysis of environmental regimes that was developed by the International Human Dimensions Programme's Project on the Institutional Dimensions of Global Environmental Change (IDGEC) (Young, 2002; IDGEC, 2005).

The IDGEC framework was developed to understand when and why specific instruments are able to change the behaviour of specific actors, given the context and drivers that influence a problem, and therefore helps to understand how these instruments can be improved. It consists of six elements: (a) institutions, (b) instruments flowing from the institutions, (c) the impact of instruments on actor's behaviour given (d) institutional drivers, (e) the impact of changed behaviour on the resource performance and social wellbeing, and (f) instrument and institutional redesign. In my thesis, adapting the framework enables me to: (i) understand the workings of HRS institutions and instruments at multiple levels of governance; (ii) evaluate the effect of the HRS instruments on actors given the drivers of poor sanitation services; and (iii) propose a redesign of HRS instruments based on an analysis of which instruments work and which do not work in achieving ID in the context of my case study (see Chapter 9). The multi-level institutional analysis provides useful insights into areas of convergence and divergence in the implementation of HRS across various levels of governance.

In adapting the IDGEC framework, I first identify the laws, policies, programmes or plans that are identified in the literature as part of HRS institutions. Next, in the theoretical chapters of my thesis (3-7) I analyse the principles, instruments and indicators which emerge from my content analysis and literature review and weigh the performance of instruments against the overarching norm of ID. This lays the background for analysing the impact of the instruments on human behaviour in my case study, given the drivers of poor sanitation services, and the outcomes on ID. My framework maintains six elements, similar to the IDGEC's: (a) institutions, (b) instruments, (c) actors, (d) drivers, (e) institutional performance, and (f) redesign.

First, institutions cover the HRS institutions and I apply Type 1 MLG in my analysis. Second, instruments cover both HRS instruments and complementary instruments from other fields like humanitarian law *inter alia*. Third, I identify the drivers of poor sanitation based on the literature (see 3.4) and my case study (see 8.2.4). Fourth, I analyse whether the existing sanitation governance instruments address the state of poor sanitation services being experienced by the various actors, given the drivers. By evaluating the performance of HRS instruments I indirectly also address institutional challenges in the realization of the HRS. Fifth, in analysing institutional performance I use both legal pluralism (to evaluate the relationship between HRS and other discourses/frameworks for sanitation governance) and my inclusive development assessment framework (to assess the impact of HRS on ID). Sixth, based on my assessment of which instruments work and which do not work in my case study, I suggest redesign. The redesign element of my conceptual framework aims to improve coherence between the HRS and ID.

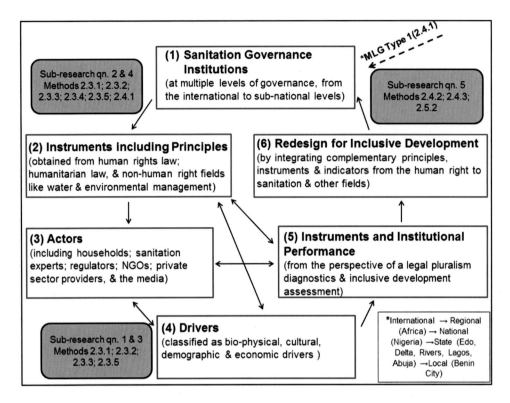

Figure 2.2 Conceptual framework outlining sub-research questions, theories and data collection methods

Instruments

In my conceptual framework, I classify various "techniques of governance, which, one way or another, involve the utilization of state resources, or their conscious limitation, in order to achieve policy goals" (Howlett & Rayner 2007, p.2) as instruments (see Box 2 in Figure 2.2). Instruments are used by government authorities "in attempting to ensure support and effect or prevent social change" (Vedung 1998, p.21). Majoor and Schwartz (2015) also argue that instruments can be employed by private non-State actors as well, and that in urban settings both private and public governance instruments interact and governance instruments operationalize public policy objectives. The functionalist or instrumentalist literature view instruments as neutral, rational, pragmatic, and technical. Conversely, political sociologists stress that instruments are neither rational nor purely technical but instead reflect the balance of power (Kassim & Le Galès, 2010), the relationship between the government and the citizen, and notions of social control and how to affect it (Kassim & Le Galès, 2010; Lascoumes & Le Galès, 2007; Majoor & Schwartz, 2015).

Instruments can be categorized based on: (a) the mechanisms or resources which the instruments apply to, (b) the purpose of the instruments, (c) the mode of application, or (d) the impact of the instruments (Majoor & Schwartz, 2015). I categorise instruments in my thesis on the basis of the mechanisms which the instruments apply because this allows me to easily integrate the wide variety of instruments that are applied by human rights practitioners and other sanitation experts in my analysis. The classification also shows that actors from different fields also tend to apply instruments that fall within the same categorisation and therefore shows further opportunities for resolving any tensions in the operations of the instruments and improving complementarity and the positive effects of pluralism.

On the basis of the mechanisms employed, I identify five categories of instruments: (a) regulatory, including administrative rules and binding policies, (b) economic, (c) suasive, (d) management, and (e) technological instruments. Economic instruments may also be part of regulatory instruments, which complicates differentiating between these two types of instruments. Regulatory instruments either compel or prohibit certain types of behaviours and include: (i) laws, (ii) administrative rules, (iii) policies, (iv) technical standards, (v) permits and licensing procedures, and (vi) liability systems like threat of prosecution to recover punitive damages.

Economic instruments induce behavioural changes through market signals, or financial incentives or disincentives and include: (i) property rights governing ownership and use of goods and services, (ii) fiscal instruments like taxes, (iii) market creation instruments like tradable quotas, (iv) charge systems like tariffs and connection fees, (v) deposit refund systems, and (vi) financial instruments like subsidies and grants (Rivera, 2002; Vedung, 1998).

Suasive instruments influence behaviours through persuasive mechanisms like: (i) education and awareness trainings, (ii) award schemes, (iii) product certification and labelling, (iv) corporate reporting requirements, and (v) disclosure requirements (Vedung, 1998).

Management instruments generally involve self-regulation and voluntary management processes by private actors (Rivera, 2002). I adapt this classification of management instruments to also include participatory instruments which directly involve the potential users in the production, operation or maintenance of sanitation goods and services, like Community-Led Total Sanitation (CLTS).

Further, the delivery of sanitation services requires some form of physical infrastructure/technology like toilets, sewers, and sludge management systems. Therefore, in Chapter 3, I analyse the technologies for improved sanitation (see 3.5). Nonetheless, technological instruments are often necessary for implementing other types of instruments for sanitation governance that I analyse in Chapters 5, 6, 7, and 8. For instance, an economic instrument like progressive pricing (see 5.5.2) would require appropriate technology for measuring the amount of services actually used.

The literature further distinguishes between the strengths and weaknesses of each of these instruments (Majoor & Schwartz, 2015) and links the capacity of authors and ability to use a mix of instruments (Howlett, 2000). Building on this, I investigate the variety of HRS and non-HRS instruments that are employed by both States and private/non-State actors for sanitation governance, as they affect the realisation of the HRS. I consider the instruments to be value laden and I therefore further investigate the underlying principles in each of the instruments and link these with the three components of ID.

Principles

The legal literature distinguishes principles from legal rules. Legal rules are legal norms to the extent that they are definitive commands which apply to defined circumstances or relationships and do not offer a general standard of judgement (Joaquín & Muñiz, 1997). Legal principles, on the other hand, are norms with general application irrespective of specific legal facts or circumstances and command the utmost compliance possible (Alexy, 2000). Hence, legal principles as basic norms reflect the general standard of judgement or basic understandings in any given society. Similarly, in the governance literature norms are defined as the "standard of appropriate behaviour for actors with a given identity" and it is the collection of norms, rules and practices that define the structure of governance institutions (Finnemore & Sikkink, 1998, p.891).

At the international level, a principle can have different meanings or connotations, including: (a) a source of international law,[18] (b) any relevant rule of international law that is to be taken into account in the relations between parties,[19] and (c) "a part of the evolving rules of international law" (de Sadeleer, 2002, p.237). And at the national level, some national constitutions like India's and Nigeria's distinguish between fundamental rights that are

---

[18] Statute of the International Court of Justice of 1945, article 38(1)(c).
[19] Vienna Convention of the Law of Treaties of 1969, article 31(3)(c).

protected through legal remedies, and Directive Principles of State Policy (which often include provisions relating to adequate standard of living, health, education, environment and other similar subjects that ordinarily fall within the scope of the International Covenant on Economic, Social and Cultural Rights of 1966) that are not enforceable in court but are nevertheless fundamental to policy making and governance.

In my analysis, I consider principles as the norms with general application which are contained in various sources of law (see 4.4) and which are operationalised through governance instruments. I therefore focus on the principles that are reflected in sanitation governance instruments and shape the structure of the institutions at multiple levels of governance. I also link each principle with the pillar of sustainable development which it is most closely related to (economic, environmental or social). Annex F expounds on the coding criteria I used for analysing the principles. Positioning principles as my unit of analysis serves two main purposes. First, it highlights the underlying plural values that influence the structure of sanitation governance institutions and thereby improves the quality of my analysis. Second, it indicates existing and potential areas of convergence and tensions in the instruments applied by various actors.

Nonetheless, I do not specifically focus on the principles for wastewater treatment or sanitation infrastructure as this falls outside the scope of my expertise and research focus and limits (von Sperling, 2007). Rather, I analyse the principles that underpin the regulatory, economic, suasive, and management instruments for sanitation governance, emerging from my literature review, content analysis and case study. I classify the resulting principles according to their sources into three main groups dealt with in separate chapters: (a) human rights principles derived from human rights law (see Chapters 4 and 5), (b) humanitarian principles derived from humanitarian law (see Chapter 6), and (c) non-human rights principles for sanitation governance which includes principles that do not directly emerge from either human rights or humanitarian law (see Chapter 7).

2.5.2 Integrated Analysis

As mentioned earlier, I used mixed methods to address my research questions within my conceptual framework. First, I applied systematic literature review and content analysis to identify the key elements of the HRS and sanitation governance institutions at different levels of governance, from international to local using the MLG Type 1 approach. This informed my analysis of institutions at multiple levels of governance, from the international to national and subnational levels during peacetime (see Chapter 5) and in humanitarian situations (see Chapter 6) (Box 1 in Figure 2.2). Further, I analysed: (a) the contestations over meaning (see 3.2, 4.2, 5.2, and 8.2), and (b) evolution of sanitation governance institutions (see 4.3, 5.3, 7.2, and 8.3).

I also used literature review and content analysis to identify the key principles, instruments, and indicators emerging from human rights, humanitarian law, and other non-human rights governance frameworks that are commonly applied for sanitation governance at multiple levels of governance (see Box 2 in Figure 2.2). In addition to the outcome of literature review and content analysis, I incorporated the opinions of experts, and other key actors working in the sanitation sector that I obtained through interviews. This enabled me to fill the gaps in the scholarly literature and integrate principles/instruments that exist *de jure*. From the foregoing, I compiled a comprehensive dataset of sanitation governance laws and principles (2.3.2) which I also linked to actors' behaviours (see Box 3 in Figure 2.2), particularly in the context of the case study (see Chapter 8).

Next, I analysed the interactions between the instruments (see Box 3 in Figure 2.2), drivers (see Box 4 in Figure 2.2), and actors' behaviours (see Box 3 in Figure 2.2). I also analysed the characteristics of the instruments/principles at multiple levels of governance, by adapting the four legal pluralism heuristic types and assessing their impacts on the three components of ID using the ID assessment framework discussed earlier (see 5.5, 6.5, 7.4, and 8.5). My case study (see Chapter 8) afforded me the opportunity to observe the operation of drivers at the national and sub-national levels. It also highlighted new drivers that are not captured in the scholarly literature and additional principles emerging from laws, policy documents, and the informal legal order. The foregoing formed the basis for my recommendations for redesign (see Chapter 9).

2.6 ETHICAL CONSIDERATIONS

For my PhD, I was affiliated with both the University of Amsterdam and IHE Institute for Water Education situated in Delft, Netherlands while my research was funded by NUFFIC's Netherlands Professional Fellowship (NFP). Further, I received funding for part of my field research from the Tertiary Education Trust Fund, Nigeria. In the course of my PhD research, I have presented part of my research findings at international conferences and always included my affiliations and funding sources in my presentations. I have also published part of my research findings in peer-reviewed papers/book chapters and I always provided details of my affiliations and funding sources. My interactions with sanitation and governance experts at the conferences I attended, as well as review comments on my papers/book chapters influenced my research design, particularly my choice of humanitarian situations as one of the three contexts for analysing the performance of the HRS. Analysing humanitarian situations eventually strengthened my analysis of the HRS and proposals for redesign by introducing new sanitation governance instruments and principles for ID.

My research design and all the steps were carried out with due consideration of the University of Amsterdam AISSR PhD Guide, including the ethical rules for research involving humans such as through obtaining prior informed consent from the respondents, maintaining confidentiality and anonymity for the respondents, and keeping the results of the interviews and surveys confidential.[20] Specifically, before the commencement of each interview and household survey, I provided each respondent with an official letter introducing me as a PhD researcher affiliated with the University of Amsterdam and the UNESCO-IHE Institute for Water Education, and informed them about the purpose of the research and their role as respondents. For the corporate bodies interviewed, I submitted my letter of introduction to the organisational heads and obtained their formal approval before scheduling an interview. Similarly, for individuals, I presented my letter of introduction in person and I obtained their verbal consent to participate in the interview/survey before administering my questions.

I did not use electronic recording devices because most interviewees declined while citing the sensitive nature of some of the information they were providing. Nonetheless, all the interviewees agreed to note taking. I continually assured each interviewee/respondent of anonymity and the confidentiality of the process and outcomes during the

---

[20] AISSR Ethical Procedure and Questions

interviews/administration of questionnaire. To ensure confidentiality and protect the respondents' identities, I have presented the outcomes in my thesis without any direct reference to the names of the individuals, households, or communities involved (see Annexes D and E).

Chapter 3. Contextualizing the Sanitation Problem

3.1 INTRODUCTION

Sanitation is a complex problem for empirical study. This chapter focuses on the secondary research question of: How does an understanding of the meaning and economic classification of sanitation, the drivers of poor domestic sanitation services, and the main technologies for domestic sanitation services affect the design of sanitation governance frameworks? I address the question through literature review and content analysis. Although the sanitation problem transcends households and permeates all spheres of life (Nygaard & Linder, 1997), this chapter focuses on personal and domestic sanitation and hygiene services as the unit of analysis. This approach is anchored on four main reasons: (a) this reflects the scope of the definition of sanitation in the development discourse (see 3.2) and the human rights framework (see 5.2.3); (b) although the human rights construct does not stipulate any economic model for sanitation service delivery, it imposes an obligation for affordability for domestic users which is affected by the economic classification of sanitation goods and services (see 3.3); (c) the human right to sanitation (HRS) is mainly focused on meeting personal and domestic sanitation needs and therefore needs to address the drivers of poor domestic sanitation services (see 3.4); (d) technologies for domestic sanitation services are not value neutral and may either hamper or improve access and therefore require further consideration (see 3.5). The chapter concludes with identifying linkages between sanitation technologies, the drivers of poor sanitation services, and inclusive development (ID) (see 3.6), and my inferences on the implications for realising the human right to sanitation (HRS) through a predominantly technocratic approach are presented in Section 3.7.

3.2 DEFINING SANITATION SERVICES

Although there was also a decade dedicated to improving access to better water and sanitation services (from 1981 to 1990), the global sanitation target was first introduced into the Millennium Development Goals (MDGs) at the World Summit on Sustainable Development (WSSD) in 2002, (Lenton, Wright & Lewis, 2005). There is no global legal definition of sanitation services. While the WSSD used the term 'basic sanitation', the Joint Monitoring Programme (JMP) of the World Health Organization (WHO) and the United Nations International Children's Emergency Fund (UNICEF) used 'improved sanitation' in monitoring and reporting on the sanitation target under the MDGs. What each of these terminologies means is still contested and the scholarly literature, often without clear definition, refers to basic sanitation (Kamga, 2013), improved sanitation (Munamati, Nhapi & Misi, 2016; van Minh & Nguyen-Viet, 2011), adequate basic sanitation (Giné-Garriga, Flores-Baquero, Jiménez-Fdez de Palencia & Pérez-Foguet, 2017), environmental sanitation (Arimah, 1996), or simply 'sanitation' (Guimarães, Malheiros & Marques, 2016). Exceptionally, Victor and Ernest (2007) define sanitation as the maintenance of hygienic conditions through garbage collection and waste disposal services. Nonetheless, the grey literature produced by international organisations like the JMP, the Water Supply and Sanitation Council (WSSCC), the World Health Organization (WHO), and the Millennium Task Force often either contain definitions of sanitation or itemise key components of sanitation from which an underlying meaning can be garnered. The definitions of sanitation focus on: (a) basic sanitation; (b) environmental sanitation; and (c) improved sanitation.

Basic

The WSSD defined basic sanitation to include: (a) improvement of sanitation in public institutions, especially in schools; (b) promotion of safe hygienic practices; (c) promotion of education and outreach focused on children, as agents of behavioural change; (d) promotion of affordable and socially and culturally acceptable technologies and practices; (e) development of innovative financing and partnership mechanisms; and (f) integration of sanitation into water resources management strategies in a manner that does not negatively affect the environment. WSSD thereby linked sanitation with safety (and improved human health, reduced childhood and infant mortality), public participation, affordability, acceptability and sustainability of sanitation infrastructure, which are some of the HRS principles discussed in Chapter 5 (Lenton et al., 2005). Unlike the scholarly literature which mainly discusses basic sanitation from the perspective of the individuals and households (for instance, Kamga, 2013), the WSSD definition has the advantage of taking into cognisance the importance of sanitation services in public places as well, and the need for financial and environmental sustainability of sanitation services.

The Millennium Development Task Force defined basic sanitation as "the lowest-cost option for securing sustainable access to safe, hygienic, and convenient facilitates and services for excreta and sullage disposal that provide privacy and dignity, while at the same time ensuring a clean and healthful environment both at home and inside the neighbourhood of users" (Lenton et al., 2005, p.30). The main strength of this definition is that it highlights the fact that the choice of sanitation instruments, particularly sanitation infrastructure is value laden and dependent on contextual factors like the conditions in the physical environment, and the available financial resources. Hence, what constitutes basic sanitation in one area, for instance in an arid low income formal settlement, may differ from the requirements in another area such as a formal settlement in a riverine area (see 3.5). It is also pragmatic in promoting the balancing of sanitation needs with ensuring environmental sustainability both in the immediate and extended surrounding of users.

The Sustainable Development Goals (SDGs) framework defines basic sanitation service as an improved facility that is not shared, limited service as "an improved facility shared with other households", and safely managed sanitation service as "a basic facility that safely disposes of human waste" (UN, 2017, p.30). It also defines hygiene coverage as the "availability of a hand washing facility with soap and water on premises" (United Nations, 2017, p.330). The basic definition under the SDG is however more restrictive than the WSSD defined which went beyond excreta management and hygiene, as elaborated above.

Environmental

There are alternative definitions of sanitation which extend beyond the focus on basic sanitation to include wider concerns for ensuring a clean and healthy environment. For instance, the World Health Organization (WHO) broadly defines sanitation, on its website, as "the provision of facilities and services for the safe disposal of human urine and faeces" and "the maintenance of hygienic conditions, through services such as garbage collection and wastewater disposal" (Sanitation, n.d.).

This definition has the advantage of integrating the safe containment of excreta with hygiene and the broader concerns for environmental sanitation in some scholarly publications. However, similar to the JMP's definition, the WHO's definition does not emphasize sewerage or any form of waste treatment after collection to ensure environmental sustainability. The WHO definition is also vague on whether facilities need to be provided for use in public places or only at the household level but it can be construed to include all spheres of human life, as human urine and faeces can be generated wherever humans exist and would then require safe disposal.

Improved

The JMP, responsible for monitoring access to sanitation under the recently concluded MDGs programme (2015), defined access to sanitation in terms of access to improved facilities which hygienically separate human excreta from human, animal and insect contact. These include flush or pour-flush toilets connected to a piped sewer system, septic tank, or pit latrine, ventilated improved pit (VIP) latrine, pit latrine with slab, and compost toilet, provided they were not public facilities (World Health Organization and UNICEF 2015). Further, only private facilities were classified as improved (Kwiringira et al., 2014; World Health Organization & UNICEF, 2015).

Although the JMP definition prioritises public health through the safe containment of human excreta by using improved sanitation facilities, improved facilities do not prevent contamination from untreated sewage that is discharged into the environment (Bain, 2014; Baum, 2013; Satterthwaite, 2016). Further, the literature suggests that there are underlying issues of poverty and low levels of education among households that rely on shared facilities (Heijnen, Routray, Torondel & Clasen, 2015a), while households that share with neighbours (presumably a relatively smaller group of users with social ties) have higher demographic status than households relying on communal facilities (open to the public use) which are also less likely to be hygienically maintained than the neighbour-shared facilities (Heijnen, Routray, Torondel & Clasen, 2015b). Communal sanitation facilities may also be poorly suited for safe use by women due to privacy and safety concerns, for instance, which raises issues of gender parity (Biran, Jenkins, Dabrase & Bhagwat, 2011). Nonetheless, shared facilities may enhance access to sanitation if hygienically maintained, culturally acceptable, and located in a safe environment (Obani & Gupta, 2016b). They may even constitute an appropriate adaptation response to limited space or resources (Rheinländer, Konradsen, Keraita, Apoya & Gyapong, 2015; Yatmo & Atmodiwirjo, 2012).

3.3   CLASSIFYING DOMESTIC SANITATION SERVICES AS ECONOMIC GOODS

The economic classification of goods and services has important implications for governance institutions that can be further illustrated by reference to Thomas Hardin's Tragedy of the Commons in the context of environmental resources (Hardin, 1968). The Tragedy of the Commons is founded on the concept of externalities, that is, the costs of production that are not reflected in the final cost of a good or service. It proposes that the environment is more of a common pool resource which is non-excludable but rivalrous and that individuals acting as rational, independent and free-enterprises realise that the cost of discharging waste directly into the environment is less than the cost of treating the waste before discharge (externalities). This creates an incentive for overexploitation or pollution of the environment in the absence of regulation. The need to internalise externalities and protect the environment from degradation is the basis for environmental law principles like the polluter pays principle which is sometimes relied upon as the justification for the use of economic instruments in governing the environment. Nonetheless, not every aspect of the environment or environmental resources can be readily classified as a common pool resource. For instance, Brölmann (2011) states that the legal model of global commons is not appropriate for global freshwater resources that cannot be localized.

The economic character of sanitation is still in issue and resolving this is important for the efficient production of sanitation services. Although the Dublin Statement of the International Conference on Water and the Environment, 1992, controversially recognised water and by extension sanitation as an economic good, it did not clarify what type of economic good. Sanitation fits into each category of economic goods at different stages, and its status changes over time and can be influenced through human actions (Mader, 2012).

On the one hand, sanitation has the characteristics of a public good (non-excludable and non-rivalrous) because the benefits of accessing and using adequate sanitation, such as reduced public health risks and improved quality of life, extend to non-users as well and are therefore non-excludable and non-rivalrous (Mader, 2012). This would support the provision of free or subsidised sanitation systems for public use. For instance, Agenda 21 frames water as essentially a 'social' good which means poor users need to be provided with free access and users can only be charged equitably for use in excess of basic human needs (United Nations, 1992). However, where funding is limited, investment may be redirected away from the direct provision of private sanitation facilities towards public good components like wastewater treatment and sewer networks (Evans, 2005), although this would not necessarily prioritise the poor.

On the other hand, sanitation can be considered a merit good (and made available to everyone on the basis of need, regardless of the ability or willingness to pay for it) because users display a preference-distortion and sometimes resort to unhygienic alternatives like open defecation instead. The provision of merit goods requires State intervention, otherwise, the private sector, including individual users, may not act in their own best interest, due to limited knowledge and resources or their value system (Mader, 2012). Still, sanitation does not entirely qualify as a merit good because the negative externalities of poor sanitation and hygiene habits are generally non-excludable. Additionally, private sanitation facilities are excludable and somewhat rivalrous depending on the technology of the sanitation system. In the case of water, there is evidence of the informal privatisation of community taps entrenching inequities in access, by improving water security for some and denying access to the poor and vulnerable, rather than promoting universal access (Udas, Roth & Zwarteveen, 2014). The sanitation system further includes public and common good components, without which the sustainability of the private infrastructure (for instance, private toilets) would not be assured (see Table 3.1).

Table 3.1 Classifying sanitation as an economic good

Components of Sanitation Systems	Economic Properties	Class of Economic Goods
Centralized sewer networks	Non-excludable Non-rivalrous	Public good
Water in aquifer, natural resources, roads, etc. required for sanitation services or used as a sink for sanitation services	Non-excludable Rivalrous	Common good
Patented sanitation technology	Excludable Non-rivalrous	Toll/club good
Physical components of sanitation systems that require connection or service fees like toilets	Excludable Rivalrous	Private good

Given that sanitation is increasingly being privatized or delivered through various public private partnerships (PPPs) models, especially in poor countries with weak infrastructure for the delivery of sanitation goods and services, there is a danger that: (a) sanitation goods and services will not be provided by the state; (b) that even if the State provides it, full cost recovery models may result in the sanitation goods and services becoming unaffordable, thereby exacerbating the existing inequities in access to sanitation at multiple levels of governance especially in the context of poor countries; or (c) that the ecological components are externalized.

3.4   DRIVERS OF POOR SANITATION SERVICES

Drivers are the causes of a problem, and can either be direct or indirect. Direct drivers influence local actors to engage in practices that result in poor sanitation services and therefore mostly operate at the local and national levels. In my analysis, I identify direct drivers by asking the following questions: (a) Does the driver reduce the capacity of users to make the necessary investments required to meet their personal sanitation needs? (b) Is the driver external to the service provider yet affecting the providers' capacity to meet the existing sanitation needs within their service areas? A driver which answers to either or both questions is direct (see 3.4.1). I classify the other drivers which occur in the literature but are not direct drivers based on the foregoing selection criteria as indirect. The indirect drivers have an incidental effect on direct drivers and can operate at multiple levels of governance, from international to local (see 3.4.2). The direct and indirect drivers, as well as their scale and context of operation are presented in Table 3.2.

3.4.1 Direct

From the literature, I identified sixteen direct drivers of poor sanitation services, including: (a) six economic; (b) seven social; and (c) four environmental drivers. The direct economic drivers are financial or fiscal factors affecting users' capacity to invest in domestic and personal sanitation services, while the direct social drivers are human, political or cultural factors hindering access to sanitation services and the environmental drivers are inherent to the physical and natural environment.

Economic drivers

First, the huge capital investments required for sanitation services (Lixil et al., 2016; Trémolet & Rama, 2012) will not be made by poor households and residents in informal settlements with a tendency to discount the future (Poulos & Whittington, 2000). Second, household poverty   limits users' ability to pay for connection and maintenance fees or flat rate subscription tariffs, and may ultimately cause service disconnections in the absence of well-targeted pro-poor instruments, like cross-subsidies (Biran et al., 2011; COHRE et al., 2008). Conversely, lack of subscription from users reduces the available financial resources for the maintenance of existing facilities, service expansion and investments in new infrastructure (World Health Organization [WHO] & UN-Water, 2015) Third, even where users subscribe for sanitation services, too low or inefficient tariff collection systems and poor revenue collection may be a driver of poor sanitation in formal and informal settlements (USAID Egypt, 2013). Fourth, the inherent preference-distortion characteristics of sanitation as a merit good (see 3.3) coupled with the existence of free though unhygienic alternatives like open defecation, hampers willingness to pay and reduces the level of demand for sanitation services (Department for International Development [DFID], 2007; Mader, 2012; Obani & Gupta, 2014a). Fifth, risk aversion affects investment in resilient systems (Saqib, Ahmad, Panezai & Rana, 2016) and some economic actors may be averse to investing in sanitation services for the poor without assurances of significant returns on their investments (Grey & Sadoff, 2007). Sixth, unaffordable tariffs and high connection and maintenance fees, coupled with low incomes, limit households' ability to access sanitation services (Biran et al., 2011; COHRE et al., 2008), and the threshold for affordability is a key issue especially in the context of informal settlements (Fonseca, 2014).

Social drivers

First, a long distance to sanitation facilities reduces accessibility, increases waiting times, and may compromise the safety of vulnerable users, like women and children who try to access the facilities at night (COHRE et al., 2008; Biran et al., 2011). Second, the exclusion of minorities from the design, operation, maintenance and use of sanitation infrastructure exacerbates inequitable access and may fuel conflicts over sanitation (Evans et al., 2009; van Stapele, 2013). Third, negative social practices like non-prioritisation of sanitation services or the location of public sanitation infrastructure based on political considerations rather than overriding public interest and efficiency hampers equitable access (Mader, 2012; Schuller & Levey, 2014). Fourth, negative cultural practices which inhibit safe waste management and the maintenance of hygiene standards (Ersel, 2015), or constrain the siting of sanitation infrastructure, without any environmental or public health basis are also drivers of poor sanitation (Akpabio, 2012; Evans et al., 2009; IRIN, 2012). Fifth, even where the infrastructure is available, poor maintenance and improper use may result in either damage or lack of use among girls especially (Biran et al., 2011; Garn et al., 2014, 2017; Simiyu, 2016). Sixth, space constraints, especially in poorly planned settlements or emergency situations, hampers the installation, operation and maintenance of facilities, technologies and infrastructures (Katukiza et al., 2010; Katukiza et al., 2012; Johannessen, Patinet, Carter & Lamb, 2012). Seventh, in informal settlements especially, tenure insecurity and the underlying power issues, limit access to formal sanitation services especially where service providers are not legally obliged to extend their coverage to informal areas, forcing the residents to resort to either open defecation and other unsanitary practices or unregulated and often more expensive informal services for their basic needs (Dagdeviren & Robertson, 2009; Murthy, 2012).

Environmental drivers

First, a challenging or inaccessible topography, compounds the technicality or cost of providing sanitation; for instance, through preventing the laying of pipes at the right depth to establish the required slope for the smooth operation of the system, requiring additional infrastructure or necessitating increased system capacity to avoid the sewage infiltration by groundwater (Cairns-Smith, Hill & Nazarenko, 2014). Paradoxically, it is the poor and marginalised groups of people who tend to settle in such physically challenging areas that are relatively cheaper to acquire or lease and during emergencies such areas are worst hit and their residents less resilient. Second, pollution, droughts and other forms of water scarcity hamper the operation of conventional sewage systems and affect self-supply options like wells, thereby reducing the availability of water for personal sanitation and hygiene uses (Johannessen et al., 2012). Third, natural hazards, which could be climatological (like extreme temperatures and wildfires), geophysical (like earthquakes and volcanoes), hydrological (like floods), or meteorological (like hurricanes and cyclones) (Watt and Weinstein 2013), may destroy non-resilient sanitation infrastructure or critical infrastructure like roads and power supply networks which are necessary for delivering sanitation services in humanitarian situations disrupt services and contaminate water sources which are required for sanitation and hygiene uses like hand washing (Bates, Kundzewicz, Wu & Palutikof, 2008; Misra, 2014). The impact of natural hazards is especially significant in poorly resilient communities (Qasim et al., 2016; Roosli & Colins, 2016; Sharifi & Yamagata, 2016). Fourth, particularly during emergencies, high temperatures and high turbidity in source water may also affect the operation of sanitation systems (Ensink et al., 2015).

3.4.2 Indirect

From the literature, I identified nine indirect drivers of poor sanitation services, including: (a) four economic; (b) four social; and (c) one environmental driver. The indirect economic drivers are financial or fiscal factors exerting an underlying influence on users' investments in sanitation, and are mainly linked to public financing. The indirect social drivers include social conditions and demographic factors, while the environmental driver is not confined to the locality.

Economic drivers

First, although there is limited evidence that foreign debt relief inevitably results in increased public expenditure on social infrastructure (Dessy & Vencatachellum, 2007; Kaddar & Furrer, 2008), foreign debts reduce the capacity of poor States for such investments (Varma et al., 2008). Second, within formal and informal settlements and humanitarian situations, insufficient funds partly due to the fragmented governance of sanitation (Isunju, 2011), and lack of targeted financing coupled with low visibility of humanitarian crises and chronic emergencies (United Nations Children's Fund [UNICEF], 2015; 2017) affects sanitation services. Third, national poverty limits the ability of poor countries to invest in sanitation infrastructure and estimate the recovery time for critical infrastructure affected by emergencies (Zorn & Shamseldin, 2015). Fourth, sanctions may also affect public investments in sanitation infrastructure; hence, General Comment No. 15 (2003), articles 31 and 32 (see 5.2) implore States to refrain from imposing embargoes and similar measures that frustrate the realisation of the right to water, or using water as a means of coercion (Obani & Gupta, 2015).

Social drivers

First, insecurity and conflicts (such as intra-communal dissensions over sanitation projects or conflicts over transboundary water resources) may hamper access to sanitation services through the destruction of sanitation infrastructure (Obani & Gupta, 2015), and limited social cohesion (often lacking in informal settlements or among people living in humanitarian situations or populations in transit) to support self-help enterprises (Isunju, 2011). Second, lack of education and awareness, minimal engagement with the relevant agencies and failure to report service problems also hinder sanitation and hygiene awareness, sustainability of services, and enforcement of civic rights (Munamati et al., 2016; Akpabio, 2012). Third, mass migration and rapid urbanisation creates additional stress for existing resources and increases the likelihood of the spread of water and sanitation related diseases in the absence of adequate infrastructure (Vuorinen, 2007). Fourth, in formal and informal settlements and emergencies, population density significantly increases pollution (Saqib et al., 2016), and affects the sustainability of sanitation infrastructure like sewer systems (Cairns-Smith et al., 2014; Schouten & Mathenge, 2010).

Environmental drivers

Although the links between climate change and human health are yet to be fully established, climate change could indirectly drive poor sanitation services by exacerbating extreme weather events and variability, leading to the destruction of sanitation infrastructure (Rabbani, Huq & Rahman, 2013). The risks are particularly high for poor countries lacking the institutions, funding, and infrastructure to invest in necessary climate mitigation and adaptation measures (Grey & Sadoff, 2007).

Table 3.2 Direct and indirect drivers of poor personal and domestic sanitation services

	Drivers	Eme.	For.	Inf.	Key References
	DIRECT				
Local — Eco.	Discounting the future				Poulos & Whittington 2000
	Household poverty				Biran et al. 2011; COHRE 2008
	Inefficient tariff collection system				USAID Egypt 2013
	Preference distortion affecting WTP				Mader 2012; Obani & Gupta 2014a
	Risk aversion				Grey & Sadoff 2007; Saqib et al. 2016
	Unaffordable tariffs & connection fees				COHRE et al. 2008; Fonseca 2014
Social	Distance to the facility				Biran et al. 2011; Mader 2012
	Exclusion of minorities from accessing services				Evans et al. 2009; van Stapele 2013
	Negative social practices				Akpabio 2012; Ersel, 2015
	Non-acceptance of sanitation facility				Ersel 2015
	Poor maintenance culture/improper use of facilities				Garn et al. 2014, 2017; Simiyu, 2016
	Space constraints				Katukiza et al. 2010; 2012
	Tenure insecurity				Dagdeviren & Robertson 2009; Murthy 2012
Env.	Challenging or inaccessible topography				Schuller & Levey 2014
	High temperatures/high turbidity in source water				Ensink et al. 2015
	Natural hazards				Labib & Read 2015; Qasim et al. 2016
	Pollution/water scarcity				Johannessen et al., 2012
	INDIRECT				
Local/Nat. — Eco.	Huge foreign debts that limit public spending				Varma et al. 2008
	Insufficient/poorly targeted funds				Isunju 2011; UNICEF 2015
	National poverty				Zorn & Shamseldin 2015
	Sanctions affecting the sanitation sector				Obani & Gupta 2015
Social	Insecurity, conflicts and poor social cohesion				Isunju 2011; Obani & Gupta 2015
	Low awareness about sanitation				Akpabio 2012; Munamati et al. 2016
	Mass migration/urbanisation				Juuti 2007; UN-Habitat 2016
	Population density/growth				Cairns-Smith et al. 2014; Saqib et al. 2016
I — Env.	Climate variability and change				Misra 2014; Rabbani et al. 2013

Nat. - National; I. – International; Eco. – Economic; Env. – Environmental; Eme. – Emergency; For. – Formal; Inf. - Informal

Source: This table builds on Table 12.1 in Obani & Gupta, 2016b

3.5 TECHNOLOGIES FOR DOMESTIC SANITATION SERVICES

This section focuses on the basic technologies used for domestic sanitation service delivery, including: (a) toilet systems (see 3.5.1); (b) on-site septic tanks (see 3.5.2); (c) sewer systems (see 3.5.3); (d) sludge treatment (see 3.5.4); and (e) the sanitation ladder and service levels commonly used for measuring and regulating the level of access to sanitation technologies and services.

3.5.1 Toilets

The first toilets in human history were simple holes in the ground. Subsequently, ancient civilizations across Africa, Asia and Europe were motivated by health, environmental, and religious values to construct both private and public toilets and washrooms for the easy and safe containment of sewage (Juuti, 2007). With the development of microscopes in the middle of the 19th century, people began to realise the health risks posed by contaminated water, leading to increased interest in water supply and toilet systems (Juuti, 2007). Toilet systems may generally be wet or dry, and in some cases bio-physical factors such as water quantity or quality may necessitate further technological innovations such as the floating toilet being developed for use in floating communities (Akpan, 2015).

Wet system

A wet toilet system, such as pour flush toilets, requires water for the evacuation of excreta from the toilet into a single leach pit, twin leach pits through a division chamber, or septic tank. While the solid wastes settle to the bottom of the tank, the organic components are decomposed by bacteria (Parkinson, Tayler, Colin & Nema, 2008). The wastewater could be treated through artificial wetlands or anaerobic filters, before discharge into a drain or watercourse, or infiltration in cases where the ground conditions permit, otherwise untreated effluents discharged into the environment poses both human health and environmental hazards (Parkinson et al., 2008). The estimated cost of a pour flush latrine is around USD 70 while a septic tank latrine costs around USD 160, including operation and maintenance costs (van de Guchte & Vandeweerd, 2004).

Dry system

A dry system, such as ecological sanitation (Ecosan) which separates faeces and urine, or the pit latrine, eliminates the need for immediate evacuation and wastewater treatment by combining the toilet and storage and is commonly used in informal settlements, humanitarian situations, and rural areas (Simha & Ganesapillai, 2016). Peepoo bags lined with sanitising agent which breaks down excreta inside the bag to be reusable as fertilizer is also an alternative dry system which presents a sustainable alternative to open defecation, especially in informal settlements and humanitarian situations (Wirseen, Munch, Patel, Wheaton & Jachnow, 2009). Ecosan requires conscious effort and places more demand on the behaviour of the users compared to some other forms of on-site systems which do not require such separation (Parkinson et al., 2008). Pit latrines can be built from local materials, and modified to suit user preferences and at very low construction and maintenance costs, hence, their prevalence in Community Led Total Sanitation (CLTS) interventions in developing countries (Chambers, 2009; Mehta, 2011; United Nations Children's Fund [UNICEF], 2012). The latrines can however cause odour nuisance (Nakagiri et al., 2016), surface and groundwater pollution due to flooding, poor drainage especially in places with a high water table, or a significantly higher risk of groundwater contamination where the walls are unlined thereby allowing nitrogen and pathogens from sewage to leach into the soil (Katukiza, 2012). The pit latrine has evolved through different design adaptations (ranging from the simple pit latrine to the ventilated pit latrine) and further technological improvements are key to improving the design of pit latrines to promote safe and sustainable use (Nakagiri et al., 2016). A simple pit latrine costs around USD 45, while ventilated improved pit latrine costs around USD 65 (van de Guchte & Vandeweerd, 2004).

3.5.2 On-Site Systems

On-site systems retain faeces and wastewater using pits, vaults, or septic tanks, until the receptacle is desludged, and are therefore appropriate for dry toilets or for use in places with adequate space for a soak pit or constructed wetland and grey water management, and a low water table with no flood risk (Katukiza et al., 2012). The receptacles of on-site systems can be manually emptied. Apart from the high haulage costs and problems of access in densely populated settlements, the public health risks of manual evacuation has led to the practise being proscribed in some countries like India and Nigeria. There are other relatively hygienic alternative technologies for pumping out the sludge, such as the Manual Pit Emptying Technology (MAPET), and the UN-Habitat Vacutug which require less skill, low operating and maintenance cost, and can manoeuvre tight spaces characteristic of informal settlements (Katukiza et al,. 2012; Thye, Templeton & Ali, 2011). However, pumping fails under weak latrine substructure; the technologies generally cover only a maximum haulage distance of

0.5 km to the treatment plant, and are unable to evacuate dry sludge and solid particles (Harvey, 2007). Alternatively, smaller vehicles such as the narrow-wheel base truck can also be used to navigate congested areas (Parkinson et al., 2008). Households typically bear at least 70% of the capital cost and 90% of the operating cost of on-site systems (Cairns-Smith et al., 2014). The capital cost of an on-site septic tank system ranges from between USD 70 and USD 360, depending on the size of the septic tank required which varies according to the size of the household using the system, average water use, and the amount of exfiltration from the system (Cairns-Smith et al., 2014; WASHCost, 2012). The annual operating cost is much less ranging from USD 4 to USD 12, depending on the distance to the disposal site, type of tank, the size of the tank relative to the size of the household, and local pricing (Cairns-Smith et al., 2014).

3.5.3 Sewer Systems

Although the water flush system has gained wide acceptance and represents the standard for improved sanitation in many formal settlements, without adequate sewage treatment it could easily lead to the spread of diseases especially in crowded conditions and to the pollution of surface waters when the untreated sewage is channelled into surface waters (Juuti, 2007). Around the twentieth century, sewers were developed to transport sewage from the toilets to the treatment facilities (Juuti, 2007). Sewer networks require a sufficient quantity of wastewater flow to convey sludge through pipes from densely populated human settlements to sewage treatment plants, and may be centralized or decentralized systems (Tilley, Zurbrügg & Lüthi, 2010).

Centralized conventional sewer systems

Centralized conventional sewer systems are ideal for high population density areas with over 30,000 people per square kilometre and are often designed to serve the maximum projected total population of the network area because of the high cost of alterations after the initial installation (Cairns-Smith et al., 2014). The projected capital cost of centralized conventional sewer systems ranges from USD 130-USD 330, with actual costs around USD 180-USD 260 per capita where the entire target population connects to the network; otherwise, the actual capital costs could be as high as USD 220-USD 940 per capita for those connected to the network (Cairns-Smith et al., 2014; Dodane, 2012; Winara et al., 2011). The operating costs for centralized conventional sewer systems range from USD 12 to USD 28 per capita, depending on the cost of energy, manpower/system automation, operation and maintenance, type of treatment, source of financing, and the topography of the service area (Cairns-Smith et al., 2014). For instance, automation may reduce manpower cost; a flat topography would require extensive pumping, thereby raising energy costs for the system, and additional financing charges for systems being funded through loans (Cairns-Smith et al., 2014).

Decentralized simplified sewers

Decentralized simplified sewers serve relatively smaller areas through reduced pipe diameters, gradients, and depths, and may be required where centralized sewer systems have become ineffective due to blockages and non-functional treatment plants (Cairns-Smith et al., 2014). Decentralized sewers are appropriate for high density urban areas with relatively high wastewater production, low soil permeability, and space constraints (see Figure 3.1) (Paterson, Mara & Curtis, 2007). They are relatively easier to upscale and cheaper to install and maintain than the conventional sewerage (Mara, 1996; Mara & Guimarães 1999). Further, such systems often involve community participation in the design and implementation which promotes households' connections to the network, thereby reducing the per capita capital costs as a result of lower number of connections to the network than projected (Cairns-Smith et al., 2014) but require periodic cleaning to prevent overflow from manholes and blockage of the sewers (Katukiza et al., 2012). The annual per capita cost of decentralized simplified sewer systems ranges from USD 105 to USD 155; the operating costs is low within the range of USD 4 to USD 10 per capita because the sewage is transported through gravity-based flow, over a shorter distance to the wastewater treatment plant (Cairns-Smith et al., 2014; van de Guchte & Vandeweerd, 2004).

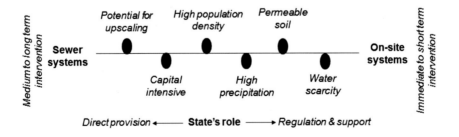

Figure 3.1 Factors influencing the choice of technologies for sewage management

3.5.4 Sludge Treatment

Although the pathogens contained in the faecal sludge undergo some natural degradation on-site, there is need for further treatment before reuse or disposal, to prevent pollution (Tilley 2008). While the type of sludge treatment required depends on the concentrations of pollutants and pathogens, legal requirements including the terms of the discharge consent, and the proposed use of the effluent, most treatment technologies often combine both physical processes, like the removal of large particles by coarse screening or the sedimentation of particles through the force of gravity, and biochemical processes such as aerobic and anaerobic degradation (Mengistu, Simane, Eshete & Workneh, 2015; Parkinson et al., 2008). The effluent from these processes may be further subjected to tertiary treatment to facilitate the removal of nitrogen, phosphorus, and heavy metals or other industrial pollutants (Fan, Zhou & Wang, 2014; Vinnerås 2007). Sludge treatment may result in a solid fraction which requires additional treatment before reuse, and a liquid fraction which requires polishing treatment in order to meet legal requirements for discharge consent or to prevent negative environmental impacts where infiltration of effluents is permissible (Parkinson et al., 2008). Short term alternatives are also available where the required treatment level cannot be achieved before reuse, including restricting the types of crops irrigated with wastewater, employing drip irrigation, and equipping farm workers with protective gear (Parkinson et al., 2008).

3.5.5 Sanitation Ladders and Service Levels

Sanitation ladders

Sanitation ladders emerged in the 1980s through participatory instruments like the Participatory Hygiene and Sanitation Transformation (PHAST) (see 7.3.3). They offer reference points for local communities to deliberate and reach a consensus on appropriate technology options for their sanitation needs (Potter et al., 2011). Sanitation ladders have also been adapted at the international and national levels of governance (see Box 3.1).

Box 3.1 Sanitation ladder for rural sanitation technology in Lao DPR

In Lao DPR for instance, six technological options were identified for rural sanitation technology using the following criteria: (a) sustainability and lasting long-term benefits, (b) immediate benefits in terms of quality, convenience, reliability, (c) capacity requirement to provide supply-side support, (d) operation and maintenance, (e) potential for up scaling, (f) cost effectiveness, and (g) accessibility. The ladder had at its lowest rung improved traditional practice, then conventional dry latrine, lid/cover latrine, ventilated improved pit latrine, pour flush latrine and at its highest the rung septic tank system (Lahiri & Chanthaphone, 2000). This encouraged the participation of local stakeholders in the policy process and offered a guide for improvements in sanitation access.

Technology-based sanitation ladder

The technology-based sanitation ladder features prominently in international sanitation governance; particularly under the MDGs framework and the 2030-bound SDGs (see 3.2). The MDGs sanitation ladder comprised of three rungs to measure progress towards: (a) improved (use of facilities which separate excreta from human contact), (b) unimproved services (use of improved facilities that are shared between two or more households), and (c) open defecation (UN, 2017). The MDGs ladder did not sufficiently address the health risks posed by poor management of excreta despite the use of improved facilities (Baum et al., 2013; Exley, Liseka, Cumming & Ensink, 2015), and downplayed the importance of shared improved facilities for the poor and people living in densely populated areas, for instance (Obani & Gupta, 2016). It also hampered innovation by imposing a predefined list of improved facilities (Kvarnström et al., 2011).

The SDGs ladder has introduced additional rungs and terminology to capture five service levels, namely: (a) safely managed (use of an improved facility which is not shared and excreta is safely treated *in situ* or transported and treated offsite), (b) basic (use of improved facility that is not shared), (c) limited (use of improved facilities that are shared with two or more households), (d) unimproved (use of pit latrines without slab or platform, hanging latrines and bucket latrines, and (e) open defecation (UN, 2017). The SDG ladder also introduces three rungs for hygiene, namely: (a) basic (hand washing facility with soap and water in the household), (b) limited (hand washing facility without soap or water), and (c) no hand washing facility.

The SDGs ladder reclassifies 'unimproved' sanitation under the MDGs as 'limited'. This is significant as there were already around 600 million people using limited service in 2015 (UN, 2017) and the reclassification supports investment in improved facilities for shared use which may be the most efficient option in densely populated informal settlements, for instance. Further, the SDGs promotes the HRS in four ways: promoting progressive realisation through prioritising improvements for people at the lower rungs; integrating environmental concerns to ensure safe handling of excreta through the use of safely managed facilities; expanding the focus on hygiene; and prioritising universal coverage. It is also important to incorporate the safe disposal and treatment of the wastewater from hygiene uses and menstrual hygiene within limited services, to advance gender equality. This is illustrated in Figure 3.2.

Function-based sanitation ladder

The function-based sanitation ladder is an alternative to the technology-based ladder that additionally incorporates user/health functions at the lower rungs and environmental functions and integrated approaches to sanitation at the higher rungs (Kvanström et al., 2011). The function-based ladder promotes safe management of different waste streams and enhances resource recovery better than the current technology-based ladder. Further, it is capable of spurring local solutions to sanitation problems and inspiring stakeholders to think beyond the provision of certain technologies, based on health and environmental considerations (Kvanström et al., 2011). Progress towards the higher rungs of the ladder may require higher capital investment. This results in poor countries focusing more on providing services at the lower rungs, while richer countries that have contained the health and microbiological risks can focus more on higher environmental functions (Keraita, Drechsel & Konradsen, 2010). It is nonetheless important to ensure that issues of accessibility, affordability, participation, and non-discrimination are addressed even at the lower rungs, to integrate HRS principles (see 5.3) in the process of increasing coverage and environmental functions (Obani & Gupta, 2016).

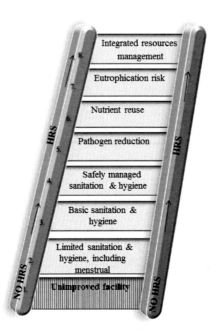

Figure 3.2 Proposed sanitation ladder that integrates technologies and functions (Adapted from Figure 2 in Obani & Gupta, 2016, building upon the new SDGs sanitation and hygiene ladder)

Sanitation service levels and service level contracts

With the participation of the private sector in sanitation service delivery, service levels are stipulated in sanitation Service Level Agreements (SLA) signed between the State and the utility/service provider, as well as in Service Level Contracts (SLC) signed between the utility/service provider and the users which stipulate the standards and terms of use of the services provided. A detailed SLA which stipulates clear standards for services promotes better coordination of the sanitation sector, and strengthens transparency, monitoring and accountability. Depending on the terms, an SLA could also improve equitable outcomes by ensuring standardized services irrespective of the location or status of the users who sign SLC with utilities/service providers (Potter et al., 2011).

Sanitation service levels like the sanitation ladder also stem from a predominantly technocratic response to the sanitation problem. The concept of sanitation service levels determine the level of access to sanitation services based on a predefined set of service parameters or indicators. Examples of indicators used to determine sanitation service levels for households include the type of technology, accessibility and ease of use, reliability, environmental impact and the levels of health concern resulting from the sanitation facility (COHRE et al., 2008; Potter et al., 2011). In line with the broad meaning of sanitation in the post-2015 development agenda (see 3.2), it is equally important for service levels to regulate not only (a) access to and standards of toilets, but additional components of sanitation such as: (b) sewage network coverage, (c) quality of sewerage and solid waste collection, treatment, disposal, reuse and recycling services, (d) efficiency of participatory mechanisms and consumer complaints handling mechanisms, (e) financial sustainability of sanitation governance, including support for the poor, vulnerable and marginalized who would otherwise be unable to access sanitation.

Further, building on the technology-based and function-based ladders, sanitation service levels may be classified into four, as follows:

(i) **No access**: Open defecation with very high health concerns

(ii) **Basic access**: Use of hygienically maintained standard toilet built with a minimum pit depth of 3.5 m and connected to a septic tank (Potter et al., 2011); located down slope and at least between 15m to 30m away from a water source, depending on the local circumstances (Potter et al., 2011; Sphere Project, 2011);[21] used by a maximum of between 10 to 20 people per drop hole, depending on the local circumstances and the needs of users (Potter et al., 2011; Sphere Project, 2011); sludge accumulation rate of 0.03 m^3/person/year (Potter et al., 2011); waste treatment and safe disposal; facilities for safely disposing menstrual products; water and soap for hand washing and hygiene, and hygiene promotion to ensure good sanitation practices (COHRE et al., 2008; Committee on Oversight and Government Reform, 2011; Moore, 2001).[22]

(iii) *Intermediate access* – In addition to (i) and (ii), use of facilities that are not shared, guaranteed privacy and continuous access day and night; mechanical emptying facilities for septic tanks; sewer connections in dense urban areas (COHRE et al., 2008).

---

[21] These values are indications at best. It is more practicable and efficient to determine the minimum technical standards for toilets and other sanitation facilities on a case-by-case basis, depending on local factors, like topography.

[22] Minimising the number of users per drop hole reduces toilet waiting times and is critical for women who often experience relatively longer waiting times than men due to physiological and cultural issues or even physical factors like having fewer toilets designated for women's use.

(iv) **Optimal access** – In addition to (i), (ii) and (iii), use of wastewater, stormwater and solid waste removal services, and the maximisation of health and environmental functions from sanitation systems (as outlined in Figure 3.2) (Kvanström et al., 2011).

3.6 TECHNOLOGIES FOR DOMESTIC SANITATION SERVICES, DRIVERS AND INCLUSIVE DEVELOPMENT

This section first provides an overview of the impact of technologies for domestic sanitation services (see 3.5) on the drivers (see 3.4) in sub-section 3.6.1, then the implications of the technologies for ID (see 3.6.2).

3.6.1 Sanitation Technologies and the Drivers of Poor Sanitation Services

Technology is crucial for the delivery of sanitation goods and services, promotes participation in the choice of instruments (through the sanitation ladder), and facilitates the monitoring of service standards for the users' protection (through SLAs and SLCs) (see 3.5.5). Depending on their design, technologies may be adapted to address direct environmental drivers related to space constraints and the local topography (Katukiza et al., 2012). Technology also provides the means of implementing other sanitation governance instruments; for instance, guaranteed free access to sanitation services (see 5.4.2) requires physical sanitation infrastructure (technology) that does not deny users' access due to non-payment. Nonetheless, technology is a relatively inflexible governance instrument (Majoor & Schwartz, 2015) and depending on the type, may require a high level of technical knowledge and expertise for operation and maintenance (see 3.5). Further, the choice, design or application of technologies (see 3.2 and 3.3) may compound the drivers of poor sanitation services and hamper ID (dos Santos & Gupta, 2017), where for instance:

(a) sanitation technology is designed with a limited public health focus on excreta management that does not address the non-health related psychosocial factors (like social norms and perceived gains for social status) which influence investment decisions and the use of sanitation services (Hulland, Martin, Dreibelbis, Valliant & Winch, 2015; Joshi et al., 2011);[23]

---

[23] cf. Gross and Günther 2014 suggest that low cost technologies rather than health, prestige, or safety will promote latrine construction.

(b) there are no public sanitation facilities and networks, and this limits access for vulnerable groups like homeless people, internally displaced people and populations in transit (de Albuquerque, 2009);

(c) there is a mismatch between conventional systems and local preferences or environmental conditions (Fatoni & Stewart, 2012; Paterson et al., 2007);

(d) patented sanitation technology is difficult or expensive to adapt locally, especially for poor countries (Viola de Azevedo Cunha, Gomes de Andrade, Lixinski & Féteira, 2013) or there are other concerns over the general cost and durability of sanitation technologies (Hulland et al., 2015); and

(e) sanitation systems are complex, or the technical expertise for the operation and maintenance of facilities is not locally available (COHRE et al., 2008; Fatoni & Stewart, 2012).

3.6.2 Sanitation Technologies and Inclusive Development

Since the 1960s, there have been concerns expressed over the impacts of science and technological innovations on various human rights. [24] In relation to the water, sanitation, and hygiene (WASH) sector, technologies ought to ensure social and relational inclusion, and environmental sustainability in order to be inclusive (see 2.4.3). To illustrate this using the sanitation ladder (see 3.5.5), the sanitation ladder can potentially improve social and relational inclusion by integrating users in the process of selecting instruments, technologies and service levels (see Q3 and Q4 in Figure 3.3). It is however important that the ladder does not only present predefined technological options but offers an opportunity for users to develop solutions that best suit their unique circumstances and maximise health and environmental functions (see Rungs 2 – 8 in Figure 3.2 and Q4 in Figure 3.3). The impact of sanitation technology also depends on the local context (for instance, are open toilets culturally acceptable? Are the facilities hygienically managed? Is the sewage safely collected, treated, disposed of, and recycled or reused?), and other related policies like whether or not

---

[24] See United Nations, Proclamation of Teheran, Final Act of the International Conference on Human Rights, Teheran, 22 April to 13 May 1968, U.N. Doc. A/CONF. 32/41 (1968). Para. 18 http://legal.un.org/avl/pdf/ha/fatchr/Final_Act_of_TehranConf.pdf; United Nations (1968b). Resolution adopted by the General Assembly. 2450 (XXIII). U.N. Doc. A/10034 (1968). Para. 1(a); United Nations (1970). Human rights and scientific and technological developments. U.N. Doc. E/CN.4/1028/Add.6 (Dec. 29, 1970), and A/8055; United Nations (1975). Declaration on the use of scientific and technological progress in the interest of peace and for the benefit of mankind. General Assembly. Resolution 3384 (XXX); United Nations, Vienna Declaration and Program of Action,} 11, U.N. Doc. A/CONF.157/23 (July 12, 1993).

informal settlements are excluded from coverage. The analysis following below is based on an inductive analysis of the technologies discusses in Section 3.5.

Social and relational inclusion

Sanitation interventions that are designed from a purely technocratic perspective may either exacerbate inequities in access to sanitation or be perceived as doing so. The 2011 toilet wars in South Africa are reminiscent of this. In the run-up to the local government elections in that year, there were many protests against unenclosed toilets in the informal settlement in Khayelitsha, Cape Town (Robins, 2014). The open toilets were widely seen as representing the political inequities, indignities and injustices of the apartheid regimes (Robins, 2014), thereby diminishing social and relational inclusion or were at least perceived to do so (that is, Q1 in Figure 3.3). Conversely, social and relational inclusion require the effective participation of all stakeholders, especially the poor, vulnerable and marginalised users, in every process for sanitation governance, for instance baseline studies on the status of sanitation services, setting of targets and service standards, design and/or selection of sanitation technologies and their location, the negotiation of SLAs and SLCs (see 3.5.5), and the implementation of monitoring and evaluation mechanisms. Social and relational inclusion also requires the equitable pricing of sanitation services at a level which stimulates providers and users to overcome the preference-distortion associated with merit goods, while also protecting access for the poor and vulnerable who may otherwise be deprived of services (see 3.3).

Ecological inclusion

Dry toilet systems like pit latrines are relatively cheap and can be built with locally sourced-materials and technology (which *prima facie* indicates social and relational inclusion within the bottom quadrants of Figure 3.3), but would fall within the lower left quadrant in Figure 3.3 (see Q3) where they are unlined and contaminate the groundwater. Ecosan contributes to environmental sustainability and therefore falls within the right quadrant of Figure 2, depending on whether or not there is equitable access in favour of the vulnerable and marginalised population (that is, Q2 or Q4 in Figure 3.3). Peepoo bags are environmentally sustainable because of their waste treatment properties and therefore fall within the right quadrants of Figure 3.3, depending on the level of accessibility for the vulnerable and marginalised population (Q2 or Q4 in Figure 3.3). Septic tanks and manual evacuation techniques, without treatment fall within the left quadrant in Figure 3.3, while evacuation technology like MAPET, VACUTUG, and narrow wheel base trucks that are suitable for informal settlements promote social inclusion would promote ecological inclusion if the sewage is treated (Q4 in Figure 3.3). Floating toilets improve accessibility for informal settlements and emergency situations like floods but can contaminate both surface and groundwater (potentially falling into Q3 in Figure 3.3). Water flush systems expend a lot of water and can cause environmental pollution, depending on the level of sludge treatment. The technology-based sanitation ladder prima facie excludes relational and environmental concerns, beyond excreta containment, therefore it falls within the left quadrant (see Q1 and Q3 in Figure 3.3), whereas the function based ladder which integrates health and environmental functions falls within the left quadrant (see Q2 and Q4 in Figure 3.3).

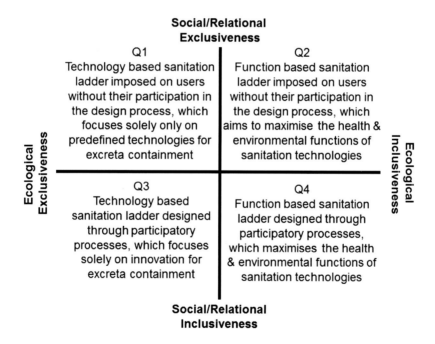

Figure 3.3 Assessing the sanitation ladder for inclusive development

3.7 INFERENCES

This chapter contributes five key messages with import for designing sanitation governance frameworks. First, it shows that there are multiple meanings of sanitation by different actors at various levels of governance, from international to local. Some definitions and actors focus on the "lowest-cost option for securing sustainable access to safe, hygienic, and convenient facilitates and services for excreta and sullage disposal that provide privacy and dignity, while at the same time ensuring a clean and healthful environment both at home and inside the neighbourhood of users" (the Millennium Development Taskforce's definition of basic sanitation). Others highlight the need for access to improved technology for excreta containment and disposal and discountenance shared facilities (for instance, JMP's definition of improved sanitation), or more broadly the need to ensure a clean and healthy environment (for instance the WHO definition) and respect for human rights (for instance the SDGs sanitation target). Nonetheless, each of these definitions are value laden and require further analysis. Taking the definition of basic sanitation as an example, among the poor, the 'lowest-cost option' may be a form of shared facility which allows for the costs of sanitation services to be spread among the users but shared facilities also raise important equity and safety concerns. Overall, the contestations show that the definition of the HRS ought to address gender equality, accessibility for vulnerable users like children, hygiene and maintenance of the facility, affordability of tariffs, and operating and maintenance costs, environmental sustainability, and social and relational equality between poor and rich users/households, to ensure inclusive outcomes (see Chapter 9).

Second, classifying sanitation or sanitation components as economic goods presupposes a quantifiable economic value and underlies the commodification and commercialisation of sanitation services and this chapter highlights an ambivalence among scholars regarding the economic nature of sanitation, with the multiple properties of a public good, merit good, or private good. A public good is non-excludable and non-rivalrous but this can be altered through commodification with the result that the poor who cannot afford to pay are excluded from accessing sanitation goods. Merit good refers to components where sanitation offers externalities that are critical to the wellbeing of both users and non-users and the environment and therefore requires State investment to augment shortfalls from the market and users. Private good refers to components of the sanitation system that allow for excludable consumption of sanitation goods and through the markets, based on the ability to pay. Further, common good components of sanitation like ecosystems that serve as natural sinks for sanitation services are non-excludable and rivalrous and therefore need to be protected from depletion as a result of unsustainable use. Toll goods components like patented sanitation technology are also excludable but non-rivalrous and may therefore be denied to poor users due to the operation of market forces. The inherent complexity of classifying sanitation goods mirrors the pluralistic foundations of sanitation governance and is an indication of the complex interactions between the different principles that converge in sanitation governance (see 5.6.3, 6.6.3, 7.4.3 and 8.6.3).

Third, there are seventeen direct drivers and nine indirect drivers of poor sanitation services covered in the literature, and they often reinforce each other within and across different scales. Although the literature mostly analyses the drivers (causes of poor sanitation services) in the context of specific vulnerable groups like residents of informal settlements, women, school children and people in detention centres or humanitarian situations, similar drivers may nonetheless apply across different settings (see Table 3.2). The challenge for sanitation governance in relation to drivers is therefore to: (a) operationalize sanitation governance principles using instruments that can address one driver (for instance, household poverty) without exacerbating another driver (for instance, pollution of natural resources and environmental stress caused by changes in the households' consumption patterns and poor management of increased wastewater), and (b) formulate instruments that can be adapted across different scales (including local, national and international) which would improve a comparison of their performance and learning.

Fourth, technologies are relatively inflexible but necessary for the operationalization of most of the other (regulatory, economic, management, and suasive) governance instruments discussed in the literature (see 5.5, 6.4, and 7.3) and in the context of my case study (see 8.5). The current predominantly technocratic response to the sanitation problem offers potential for enhancing participation (like the sanitation ladder) and instruments for containing human excreta. However, it does not always adequately address the complex interconnectedness of the drivers of poor sanitation services, definitional issues and the complexity of classifying sanitation goods and services in economic terms due to underlying factors like the design and suitability of the technologies to the local context; access problems for vulnerable groups; the cost and durability of the available technologies; and the poor integration of psychosocial considerations like social norms and the status of users, which affect the adoption of sanitation technologies (see 3.6.1). .

Fifth, sanitation technologies differ in the extent to which they promote ID depending on whether or not the technologies improve affordable access to sanitation services and participation in the sanitation governance processes for the poor, vulnerable and marginalised groups (social and relational inclusion), and whether or not the technologies promote environmental sustainability and minimise the harmful effect of sanitation service delivery on the environment (ecological inclusion). Conversely, even where a sanitation technology *prima facie* ensures environmental sustainability and is designed to be affordable, excluding the poor or residents of informal settlements from accessing the technology would still hamper ID. Hence, the truth about the impact of sanitation technologies on ID lies in the details of the principle or (regulatory, economic, management or suasive) instrument which the sanitation technology is designed to deliver. As a result, sanitation technologies are not discussed as a stand-alone theme in the subsequent discussion of instruments in Chapters 5, 6, 7 and 8.

The above five conclusions are critical for further elaborating on the HRS as will be shown in this thesis.

Chapter 4. Human Rights Principles

4.1 INTRODUCTION

The human right to sanitation (HRS) norm, which initially evolved through a bottom-up approach, has been enriched through principles contained in various sources of human rights law. Therefore, as a precursor to analysing the HRS in further detail, this chapter addresses the secondary research question: What are the human rights principles for addressing key human development challenges; and how do the principles promote inclusive development? The scope of the chapter is limited to analysing four aspects of international human rights (HR) that are directly relevant to understanding the HRS norm, subsequently discussed in Chapter 5, namely: the sources and meaning of HR (see 4.2); an overview of the HR principles (see 4.3); and the indicators for evaluating the performance of the HR principles (see 4.4). Section 4.5 presents the inferences from the chapter. However, this chapter does not cover other wider aspects of international HR law, such as the (non-)existence of a hierarchy in international legal norms, in further detail.

4.2 SOURCES AND MEANING OF HUMAN RIGHTS

This section analyses the sources of international HR law (see 4.2.1), as a precursor to understanding the meaning of HR (see 4.2.2).

4.2.1 Sources

The Statute of the International Court of Justice of 1945, article 38(1) recognises five sources of international law, including: customs, treaties between States, judicial decisions and the writings of publicists, soft law and principles. Principles are treated under the broader category of instruments (Box 2 in Figure 2.2) in the conceptual framework for this thesis and are therefore not specifically discussed below, in order to avoid redundancy.

Customary international law

Customs predate treaties as the oldest source of international law (Alston & Goodman, 2012; Smith, 2010), and are generally established where the following elements exist: (a) concordant practise by a number of States in relation to a particular situation; (b) continuity of that practise over a considerable time period; (c) a sense of legal obligation to follow the practise, *opinio juris*; and (d) general acquiescence of the practice by other States (see Nicaragua v. US (Merits) (1986). It is difficult to establish whether *opinio juris* reflect *lex lata* (the law as it is) or *lex ferenda* (the law as it ought to be), just as it is controversial whether treaties and declarations constitute State practice or *opinio juris* (Roberts, 2001). As a result, some scholars propose that customs have evolved into a less rigid source of international law, the so-called 'modern customs' deduced from treaties and declarations rather than State practice; this portends to be more practical and democratic law (Alston & Goodman, 2012). Nonetheless, a few contradictions do not undermine consistent State practise, provided there is sufficient similarity in the practise among States and the inconsistency is treated as a breach by States.[25]

Customs apply *erga omnes*. However, States can opt out or alter CIL through treaties or varied State usage, though the latter is more difficult to ascertain (Smith, 2010). However, the ratification of a treaty by a majority of States does not automatically convert all the treaty provisions into CIL without proof of States practise. For instance, though the Convention on the Rights of the Child 1989 (CRC) has been ratified by all the UN member states except the US, some provisions are merely aspirational and not all States comply with all the norms, but the principle of the best interest of the child which runs through several provisions of the CRC,[26] is now considered as CIL (Smith, 2010).

Although economic, social, and cultural rights are not commonly listed as part of customary international law, perhaps because the principle of progressive realisation limits evidence of a consistent pattern of gross violations, the gross violation of economic, social and cultural rights like food, health, adequate standard of living and related derived rights that are fundamental to human dignity could qualify as a violation of customary international law (Smith, 2010). Hence, I explore the customary international law (CIL) status of the HRS in Chapter 5.

Treaties

The Vienna Convention on the Law of Treaties of 1969 (VCLT) covers most of the fundamental aspects of the treaty making process (cf. Brölmann, 2012) and its article 2(a) defines a treaty as "an international agreement concluded between States in written form and governed by international law, whether embodied in a single instrument or in two or more

---

[25]   ICJ, Fisheries case (United Kingdom v. Norway), Judgement, 18 December 1951, ICJ Reports 1951, p. 131; ICJ, Continental Shelf case (Tunisia v. Libyan Arab Jamahiriya), Judgement, 24 February 1982, ICJ Reports 1982, p. 74; Continental Shelf case, p. 33; ICJ, Case concerning Military and Paramilitary Activities in and against Nicaragua, p. 98.

[26]   Articles 3(1), 18, and 40

related instruments and whatever its particular designation." Compliance with most treaties is monitored by a designated treaty body mostly through regulatory instruments including a system of regular self-evaluative reports and an individual complaints procedure. States exercise their discretion on the application of treaty provisions through declarations, derogations, denunciations and reservations, where permissible (Shelton, 2014). Where a treaty conflicts with a peremptory norm of general international law or *jus cogens*, such a treaty is void *ab initio*. A peremptory norm in itself "is a norm accepted and recognized by the international community of States as a whole as a norm from which no derogation is permitted and which can be modified only by a subsequent norm of general international law having the same character" (VCLT, article 53).

The International Bill of Rights is comprised of the Universal Declaration of Human Rights 1948 (UDHR), the International Covenant on Civil and Political Rights 1966 (ICCPR) and the International Covenant on Economic, Social and Cultural Rights 1966 (ICESCR). The rights contained in the International Bill of Rights have been expanded through subsequent international HR treaties, mainly during the 1960s and the 1980s (Ramcharan, 1991), including: the International Convention on the Elimination of all Forms of Racial Discrimination 1965 (ICERD), Convention on the Elimination of Discrimination Against Women 1979 (CEDAW), Convention Against Torture and other Cruel Inhuman or Degrading Treatment or Punishment 1984 (CAT), Convention on the Rights of the Child 1989 (CRC), International Convention on the Protection of the Rights of All Migrant Workers and Members of their Families 1990 (ICMW)), Convention on the Rights of Persons with Disabilities 2006 (CRPD), and International Convention on the Protection of All Persons from Enforced Disappearances 2006 (CPED) which is not yet in force. Certain rights are prioritised in some cultures and regions; these results in regional HR treaties which may either promote the cultural diversity of the region,[27] or proscribe cultural practices which offend international public policy.[28]

---

[27] UN Convention on the Rights of the Child 1989, Article 20(3) recognises alternative forms of care for children including kafalah of Islamic law.

[28] For instance, the international community has repeatedly condemned and called for the eradication of the practise of female genital mutilation which is widely practised in some parts of Africa, the Middle East and Asia (WHO 2017). The eradication of the practise would depend on a change of cultures of the affected countries and peoples before legal measures can be successfully established for the purpose.

States exercise their discretion on the application of treaty provisions through reservations,[29] declarations,[30] derogations, and denunciations, where permissible.[31] Treaties may subsequently be amended mostly through an optional protocol containing either additional rights and freedoms,[32] or an optional enforcement mechanism which may have been initially excluded.[33] Where a treaty conflicts with a peremptory norm of general international law or *jus cogens*, such a treaty is void *ab initio* (Smith, 2010).[34] A peremptory norm in itself "is a norm accepted and recognized by the international community of States as a whole as a norm from which no derogation is permitted and which can be modified only by a subsequent norm of general international law having the same character".[35] The conditions for *jus cogens* are that the norm must be: (a) a norm of general international law, (b) expressly or implicitly accepted and recognised by the international community of States as a whole, (c) one from which no derogation is permitted and which can only be modified by a subsequent norm of the same character (Shelton, 2014).

Compliance with most treaties is monitored by a designated treaty body mostly through regulatory instruments including a system of regular self-evaluative reports and individual complaints procedure (see Table 4.1). The functions of the treaty bodies include: considering complaints of treaty violations by States; conducting inquiries into allegations of grave, serious, or systematic violations of HR treaty provisions;[36] considering State-to-State

---

[29]   The Vienna Convention on the Law of Treaties 1969, Article 2(1)(d) defines reservation as: "... a unilateral statement, however phrased or named, made by a State, when signing, ratifying, accepting, approving or acceding to a treaty, whereby it purports to exclude or to modify the legal effect of certain provisions of the treaty in their application to that State."

[30]   An example is the Kingdom of Saudi Arabia that ratified the Convention on the Elimination of All Forms of Discrimination against Women (CEDAW) in 2000 but declared that "In case of a contradiction between any term of the Convention and the norms of Islamic law, the Kingdom is not under obligation to observe the contradictory terms of the Convention." See Concluding Observations of the CEDAW Committee, February 2008, retrieved from https://www.hrw.org/reports/2008/saudiarabia0408/5.htm. Although many States have objected to this declaration, similar reservations and declarations have also been lodged by other Islamic states. The result is that despite the ratification of the treaty, such States can continue to pursue and endorse practices that are prejudicial to the rights of women guaranteed in the CEDAW, within their respective territories.

[31]   The Optional Protocol to the Convention on the Elimination of All Forms of Discrimination against Women, article 17, for instance, does not permit reservations.

[32]   Examples include the Protocols to the United Nations Convention on the Rights of the Child, on the involvement of children in armed conflicts and on the sale of children, child prostitution, and child pornography.

[33]   For instance, see the First Optional Protocol to the International Covenant on Civil and Political Rights which provides for a system of individual petitions.

[34]   Vienna Convention on the Law of Treaties 1969, article 53.

[35]   Vienna Convention on the Law of Treaties 1969, article 53.

[36]   CAT, article 20; Optional Protocol to the CEDAW, articles 8 - 10.

complaints over treaty violations;[37] and convening periodic meetings of States parties to discuss matters of mutual concern for treaty implementation and monitoring (Alebeek & Nollkaemper, 2012). Some treaty bodies also organise general discussions on themes relevant to their work, the outcomes of which may inform the drafting of a new general comment on the interpretation of treaty provisions.[38] The concluding observations of the monitoring body on States' reports may be indicative of the scope and practical nature of the rights contained in the treaty, as well as the direction in which the rights are evolving (Smith, 2010). Other HR treaty monitoring and enforcement mechanisms exist in the form of country reports and thematic procedures, including investigations of HR situations by a special rapporteur concerning specific HR,[39] and world conferences on HR such as the World Conference Against Racism, Racial Discrimination, Xenophobia, and Related Intolerance (2001) and the Rio+20 United Nations Conference on Sustainable Development (2012). At the regional level, HR bodies (RHRB) are also generally established by Member States to promote and protect HR, and in the case of HR courts to resolve disputes over alleged HR violations within the respective regions.

---

[37] The CAT (article 21), ICMW (article 76), ICERD (articles 11-13), and ICCPR (articles 41-43) respectively specify the procedures for the relevant treaty body to consider complaints from one state party against another, on the grounds that the latter is not in compliance with the treaty.

[38] ICERD and CEDAW use the term "general recommendations".

[39] United Nations Charter of 1945, articles 55 and 56.

Table 4.1 Examples of international human rights treaties, treaty bodies and enforcement mechanisms

Year	Treaties	Treaty bodies	Monitoring & enforcement mechanisms
1965	International Convention on the Elimination of all Forms of Racial Discrimination (ICERD)	Committee on the Elimination of Racial Discrimination	Reports; individual complaints under article 14
1966	International Covenant on Economic, Social and Cultural Rights (ICESCR)	Committee on Economic, Social and Cultural Rights	Reports; the United Nations Economic and Social Council (ECOSOC)
1979	Convention on the Elimination of All Forms of Discrimination against Women (CEDAW)	Committee on the Elimination of Discrimination against Women	Reports, individual complaints & inquiries
1984	Convention against Torture and Other Cruel, Inhuman or Degrading Treatment or Punishment (CAT)	Committee against Torture; Subcommittee on Prevention of Torture	Visits, reports, individual complaints under article 22 & inquiries
1989	Convention on the Rights of the Child (CRC)	Committee on the Rights of the Child	Reports
1990	International Convention on the Protection of the Rights of All Migrant Workers and Members of their Families (ICMW)	Committee on the Protection of the Rights of All Migrant Workers and Members of their Families	Reports, individual complaints under article 77

In Chapter 5, I analyse the recognition of the HRS in treaties.

Judicial decisions and the writings of "the most highly qualified publicists"

There are two views in the literature on judicial decisions and the writings of the most highly qualified publicists as a source of international law. One view is that they constitute a direct source of law (Jennings, 1996; Shahabuddeen, 2011). A contrary view, which resonates with the decisions of international courts and tribunals, only regards judicial decisions and writings of the most highly qualified publicists as subsidiary sources but this is not an indication of less importance (see Prosecutor v. Zoran Kupreskic, Mirjan Kupreskic, Vlatko Kupreskic, Drago Josipovic, Dragan Papic, Vladimir Santic, Judgment, IT-95-16-T, ICTY Trial Chamber, 14 Jan. 2000, at 540 (International Criminal Tribunal for the Former Yugoslavia); Prosecutor v. Issa Hassan Sesay, Morris Kallon, Augustine Gbao, Judgment, Case No. SCSL-04-15-T, SCSL Trial Chamber, 2 Mar. 2009, at 295 (Statute of the Special Court for Sierra Leone; Wrońska 2011). In line with this, judicial decisions and writings of the most highly qualified publicists may be regarded as subsidiary or secondary to the preceding 'principal' means in Article 38(1)(a)-(c) (international conventions, international customs, and the general principles of law recognized by civilized nations). Nonetheless, both the 'principal' and 'subsidiary' means for the determination of rules of law supplement each other (Borda 2013). In line with this, I explore judicial decisions on HRS principles and instruments in Chapter 5.

Soft law

Soft law refers to regulatory measures such as the declarations and resolutions of the UN General Assembly,[40] which are normative and bear considerable influence on States' practices without being legally binding *per se* (Kennedy, 1987; Shelton, 2003). The positivist binary distinction between law (because it is justiciable in a court or tribunal) and non-law (because it is not justiciable or legally enforceable) instruments does not recognise soft law (Weil, 1983), even though some positivists recognise that non-binding instruments could provide significant evidence of State practise or consensus on particular rules (Brownlie, 2003).

In the literature, there is a further distinction between two forms of soft law, when an instrument negotiated between parties either: (a) does not take the form of a treaty or any other form recognized as a binding legal instrument under international law (*soft instrumentum*), or (b) is in a legally binding form but does not contain any binding normative commitments e.g. because its provisions are vague (*soft negotium*) (D'Amato, 2009; d'Aspremont, 2008, 2009). The *soft instrumentum* can be further categorised into three: (a) non-obligatory agreements and declarations developed by States (for instance, non-binding political declarations and conference declarations), (b) non-binding decisions of international organisations (including guidelines and programmes), and (c) recommendations adopted by NGOs.

Nonetheless, soft law may be a more appropriate option than hard law when there is need to take urgent action in the face of scientific uncertainty, where legal pluralism prevents consensus on legally binding norms, or where there is need for participation by non-State actors who are not traditionally parties to the formulation of hard law like treaties (Shelton, 2003). Soft law measures like the Declaration on the Rights of the Child 1959, the Declaration on the Elimination of all forms of Racial Discrimination 1963, and the Declaration on the Elimination of Discrimination against women have been important precusors to treaties. In this regard, the UN General Assembly and Human Rights Council

---

[40] See Advisory Opinion, Legal Consequences for States for the Continued Presence of South Africa in Namibia (South West Africa) notwithstanding Security Council Resolution 276 (1970), ICJ, 21 June 1971, p. 50.

resolutions on the rights to water and sanitation have significantly contributed to the status of the HRS as a distinct right as discussed in Chapter 5.

4.2.2 Meaning of Human Rights

The contestations overs the meaning of HR predate the codification of HR principles and standards (Sen, 2004). HR evolved through the different periods of human civilization as a concept for addressing pressing development issues through the tripartite obligations of respect, protection and fulfilment that it imposes on States as the primary duty bearers, as well as non-State actors whose operations affect the realisation of the human rights obligations (see Table 4.2). HR includes economic, social and cultural, and civil and political rights, and are universal, inalienable, interdependent, indivisible and non-discriminatory. It generally imposes three types of obligations on States as the primary duty bearers, namely: to *respect* the right by refraining from any form of interference with the right; to *protect*, by preventing interference by third parties; and to *fulfil*, by facilitating the provision of services where necessary or directly providing the resources to realise required by the right (Schenin, 2003). With such clearly defined principles (see 4.3.3), it is therefore counter-intuitive to propose that "[H]uman rights is facing a crisis of meaning and legitimacy" (Mooney, 2012, p.169). The major challenge is to formulate a definition which reflects the historical aspects of HR without compromising the dynamism of the emergent global practice of a proliferation of HR to address emerging human development challenges (Flynn, 2012; Griffin, 2008).[41]

---

[41]   Flynn uses the term "rights inflation" to describe the increasing number of human rights (Flynn, 2012:3).

Table 4.2 Evolution of human rights principles across civilizations[42]

Period	Proponents & source documents	Main tenets
Antiquity	King Hammurabi's Code (1795-1750 B.C.)	Ethical standards for communal life
	Judaism, Christianity & Islamic traditions	Notions of human dignity; non-discrimination
	Pericles (490-429 B.C.)	Non-discrimination
	Sophocles (469-406 B.C.)	Right to expression & civil disobedience
	Cicero	Brotherhood of man; justice
	Marcus Aurelius	Brotherhood of man; justice; notion of duties
Middle Ages	St. Augustine (396-430)	Precursor of solidarity rights
	Thomas Aquinas (1225-1274)	Human dignity; intellectual rights
	John of Salisbury (1120-1180)	Notions of the common good
	Magna Carter (1215)	Property & family rights
Renaissance	Petrarch (1304-1374)	Education for morality
	Giovanni Pico (1463-1546)	Equality of all persons
	Desiderius Erasmus (1466-1536)	Importance of proper education & non-elitist approach to learning the scripture
	Martin Luther (1483-1546)	Freedom of conscience in the pursuit of religion
	The Vindiciae Contra Tyrannos (1579)	Justice & the rule of law
Enlightenment	John Locke (1632-1704)	Rights to life, freedom & property; poverty as a result of moral failure
	Marie Voltaire (1694-1778)	Freedom of press & religious thoughts
	Thomas Paine (1737-1809)	Poverty as a result of social disorder
	Jean Jacques Rousseau (1712-1778); Gracchus Babeuf (1760-1797)	Equitable allocation of resources
	US Declaration of Independence (1776)	Equality of all men; did not recognise women; referred to the indigenous people as "Indian savages"
	US Constitution (1789) & Bill of Rights (1791)	Freedom of thought & conscience
	French Declaration of the Rights of Man & Citizen	Civil, political & economic rights
Industrialization	Karl Marx (1818-1883); Friedrich Engels (1820-1895)	Civil & political rights as a façade of capitalism
	Malcolm X	Civil rights diminish the notion of human rights
	Papal Encyclicals	Indivisibility of human rights
	Soviet Constitution (1936)	Economic & social rights

Although the term 'human rights' was mentioned only in the UN Charter of 1945, the origins of the concept can be traced far back and has evolved with humanity's constant strive for

---

[42] I compiled Table 4.2 from my literature review of scholarly literature on human rights law, published in the English Language. Nonetheless, literature available in other languages may further provide useful insight into the evolution of human rights principles.

self-actualisation over different periods of human history (see Table 4.2) (Wronka, 2008). The debate over the meaning of HR predates codification (Sen, 2004) and is traceable to a lack of a unified foundation (Mooney, 2012). This is not purely a theoretical question, but one that requires a consideration of the political processes for HR implementation as well (Flynn, 2012). On the political level, after the Cold War, the lack of consensus between the Soviet Bloc and the United States, including its allies, led to the division of indivisible, universal, interdependent, interrelated, and presumably equally important,[43] list of HR into civil and political, and economic, social and cultural, without any formal hierarchy, leading to scepticism over the value of HR.

In the legal literature, there is a disagreement between the natural law theory of the rights of man characterised by a strong individualistic bias and contemporary HR definitions characterised by a strong egalitarian and international orientation (see Table 4.3) (Flynn, 2012). HR are represented both as: (a) a moral theory for mutual recognition and reciprocity among States in international relations and an inherent part of constitutional democracies (Rawls, 1999), and (b) the threshold for the legitimation of State intervention in the defence of individual autonomy (Flynn, 2012). The legal literature also distinguishes between legal rights derived *de jure* from the formal rules of law that are dependent on the relationship between the individual and the state, and non-legal rights or the bare HR which are those *de facto* conditions that make human existence possible and are therefore fundamental to the individual's existence, irrespective of the relationship with the state (Mooney, 2012; Parekh, 2007; van Boven, 1982). The distinction highlights the inherent limitations of a purely legalistic (and narrow) definition of HR, and the paradox that arises where it is the vocal 'rightsholders' who are able to demand enforcement of their rights and become the main beneficiaries, rather than the marginalized 'rightsholders' who require HR protections more (Douzinas, 2000; Gearty, 2006; Mooney, 2012).

---

[43]   Vienna Declaration and Programme of Action 1993, para 5. Nonetheless, the right to life is regarded as being important and necessary for the fulfilment of all other rights but its supremacy is contestable and can be legitimately denied under certain circumstances. For instance, in the execution of a court sentence following conviction for a crime. See European Convention on Human Rights 1950, Article 2; Constitution of the Federal Republic of Nigeria 1999, Section 33.

Table 4.3 Perspectives on the meaning of human rights

Meaning of Human Rights	Brief Description	Key References
Moral theory of international relations & condition for reciprocity among States	Human rights generally determine the standard of behaviour for both liberal & non-liberal States, & regulate the relations between States	Rawls, 1999
Moral threshold for the legitimation of State's extra-territorial interventions	Human rights primarily aim at protecting individual autonomy & form part of political morality	Flynn, 2012
Dependent on the relationship between the State and an individual	Human rights are determined by formal laws which are products of local culture and practice & therefore lacks a universal foundation	Mooney, 2012; Parekh, 2007; Douzinas, 2000; van Boven (1982)
Fundamental to human existence, irrespective of the relationship with the State	Human rights are necessary for human existence, universal & therefore have a defensible foundation globally	Mooney, 2012; Parekh, 2007; van Boven (1982)
Universal, unchanging and applicable to all humans	Human rights are a given and integral to human existence; perspective of natural law scholars	Rogers & Kitzinger (1986); Dembour (2006); Stenner (2010)
Outcome of negotiations between rightsholders and the State	Human rights are negotiated & agreed by stakeholders; perspective of discourse scholars	Rogers & Kitzinger (1986); Dembour (2006); Stenner (2010)
Offers protection for the vulnerable members of the society	Human rights are fought for; perspective of protest scholars	Rogers & Kitzinger (1986); Dembour (2006); Stenner (2010)
Subject for discussion or deliberation	Human rights are to be discussed; perspective of deliberative scholars	Rogers & Kitzinger (1986); Dembour (2006); Stenner (2010)

Remarkably, philosophy scholars define HR broadly as the rights to which all human beings are entitled literally because they are human, and there can be no justification for violation under any circumstances (Donnelly, 2013; Wiseberg, 1996). A mosaic of discourses support the broad definition of HR, including: grounded universals; radical activist politics; socio-political construction; rights and responsibilities; and rights and democracy (Rogers & Kitzinger, 1986; Stenner, 2010). In line with these alternative perspectives, HR may be considered to be: (a) the product of natural law and therefore universal, unchanging, and applicable to all human beings without discrimination; (b) an important instrument for protecting the vulnerable members of society; (c) a subject for deliberation between the rightsholders and the State; (d) the outcome of negotiations and agreements on rights, responsibilities, and democratic principles; and (e) a product of notions of community belonging and religious foundations (Dembour, 2006; Stenner, 2010).

Douzinas (2013, p.52) proposes that "no global 'theory' of rights exists or can be created. Different theoretical perspectives and disciplinary approaches are therefore necessary." This is because HR are the result of multiple discourses, practices, institutions and campaigns (Douzinas, 2013; Mooney, 2012). The lack of consensus on meaning makes it difficult to find consistency in the implementation of HR at the level of individuals. As a result, at present, any inconsistency in the understanding of HR is mainly to be found at the social and institutional level, rather than with individuals (Stenner, 2010). This poses a challenge to the continuous relevance of HR as a legitimate approach for social regulation. Hence, while HR scholars are mainly preoccupied with codification and implementation, the meaning of HR is still in issue; HR is open to either a broad or a narrow interpretation.

Although appreciating the finite diversity of the distinct ways of understanding HR is essential to understanding the meaning and content of HR within different social and cultural settings (Stenner, 2010), it is equally important to have a common understanding of HR among stakeholders in order to strengthen the normative framework, enhance legitimacy and ensure consistency. International HR can regulate State relations and establish the standard for State intervention and extraterritorial actions to protect individual autonomy. At the national and sub-national levels, HR can: (a) counter significant power imbalances (Sinden, 2005; 2009); (b) protect the weakest and all persons who suffer various forms of inequities (de Albuquerque, 2014; London & Schneider, 2012); (c) limit State autonomy and sovereignty (Limon, 2010); (d) prioritise human protection in development policies (White, 2011); and (e) lend a sense of urgency to human development challenges such as the lack of sanitation (Qerimi, 2012). However, these outcomes are mostly achievable where HR are broadly defined to include both rights *de jure*, as contained in the formal sources of international law (see 4.2.1), and *de facto*, in recognition of emerging categories of HR like third generation solidarity rights (Alston, 2009), the fourth generation right to biotechnology (Gupta, 2014), and by extending HR to all persons without discrimination (Mooney, 2012).

4.3   HUMAN RIGHTS PRINCIPLES

General principles of law consist of "logical propositions resulting from judicial reasoning on the basis of existing pieces of international law" (Wrońska, 2011, p.38). International courts have relied on general principles of law applied in national legal systems to determine legal concepts like the legal status of corporations.[44] The HR principles analysed below are mostly derived from treaties (see 4.2.1), and establish a minimum standard of conduct for both States and non-State actors particularly when integrated with substantive HR (Haugen, 2011). I categorise the principles that I derive from my literature review and content analysis along the three pillars of sustainable development which are reflected in my conceptual framework (see 2.4.3), namely: social, economic and environmental principles. While this approach shows that the HR principles are mostly social, I concede that in practise the impact of the principles may extend to the other pillars of sustainable development. For instance, empowerment is a social principle that may nonetheless require the equitable distribution of economic resources and thereby impact economic growth.

Accountability

Accountability is enshrined in the ICESCR, article 2(1) and ICCPR, articles 2(1) and (2). Accountability affords rightsholders with an opportunity to obtain an explanation from duty-bearers on how or why they failed to discharge their duties and a remedy or reparation for any losses they suffer as a result; it also affords the duty-bearers an opportunity to explain their actions or omissions without necessarily implying punishment for the violation of duties. Accountability can be achieved through a variety of mechanisms, including: (a) judicial (review of executive acts by the courts), (b) quasi-judicial (e.g. international HR treaty bodies, ombudsmen), (c) administrative (e.g. HR impacts assessment), and (d) political (e.g. parliamentary processes).[45]

Dignity

Human dignity is a core principle of international HR law based on the premise that every human being deserves respect, which supports a broad definition of HR (see 4.3) and inspires the universal application of the International Bill of Rights.[46] The Vienna Declaration and Programme of Action equally "affirms that extreme poverty and social exclusion constitute a violation of human dignity."[47] However, the source of human dignity and its links with HR is still contested. One view, expressed in the political science literature, is that: (a) regard for human dignity is central to non-Western cultural traditions (like the African, Chinese, Islamic and Indian), (b) HR as understood by Westerners (entitlements inherent to human beings) are

---

[44] See, for instance, the Barcelona Traction Co. Case (1970).
[45] Office of the United Nations High Commissioner for Human Rights, Principles and Guidelines for a Human Rights Approach to Poverty Reduction Strategies HR/PUB/06/12.
[46] Preamble ICESCR and ICCPR; UDHR, article 1.
[47] Doc. A/CONF.157/23 of 12 July 1993, Vienna Declaration and Programme of Action, para. 25.

alien to the non-Western approaches to dignity, and (c) there are other ways to protect human dignity besides HR (Donnelly, 1982; 2013). The ethical theory literature further argues for separating HR and human dignity on three grounds: (a) human dignity aggravates the justification problem for HR in secular societies; (b) HR hinged on human dignity loses universal validity, and (c) human dignity has become more controversial than HR (Schroeder 2012). Conversely, the legal literature argues that the meaning of human dignity is context specific and varies significantly both within and across jurisdictions; in the absence of a common understanding of the substantive meaning of human dignity it still plays a significant role in HR adjudication and provides "a language in which judges can appear to justify how they deal with issues such as the weight of rights, the domestication and contextualization of rights, and the generation of new or more extensive rights" (McCrudden, 2008, p.724).

Equality and non-discrimination

Equality and non-discrimination are enshrined in various treaty provisions including the International Bill of Rights, CEDAW, CRC and ICERD (see Table 4.4).[48] Equality specifically guarantees equal treatment of all persons by the law and protection against arbitrary treatment and discriminatory practices on the unlawful grounds of "race, colour, sex, language, religion, political or other opinion, national or social origin, property, birth, disability and health status, including HIV/AIDS, age, sexual orientation or other status."[49] Inequalities and discrimination may either occur as overt "legal inequalities in status and entitlements, deeply rooted social distinctions and exclusions", or in more covert forms.[50] Nonetheless, not every case of differential treatment amounts to a violation of the principle, except where: (a) equal cases are treated differently, (b) the difference in treatment is not reasonably justifiable or objective, or (c) if the difference in treatment is not proportional to the purpose of the action.[51]

Public participation

The CEDAW protects women's right to participate in political and public life, represent their government at the international level, and participate in education, economic and social

---

[48]   See UDHR, articles 1, 2, 4, 7; ICESCR, articles 2(2), 3; ICCPR, articles 291), 3, 24(1), 26; CRC, article 2; and the UN Charter, articles 1(3), 55(c), 56.

[49]   Office of the United Nations High Commissioner for Human Rights, Principles and Guidelines for a Human Rights Approach to Poverty Reduction Strategies HR/PUB/06/12, p. 9.

[50]   Ibid.

[51]   See, the Inter-American Court, *Advisory Opinion No. 4*, para. 57; *Marckx v. Belgium* (European Court); Jacobs v Belgium (the Human Rights Committee)

benefits accruing in their society, and for rural women to participate in development planning and community activities.[52] The ICESCR also provides under the right to education "that education shall enable all persons to participate effectively in a free society, promote understanding, tolerance and friendship among all nations and all racial, ethnic or religious groups, and further the activities of the United Nations for the maintenance of peace."[53] The ICESCR also recognizes the right to participate in cultural life.[54] The ICCPR and the CRC also enshrine participation.[55] Public participation requires active engagement with the public through four stages: (a) preference revelation (this occurs before policy formulation and allows people to express the objectives they want to achieve through the policy); (b) policy choice (this enables people to participate effectively in the policy formulation process); (c) implementation (this is achieved through decentralization and subsidiarity); and (e) monitoring, assessment and accountability (which ensures that people affected by policies take part in the monitoring and evaluation process and can also hold the duty-bearers to account).[56]

Transparency and empowerment

Transparency is explicitly recognized in the CRPD, article 4(3) which provides that State Parties shall "closely consult with and actively involve" persons with disabilities in the process of formulating and implementing the legal framework for implementing the CRPD. Such close consultation requires transparency on the part of the State (Haugen, 2011). Transparency empowers rightsholders through ensuring public access to information held by the State about public processes, decisions, and outcomes that affect their lives. It thereby supports effective participation, accountability and monitoring.

---

[52]   CEDAW, articles 7(b), 8, 10(g), 13(d), 14(2)(a)(f). The participation principle is also enshrined in the Rio Declaration, principles 10, 20 and 22.
[53]   ICESCR, article 13(1)
[54]   ICESCR, article 15(1)(a)
[55]   ICCPR, articles 19, 21, 22(1), 25; CRC, articles 13, 15, 31.
[56]   Office of the United Nations High Commissioner for Human Rights (n. 49).

Table 4.4 Overview of human rights principles and their legal status

Human Rights Principles	Examples of Source Documents	Legal Status	Key References
Dignity	1948 Universal Declaration of Human Rights 1993 Vienna Declaration and Programme of Action	Established in hard and soft law	Donnelly, 1982, 2009, 2013; McCrudden, 2008
Rule of Law	1948 Universal Declaration of Human Rights 2000 United Nations Millennium Declaration 2005 World Summit Outcome 2012 Declaration of the High-level Meeting on the Rule of Law	Established in hard and soft law	Farrall, 2009; Janse, 2013
Participation	1966 International Covenant on Civil and Political Rights 1966 International Covenant on Economic, Social and Cultural Rights 1979 Convention on the Elimination of Discrimination Against Women 1989 Convention on the Rights of the Child 1992 Rio Declaration	Established in hard and soft law	Donnelly, 2013
Equality and Non-discrimination	1965 International Convention on the Elimination of all Forms of Racial Discrimination 1966 International Covenant on Civil and Political Rights 1966 International Covenant on Economic, Social and Cultural Rights 1979 Convention on the Elimination of Discrimination Against Women	Established in hard and soft law	Donnelly, 2013
Transparency and Access to information	1992 Rio Declaration 1998 Aarhus Convention 2006 Convention on the Rights of Persons with Disabilities	Established in hard and soft law	Donnelly, 2013
Accountability	1966 International Covenant on Civil and Political Rights 1966 International Covenant on Economic, Social and Cultural Rights	Established in hard and soft law	Donnelly, 2013

4.4 INDICATORS FOR MEASURING AND EVALUATING HUMAN RIGHTS PRINCIPLES

There is a pressing need for improved methodologies for objectively measuring and evaluating HR violations, compliance and general progress, and HR indicators are among the most powerful quantitative tools available. The realisation of HR may be constrained by any of the following five factors. First, compliance largely depends on the internalization of international law norms within the domestic legal framework and inherent right of self-determination constrains enforcement through international law mechanisms (Koh, 1999). Second, States may be directly or indirectly complicit in HR violations (Collingsworth, 2002) and the local political circumstances are crucial to HR implementation (Zhou, 2012). Third, the lack of capacity and resources, especially for poor States, may hamper the provision of basic services required under the ICESCR, for instance. Fourth, HR enforcement raises the issue of the principal's moral hazard as State agencies are responsible for the protection of rightsholders, irrespective of the best interest of their principal (the State) (Miller, 2005). Non-State actors, like NGOs, also face the challenge of balancing accountability to donors with accountability to communities (Mutua, 2013). Fifth, there may be inherent power imbalances within the domestic framework which hamper voluntary compliance with HR standards and therefore require stronger regulatory mechanisms, just like commercial and property rights are usually protected by enforceable contracts. It is therefore important to support the HR framework with indicators for measuring and evaluating performance. This section therefore presents a brief overview of HR obligations and indicators, based on a literature review and content analysis.

HR indicators are the pieces of information used to measure the degree of fulfilment or compliance of a right in any given context (Green 2001). HR indicators also include information on events, as well as outcomes, objects or activities through which HR principles and instruments can be assessed and monitored.[57] There are three functions of HR indicators, namely: (a) clarity in monitoring compliance with obligations and progressive implementation, (b) an objective assessment of the impacts of human development programming based on HR standards, and (c) evidence of HR violations (de Beco 2010; Rosga and Satterthwaie 2009; Welling 2008).[58]

---

[57] See The Use of Indicators in Realizing ESCR, Report of the High Commissioner for Human Rights UN Doc. E/2011/90 (13 May 2011); Report on Indicators for Promoting and Monitoring the Implementation of Human Rights UN Doc. HRI/MC/2008/3 (6 June 2008).
[58] Ibid.

The formulation of HR indicators involves multiple disciplines, methodologies and standardisation of processes which transcends the international HR law sphere (Rosga & Satterthwaie, 2009; Welling, 2008). Relevant considerations in the selection of HR indicators include: a) a clear link to the relevant HR normative framework; b) disaggregation to capture minorities; c) balance between context specific and universal indicators; d) practicality of use and comparison of data; e) compliance with legal, ethical and HR safeguards in the collection, processing and dissemination of data; and f) data reliability and validity.[59] Indicators may be quantitative, qualitative, structural, process related or outcome related.[60] Structural indicators capture the necessary institutional mechanisms for HR implementation; process indicators link HR policies with milestones that translate into outcome indicators reflecting the status of HR realization.[61]

Compliance monitoring through the use of indicators involves the selection and contextualisation of indicators, selection of benchmarks and targets as required by the specific context, reporting on the indicators, benchmarks and targets, and monitoring reported indicators to supplement recommendations made by HR mechanisms. This involves highly technical processes but at the same time, human judgement and local actors play a crucial role which ought to be reflected in the HR monitoring process for two main reasons. First, sole reliance on quantitative indicators could reduce the measurement and evaluation of HR into a technical exercise and erode human judgment and democratic accountability (Rosga & Satterthwaie, 2009). Second, there is a tension between the rationalist approach to the development of a global set of indicators and the decentralisation of the process (McGrogan, 2016). Hence, while a global set of indicators allows for monitoring HR over different timescales and across countries, it is further necessary to ensure that the local actors can contextualise the indicators to meet local circumstances and thereby enrich the practical relevance of the HR indicators locally.

4.5 INFERENCES

This chapter draws four main conclusions that lay the foundation for further analysis of the HRS institution and instruments in the remaining chapters. First, human rights have evolved over centuries and suffer from two contestations: (a) the structure of some HR systems may create a paradox where it is the vocal 'rightsholders' (like the population in formal

---

[59] The Use of Indicators in Realizing ESCR (n. 57).
[60] Report on Indicators (n. 57).
[61] Ibid.

settlements) who are able to demand enforcement of their rights rather than the marginalized 'rightsholders' (like the population in informal settlements) for whom these rights are meant to protect; and (b) conflicts could occur between human rights held by individuals or groups at different levels of governance; for instance, at the international level there are inherent conflicts between the right to development and environmental protection heralded under the emergence of the right to promote sustainable development in the United Nations Framework Convention on Climate Change (Gupta & Arts, 2017), while at the local level, individual rights like the freedom of conscience and religion may inadvertently limit equality and universality (Bayefsky (2001) discusses universality in the UN human rights system). Considering the contestations and their effects on the human rights institution is particularly relevant for the implementation of the HRS.

Second, there are seven core human rights principles, namely: accountability, dignity, equality and non-discrimination, participation, rule of law, transparency and empowerment. These principles are mostly established in treaties, but have also evolved from national laws. The human rights principles mainly relate to the social pillar of sustainable development and focus on improving transparency and the effectiveness of participatory processes while countering power politics that lead to marginalisation or exacerbate inequalities in human development, thereby advancing social and relational inclusion. The human rights principles are much less directly linked to the ecological component of ID, although the right to a healthy environment exists with varying legal status in various human rights instruments and the right to promote sustainable development is in a legally binding Convention. Nonetheless, human rights principles such as participation play a prominent role in the development of environmental rights.

Third, the predominance of social and relational HR principles and their limited relevance to ecological inclusion further points to the need to clarify how and under what circumstances the HR normative framework can address ecological concerns, in the interest of ID, and where necessary integrate complementary environmental principles (see 5.3.2 and 7.2.3).

Fourth, HR institutions often impose mainly State-centric human rights obligations without adequately addressing the liability of non-State actors. It is therefore equally important to advance HR principles as a guiding norm for the activities of non-State actors who are increasingly involved in the delivery of services and other activities affecting the fulfilment of HR. This can be linked to the emerging discourse on the liability of non-State actors for HR violations (Obani & Gupta, 2016). While HR indicators are useful for monitoring compliance by the duty bearers, the indicators by themselves do not guarantee the best outcomes for the most vulnerable rightsholders without opportunities for human judgement and downward accountability to the rightsholders. This underscores the need to decentralise HR formulation and empower the weakest rightsholders to demand accountability from the relevant duty bearers.

Chapter 5. Human Right to Sanitation Principles

5.1 INTRODUCTION

As shown in Chapter 4, international human rights (HR) law mostly prioritises social and relational inclusion which has implications for human development and issues of access to resources like water, sanitation and hygiene facilities, characteristically falling within the domestic jurisdiction. The subsidiary research questions for this chapter are therefore: (a) How did the international human right to sanitation emerge and what does it mean across multiple levels of governance? (b) What are the principles of the human right to sanitation and how can progress towards realising the right be monitored? (c) How do the principles of the human right to sanitation address the drivers of poor sanitation services to promote inclusive development? The chapter adopts a multi-level governance approach because the human right to sanitation (HRS) initially evolved internationally through a bottom-up approach, as a result of local advocacy efforts and HRS norms now transcend multiple levels of governance. It analyses the emergence, legal bases and status of the international HRS (see 5.2); the analyses is enriched through content analysis of HRS legislations in 67 States across Africa, Asia, Europe, North America, Oceania and South America (see Annex I), and decisions of national, regional and international courts. The remaining sections address the principles of the HRS (see 5.3), and the instruments through which the HRS principles are operationalised (see 5.4). Section 5.5 discusses the structural, process and outcome indicators for monitoring the HRS, while Section 5.6 discusses the HRS principles in relation to drivers and ID. Further, Section 5.7 articulates the key inferences derived from this chapter.

5.2 EMERGENCE, LEGAL BASIS, AND MEANING OF THE HUMAN RIGHT TO SANITATION

This section analyses the emergence of the HRS in international law (see 5.2.1) before exploring the legal bases (see 5.2.2) and status (see 5.2.3), and the meaning of the right (see 5.2.4). In exploring the emergence, I make relevant references to the closely linked human right to water (HRW) because of the similarities in the evolution of both rights despite their differences illustrated in Table 1.1 in Chapter 1. In exploring the legal status and meaning, I also examine regional and national instruments which support better understanding of the normative elements of the international HRS.

5.2.1 Emergence

I begin this section with an overview of developments in the emergence of the international HRW that are relevant for the emergence of the HRS. The HRW itself was initially not expressly recognized in the International Bill of Rights (see Chapter 4). Prior to the evolution of the HRW in its current understanding as an individual right, the need for equitable access to clean water was central to the international environmental policy and sustainable development discourse since the 1970s. The incorporation of human rights principles like availability, safety and affordability in the 2015 Sustainable Development Goals (SDGs) water and sanitation target is evidence of the synergies between international environmental policy, sustainable development and human rights discourse. Table 5.1 illustrates key international environmental, water and (sustainable) development law and policy instruments and political declarations with provisions relevant to the rights to water and sanitation, without distinguishing the legal status of the instruments.

Table 5.1   International environmental, water and (sustainable) development instruments and political declarations relevant to the human rights to water and sanitation principles

Year	Instruments	HRWS Principles				
		Acceptable	Accessible	Affordable	Safe	Sufficient
1972	Stockholm Declaration				■	
1977	Mar del Plata Convention		■		■	■
1992	Rio Declaration*		■		■	
1992	Agenda 21*		■	■	■	■
1997	UN Watercourses Convention		■			■
1999	UNECE Protocol on Water and Health*		■		■	■
2000	UN Millennium Development Goals*		■		■	■
2001	Bonn International Conference on Freshwater*		■	■	■	
2002	Johannesburg Plan of Implementation*			■	■	
2004	ILA Report	■	■	■	■	■
1992	Dublin Statement*			■	■	
1994	International Conference on Population and Development*		■		■	
2001	Recommendation (2001) of the Committee of Ministers to Member States on the European Charter on Water Resources		■		■	■
2006	Abuja Declaration*		■		■	
2007	Message from Beppu, First Asia-Pacific Water Summit*		■		■	
2008	Delhi Declaration*		■		■	
2015	Sustainable Development Goals*		■	■	■	■
*The instrument also explicitly addresses human sanitation needs.						

Remarkably the instruments mainly focus on accessibility (including physical accessibility), affordability, safety and sufficiency, but rarely on acceptability. Acceptability is crucial to the HRS, and this is a salient difference between water and sanitation because the latter also needs to be culturally acceptable to ensure use (see 3.4.1). Other recent political declarations explicitly recognising the HRS include the 4th and the 5th South Asian Conference on Sanitation (SACOSAN), held in 2011 and 2013, respectively; and the Ngor Declaration on Sanitation and Hygiene adopted at the Fourth African Conference on Sanitation and Hygiene, held in 2015.

Within the human rights framework, some scholars and water rights advocates argued that a HRW can be implied from specific rights in the International Covenant on Economic, Social and Cultural Rights 1966 (ICESCR) (see for instance, Gleick, 1998; McCaffrey, 1992; Kiefer & Brölmann, 2005). Similarly, the HRS is equally essential for the attainment of relevant ICESCR rights (Ellis & Feris, 2014; Meier et al., 2014; Obani & Gupta, 2015). The HRW was initially recognized as subordinate to economic, social and cultural rights like the rights to an adequate standard of living,[62] food,[63] housing,[64] health[65] and development[66] *inter alia*. Sanitation is also necessary for the realisation of various economic, social and cultural rights, illustrated in Table 5.2, and it is this realisation that has culminated in the recognition of the HRS. Nonetheless, in 2002, the Committee on Economic, Social and Cultural Rights (CESCR) adopted General Comment No. 15 (GC 15) which remarkably recognises the HRW as self-standing and essential for realising an adequate standard of living and other related ICESCR rights (Obani & Gupta, 2015). The HRW has also been recognised in treaties *ratione personae* like the Convention on the Elimination of All Forms of Discrimination against Women 1979 (CEDAW), and the Convention on the Rights of the Child 1979 (CRC). Further, the HRWS have been upheld in various resolutions of the UNGA and the UN HRC,

---

[62] International Covenant on Economic, Social and Cultural Rights (New York, 16 December 1966; in force 3 January 1976) ('ICESCR') Article 11;

[63] UN Committee on Economic, Social and Cultural Rights, General Comment No. 15, The Right to Water (Articles 11 and 12 of the Covenant) (UN Doc. E/C.12/2002/11, 20 January 2003) ('General Comment No. 15'), at paragraph 6.

[64] UN Committee on Economic, Social and Cultural Rights, General Comment No. 4, The Right to Adequate Housing (Article 11 (1) of the Covenant) (UN Doc. E/1992/23, 13 December 1991) ('General Comment No. 4'), at paragraph 8(b).

[65] UN Committee on Economic, Social and Cultural Rights, General Comment No. 14, The Right to the Highest Attainable Standard of Health (Article 12 of the Covenant) (UN Doc. E/C.12/2000/4, 11 August 2000) ('General Comment No. 14'), at paragraphs 11, 12 (a), (b) and (d), 15, 34, 36, 40, 43 and 51.

[66] Declaration on the Right to Development, 4 December 1986, (UNGA Resolution A/RES/41/128, 4 December 1986), Article 8.

including the UNGA Resolution 70/169 of 2015 which clearly recognized sanitation as a self-standing right, without vote.

Prior to the 2015 UNGA Resolution, the HRS evolved through UN human rights mechanisms like the Sub-Commission on Human Rights, while also being considered in relation to other rights as illustrated in Table 5.2 (Kamga, 2013; Obani & Gupta, 2015; Winkler, 2016). The GC 15 only considered sanitation marginally, highlighting the importance of the HRS for water quality and human dignity and privacy. Nonetheless, in 2010, the CESCR issued a "Statement on the Right to Sanitation"[67] reaffirming the interrelatedness of the HRS with the HRW (Obani & Gupta, 2015). The normative content of the HRS has been enriched at the international level through the work of the Special Rapporteur on the Human Right to Safe Drinking Water and Sanitation. The HRC first established the mandate of the Independent Expert on the Issue of Human Rights Obligations Related to Access to Safe Drinking Water and Sanitation in 2008 and appointed Catarina de Albuquerque to the position that same year. The position was however renamed Special Rapporteur on the Human Right to Safe Drinking Water and Sanitation in 2011 and the tenure of Ms. de Albuquerque was extended for another three years until 2014, when Mr. Léo Heller was appointed as the new Special Rapporteur.

---

[67]   UN Committee on Economic, Social and Cultural Rights, Statement on the Right to Sanitation (UN Doc. E/C.12/2010/1, 19 November 2010) ('Sanitation Statement').

Table 5.2   Evolution of a distinct international human right to sanitation through interrelated rights

Human rights	Source Documents	Key References
Life; Healthy environment	1982 CCPR/C/21/Rev.1 2009 A/HRC/12/24	Kamga, 2013; Winkler, 2016
Adequate housing	1992 E/1992/23 Annex III 2002 E/CN.4/2002/59 2009 A/HRC/12/24 2009 E/C.12/2008/2	Kamga, 2013; Winkler, 2016
Adequate standard of living	1998 E/CN.4/1998/53/Add.2 2008 E/CN.4/Sub.2/2005/25 2009 A/HRC/12/24	Kamga, 2013; Winkler, 2016
Health	2000 E/C.12/2000/4 2005 CRC/C/GC/7/Rev 2008 E/CN.4/Sub.2/2005/25 2009 A/HRC/12/24 2009 CRC/C/GC/11	Kamga, 2013; Winkler, 2016
Work	2000 E/C.12/2000/4 2009 A/HRC/12/24	Kamga, 2013; Winkler, 2016
Water	2002 E/C.12/2002/11 2009 A/HRC/12/24	Kamga, 2013; Obani & Gupta, 2015; Winkler, 2016
Education	2003 CRC/GC/2003/4 2009 A/HRC/12/24	Kamga, 2013
Water and Sanitation	2006 E/C.12/GC/19 2009 A/HRC/12/24 2010 A/HRC/RES/12/8 2011 A/HRC/RES/18/1 2012 A/HRC/RES/21/2 2013 A/HRC/RES/24/18 2014 A/HRC/RES/27/7	Obani & Gupta, 2015; Winkler, 2016
Sanitation (as a distinct right)	2007 A/HRC/6/3 2015 A/RES/70/169	Obani & Gupta, 2015; Winkler, 2016
Physical security	2009 A/HRC/12/24	Winkler, 2016
Prohibition of inhuman treatment	2009 A/HRC/12/24	Winkler, 2016
Equality of men & women	2009 A/HRC/12/24	Winkler, 2016
Prohibition of discrimination	2009 A/HRC/12/24	Winkler, 2016

5.2.2 Legal Basis

The literature on the HRS recognises it as a distinct right and analyses its legal foundations (Winkler, 2016), but the question of the legal status of the right remains unanswered. It is the legal status, based on the legal foundations, that determines the nature of obligations established by the right, the relationship between the duty bearers and rightsholders, and the available remedies in case of violation of the right. This is all the more significant because the HRS is mostly implied in treaties (see 5.2.2) and only recently emerged as a self-standing right (see 5.2.1). In this section, I therefore analyse the legal bases for the HRS together with the legal status of the right.

Customary international law

The evolution of HRS into customary international law (CIL) requires evidence of: (a) concordant practise by a number of States in relation to a particular situation; (b) continuity of that practise over a considerable time period; (c) a sense of legal obligation to follow the practise, *opinion juris*; and (d) general acquiescence of the practice by other States (see 4.4.2). I now analyse political declarations of States and the recognition of the HRS in national constitutions and laws, as evidence of State practise (Bates (2010) analyses the customary international law status of the HRW and argues that it is a principle of customary international law based on usage and *opinio juris*, but that this further requires formal recognition by an international court.

States have increasingly made declarations and committed to promoting the HRWS, and the recent SDGs commitment which incorporates HRWS principles supports the emergence of custom (Obani & Gupta, 2016). However, taking HRS as a distinct right, States have not shown consistency in implementing the commitments reached on sanitation. A Government Spending Watch report on the financing of the MDGs sectors in 67 low and lower-middle income countries, shows that only three out of thirty-one countries with data on MDGs financing actually allocated the recommended 1.5% of GDP to water, sanitation and hygiene services (WASH) in 2014 (Martin & Walker, 2015). And on an average, since 2011, WASH has accounted for only 0.9% of GDP and 2.3% of government annual budget (Martin & Walker, 2015). The figures highlight both the underfunding of WASH and the difficulty of tracking sanitation financing as a result of the fragmented implementation and linked funding with water. Other declarations like the eThekwini commitments on sanitation in Africa, under which State Parties committed to a national budgetary allocation of a minimum of 0.5% of GDP for sanitation and hygiene programs *inter alia* are also poorly implemented.

States have also increasingly recognised the HRS in their national constitutions and laws, which indicates State practice to some extent. However, the numbers of HRS legislations are still few and fairly recent. The UN-Water global analysis and assessment of sanitation and drinking-water (GLAAS) 2014 report indicates that 63 countries out of the 94 respondents recognise the HRS in their legislations. In addition, countries like Angola, Algeria and Nicaragua, also have HRS legislations while Burkina Faso recognised the right in 2015. Hence, in total, there are about 67 States with HRS legislations globally. The States are specified in Annex I, according to their continents. However, the Tables in Annex I only show States that were recorded in the 2010 UNGA Resolution A/64/292 on the human right to water and sanitation, and therefore exclude Cook Island, South Sudan, Tonga, West Bank and Gaza Strip even though they have legislation on the HRS. Nonetheless, the uniformity and consistency in the HRS legislations that is required for CIL status is lacking. For instance, the United Kingdom while it recognised the HRS in the Statement on the Human Right to Sanitation of 2012 also declared that the right does not require "the collection and treatment of human waste" (see 5.2.3).[68] This is a very limited conception of the HRS which falls short of basic sanitation both under the HRS (see 5.2.3) and the SDGs (see 3.2 and 3.5.5).

Hence, the HRS is yet to fully evolve into CIL due to variations in actual State practices (and so I do not analyse the need for formal recognition of the CIL status by an international court as suggested by Bates (2010) in the case of the HRW). Nonetheless, the HRS is nonetheless binding where it is either expressly or implicitly contained in legally binding instruments like treaties, constitutions and laws.

Treaties

As already mentioned, the HRS is implied from the ICESCR, particularly articles 11 and 12, which respectively guarantee "an adequate standard of living" and "the enjoyment of the highest attainable standard of physical and mental health", although the ICESCR and the General Comments on these articles did not expressly address sanitation.[69] The ICESCR provisions are subject to progressive realisation and require States to cooperate and apply the maximum of their available resources for this purpose (see 5.2.3) (Kiefer & Brölmann, 2005).[70] The HRS is also implied from other human rights treaties of limited scope either *ratione loci* or *ratione personae* like the CEDAW,[71] CRC,[72] and the 2006 Convention on the Rights of Persons with Disabilities (CRPD). The CEDAW imposes an obligation on States to ensure the right of women in rural areas to an adequate standard of living, including sanitation and water;[73] this falls short of universal recognition of the HRS for all women in other vulnerable settings. The CRC obliges States to conduct hygiene and environmental

---

[68] United Kingdom, Statement on the human right to sanitation, Paragraph 2. Retrieved from https://www.gov.uk/government/uploads/system/uploads/attachment_data/file/36541/human-right-sanitation270612.pdf.

[69] Human Rights Council, Report of the independent expert on the issue of human rights obligations related to access to safe drinking water and association, A/HRC/12/24, 1 July 2009.

[70] ICESCR, article 2(1).

[71] Article 14(2)(h).

[72] Article 24(2)(e).

[73] CEDAW, article 14(2)(h).

sanitation education to promote children's right to health; this is presumably broader than equating sanitation to excreta management (see 3.2) but does not oblige States to provide sanitation services but only sanitation education. The CEDAW and CRC significantly delink the rights to water and sanitation.

The HRS can be implied from regional human rights treaties. For instance, in Africa, the HRS can be implied from the right to health in the African Charter of Human and People's Rights 1981 and the African Charter on the Rights and Welfare of the Child 1990, the Senegal River Water Charter 2002; the rights of women to a healthy and sustainable environment and States' duty to provide women with access to clean drinking water as part of measures to promote their right to nutritious and adequate food in the Protocol to the African Charter on Human and Peoples' Rights on the Rights of Women in Africa 2003. The Senegal River Water Charter 2002 recognises the HRW and authorises free access for domestic uses. The African Convention on the Conservation of Nature and Natural Resources 2003 further obliges States to guarantee "a sufficient and continuous supply of suitable water" for their populations; and the Niger Basin Water Charter 2008 guarantees the HRW as well as the right to access information, and participate in water governance. While these provisions do not expressly guarantee the HRS, they can be considered to imply the HRS and are especially salient on the development of the normative content of the right like affordability, access to information and participation.

The implied recognition of HRS imposes limitations on the normative development of the right that are easily illustrated by the limitations on a subordinate HRW: (a) the HRW became solely dependent on the legal remedies for the primary right, (b) each State has a different set of obligations for each primary right,[74] and (c) the amount of water needed to fulfil the primary right and the HRW varies (Obani and Gupta, 2015). In the South African case of *Johnson Matotoba Nokotyana and Others v Ekurhulen Metropolitan Municipality and Others*,[75] the court held that the right of access to adequate housing which is guaranteed under national law cannot be interpreted to include basic sanitation in the absence of minimum standards for housing. Nonetheless, in the absence of a treaty on the HRS, the right

---

[74]   Despite increasing constitutional convergence, national constitutions still grant varying legal status to human rights. For instance, while economic, social and cultural objectives are classified as non-justiciable fundamental objectives and directive principles of State policy in the Constitution of the Federal Republic of Nigeria 1999, similar rights are guaranteed and justiciable under the Constitution of the Republic of South Africa (Act No. 108 of 1996).

[75]   [2009] Constitutional Court CCT31/09, reported in WaterLex 2014.

can be implied from existing binding legal rights in courts at various levels of governance and benefit vulnerable groups like women (through the CEDAW) and children (through the CRC), especially where lack of sanitation results in violation of other HR guarantees. This also allows HRS to benefit from the existing enforcement mechanisms for the respective HR from which HRS may be implied (see 4.2.1).

Judicial decisions

Courts at different levels of governance have been instrumental to developing the normative content of the HRS and promoting implementation (Obani, 2015). Even in the absence of formal recognition of HRS in national legislations, national, regional and international courts have relied on other national and international human rights guarantees, including the rights to life, health, adequate housing, and education, and the guarantees of dignity and freedom from inhuman and degrading treatment, as a basis for recognising the rights to water and sanitation. Some examples of such cases are summarised in Table 5.3.

Soft law

Although the resolutions of the UNGA and HRC on the HRS indicate increasing State commitment and support for the HRS as a distinct right (see 5.2.1), the resolutions are not legally binding *per se*. It is however important to note that the reservations which States expressed about the HRWS while voting the UNGA Resolution A/64/292 did not consider HRS separately otherwise, there may have been less concerns over the implications of the HRWS for State sovereignty, for instance (as expressed by States like Ethiopia). In total, 33 African, 35 Asia, 22 European, 18 North American, 4 Oceanian and 10 South American States voted in favour of the resolution. The details of the States that abstained or were absent are also presented in Annex I, according to their continents. The HRC resolutions are significant because they are based on the UN Charter and therefore apply generally to all UN Member States, while GCs are only relevant to ICESCR State Parties. Remarkably, the United States of America was among the States that adopted the HRC Resolution even though it viewed the rights as being a derived right from the ICESCR and the United States is not a party to the ICESCR. It is equally remarkable that many national HRS legislations followed in the wake of the UNGA and HRC resolutions (like Kenya, Morocco, Zimbabwe and Tunisia; see Annex I), and more can be expected with the distinct recognition of HRS by the UNGA in 2015.

Table 5.3 Judicial decisions on the human right to sanitation

Case	Summary
Municipal Council, Ratlam v. Shri Vardhichand & Others (1980)[d] India	A court can through affirmative action compel a statutory body to provide sanitation services.
Carlos Alfonso Rojas Rodríguez c/ ACUAVENORTE y Otros (1992)[d, e] Colombia	Connection to the sewage network is a fundamental right.
Government of the Republic of South Africa and Others v Grootboom and Others (2000)[d, e, g] South Africa	The State is obliged to fulfil the basic needs of extremely poor people, including access to sanitation for the homeless.
Case of the Sawhoyamaxa Indigenous Community v. Paraguay (Merits, Reparations and Costs) (2006)[d, e] Paraguay	Lack of sanitation services is a threat to the right to life of the indigenous community.[76]
Carolina Murcia Otálora c/ Empresas Públicas de Neiva ESP (2009)[f] Colombia	Service disconnections for non-payment, particularly affecting vulnerable groups, due to involuntary reasons violates human rights.
Hernán Galeano Díaz c/ Empresas Públicas de Medellín ESP, y Marco Gómez Otero y Otros c/ Hidropacífico SA ESP y Otros (2010)[c, d, e] Colombia	Failure to connect a property to the sanitation network is a human rights violation.
Ibrahim Sangor Osman v Minister of State for Provincial Administration & Internal Security eKLR (2011) Kenya [b, g]	Forcible eviction of an informal settlement and cutting off their access to sanitation, *inter alia*, violated their human rights.
Beja and Others v Premier of the Western Cape and Others (2011)[a, g, h] South Africa	Providing unenclosed toilets to a poor community violates the rights to dignity and adequate housing; failure to ensure meaningful participation is also a human rights violation.
Habeas Corpus Colectivo presentado por Víctor Atencio c/ el Ministerio de Gobierno y Justicia, Director General del Sistema Penitenciario (2011)[e] Panama	Failing to provide detainees with adequate access to sanitation constitutes amounts to inhuman or degrading treatment and violates dignity.[77]
Dagoberto Bohórquez Forero c/ EAAB Empresa de Acueducto y Alcantarillado de Bogotá y Otros (2012)[b, e] Colombia	Public authorities and service providers are obliged to provide efficient and timely sanitation services to legalised settlements.
Environment & Consumer Protection Foundation v Delhi Administration and Others (2012)[a, e] India	The State has an immediate obligation to provide toilets for boys and girls in schools, in order to guarantee the right to education.
Human right to sanitation principles expressed in the court's decisions: a – Acceptability; b – Access (physical); c – Access to information; d - Accountability; e - Availability; f - Affordability; g – Equality and non-discrimination; h – Participation	

Source: Compiled by the author, from case summaries reported in WaterLex 2014

---

[76]   The Inter-American Commission on Human Rights/Paraguay reached a similar decision in the case of *Yakye Axa Indigenous Community v Paraguay* Inter-American Court of Human Rights Series C no 125 (17 June 2005), reported in WaterLex 2014.

[77]   Similar decisions on access to sanitation for detainees has been reached in national courts (in Angola, France, Fiji and Ireland, for instance), as well as regional courts (like the European Court of Human Rights and the Inter-American Commission on Human Rights), reported in WaterLex 2014.

5.2.3 Meaning

The HRS generally imposes tripartite obligations on States, to respect, protect and fulfil the right, and soft law instruments like the United Nations "Protect, Respect, Remedy" Framework and Guiding Principles on Business and Human Rights (Ruggie, 2011) further expand the possibility of holding businesses liable for HRS violations (Obani & Gupta, 2016). The framework and guiding principles are set on three pillars: (a) State's duty to protect against human rights abuses by third parties, including business; (b) corporate responsibility to respect human rights; and (c) improved access by victims to an effective remedy, in case of violation. The enforcement of HRS obligations against either the State or businesses however requires a clarification of the meaning of the HRS.

The HRS is defined differently by actors at various levels of governance. The former Special Rapporteur, de Albuquerque, defined sanitation in human rights terms: "as a system for the collection, transport, treatment and disposal or reuse of human excreta and associated hygiene," including "domestic wastewater, which flows from toilets, sinks and showers, … insofar as water regularly contains human excreta and the by-products of the associated hygiene," and "in some places, existing solutions for human excreta management make it inseparable from solid waste management" (de Albuquerque 2009, paragraph 63). This definition extends to "all spheres of life" (de Albuquerque 2009, paragraph 63), and protects other human rights, like health, which could otherwise be violated due to unsafe sanitation or hygiene practices (Winkler, 2016; Zimmer, Winkler & de Albuquerque, 2014). It thereby places a corresponding duty on the rightsholders to use and hygienically maintain the available sanitation facilities, and may require those who can afford to, to cross-subsidise access for the poor, in order to advance universal access (Obani & Gupta, 2014a). Nonetheless the definition does not integrate environmental sustainability (Feris, 2015). This is evident from the narrow list of sanitation services excluding non-excreta related waste streams, like solid waste management, which nonetheless poses a threat to human wellbeing and the environment if poorly managed. Similarly, the UNECE Protocol on Water and Health defines sanitation as: "the collection, transport, treatment and disposal or reuse of human excreta or domestic waste water, whether through collective systems or by installations serving a single household or undertaking",[78] and a collective (sanitation) system as: "[A] system for the provision of sanitation which serves a number of households or undertakings and, where appropriate, also provides for the collection, transport, treatment and disposal or

---

[78] Article 8.

reuse of industrial waste water, whether provided by a body in the public sector, an undertaking in the private sector or by a partnership between the two sectors".[79] Further, the Message from Beppu 2007, emphasises the need to "[P]romote thinking of sanitation as the full cycle of proper arrangements, safe conveyance and sanitary disposal/re-use of liquid and solid wastes (including solutions that do not adversely impact the quality of land and water resources), and associated hygiene behaviour".

More recently, the UNGA Resolution 70/169 of 2015 recognised *inter alia* that the HRS "entitles everyone, without discrimination, to have physical and affordable access to sanitation, in all spheres of life, that is safe, hygienic, secure, socially and culturally acceptable and that provides privacy and ensures dignity, while reaffirming that both rights are components of the right to an adequate standard of living".[80]

At the national level, there have been three main approaches to defining the HRS. One emphasises the relational focus of the HRS and the interests of vulnerable groups. For instance, in South Africa, the Water Services Act, Act 108 of 1997, provides *inter alia* that everyone has a right to access basic sanitation services,[81] defined as "...the prescribed minimum standard of services necessary for the safe, hygienic and adequate collection, removal, disposal or purification of human excreta, domestic waste-water and sewage from households, including informal households".[82] Hence, basic sanitation which is the minimum standard is defined by the Act to include the management of human excreta as well as wastewater and sewage and also applies to those in informal settlements. Another approach integrates environmental protection. For instance, in Guinea Bissau the aim of sanitation is broadly stated as being "to ensure the immediate removal of domestic and industrial wastewater likely to cause harm, and of rivers likely to flood inhabited places, subject to public health requirements and environmental protection".[83] The third approach is more restrictive than the definition by the former Special Rapporteur. For instance, the United Kingdom defines HRS to include: "... achieving the outcome of providing a system for the treatment and disposal or re-use of human sewage and associated hygiene. It entitles right holders to reasonable access to the elements of the right, but allows for recovery of the cost of providing such access, including any environmental and resource costs. The right does not

---

[79]   Article 9(b).
[80]   Paragraph 2.
[81]   Section 3.
[82]   Chapter 1.
[83]   Guinea Bissau, *Water Code, Law No. 5-A/92* (unofficial translation), article 29.

prescribe any particular model of delivery for public and private sectors. It also does not require the collection and transport of human waste..."[84] This further threatens environmental sustainability due to the risk of contamination of the environment from uncollected or poorly managed human waste (Zimmer et al., 2014).

The meaning of the HRS can be further expanded on the basis of the ICESCR, from which the right is mainly derived in international law, to require that States use their maximum available resources to progressively achieve full realisation. The ICESCR, article 2(1), provides that each State Party "undertakes to take steps, individually and through international assistance and co-operation, especially economic and technical, to the maximum of its available resources, with a view to achieving progressively the full realization of the rights recognized in the present Covenant by all appropriate means, including particularly the adoption of legislative measures." States are also obliged by the ICESCR, article 2, to guarantee human rights without discrimination,[85] while "[D]eveloping countries, with due regard to human rights and their national economy, may determine to what extent they would guarantee the economic rights recognized in the present Covenant to non-nationals."[86] Progressive realisation reflects the reality that States may be unable to immediately attain full realisation of the ICESCR provisions, depending on their available resources, for instance.[87] Nonetheless, the HRS imposes a minimum set of obligations, the core content of the right (see 5.3), from which States are not allowed to deviate, not even due to lack of economic resources, otherwise, the right essentially becomes meaningless (Martin, 2011).

The HRS therefore means that additionally, States can apply the "maximum of its available resources" to fulfil the core content of the right (see 5.3) and establish safeguards against retrogression from any progress achieved in the implementation process (Obani & Gupta, 2016). The available resources include both financial and non-financial resources (including human capital, natural resources like land and water, information, and technology) on which the realisation of the HRS depends. Such a broad definition of resources prevents States from relying on the scarcity of finances as an excuse for non-implementation of the HRS (Skogly, 2012). It also means that even a developing country with abundant non-financial resources, like natural resources, bears a duty towards assisting other countries with less resources to

---

[84]   United Kingdom Statement (n. 68).
[85]   ICESCR, article 2(2).
[86]   ICESCR, article 2(3)
[87]   The CESCR describes this as a "necessary flexibility device" in the ICESCR. See Committee on Economic, Social and Cultural Rights, 1990, paragraph 9.

realise the HRS based on the principle of extraterritorial obligations (see 5.3.2). Applying the tripartite obligations to respect, protect and fulfil further means that States are to take steps to: (a) protect rightsholders against arbitrary disconnection, exclusion from services, unaffordable tariffs and discriminatory practices which interferes with their access to sanitation services; (b) respect the HRS both within their jurisdictions and abroad, and establish safeguards against violations; and (c) take appropriate legislative and other measures necessary to fulfil the HRS.

5.3   PRINCIPLES OF THE HUMAN RIGHT TO SANITATION

This section presents eleven principles of the HRS that I derive through content analysis and literature review, based on the definition of principles in my conceptual framework (see 2.5.1). I derive the principles from the core content of the HRS (availability, physical accessibility, safety, acceptability, affordability, and equality and non-discrimination),[88] and cross-cutting HR (accountability, extra-territorial obligation, participation, sustainability, and transparency and access to information) (see 4.3) (de Albuquerque, 2009; 2014). A clear definition of the HRS principles is essential for limiting the administrative discretion and ensuring predictability in HRS enforcement by the courts (McIntyre, 2012). One of the consequences of the historical linking of the HRWS (see 1.2.2 and 5.2.1) is that HRS principles have largely been subsumed under the development of the HRW principles; hence my discussion of the HRS principles in this section draws heavily from the HRW where relevant.

Accessibility

Access is mentioned in the various UNGA and HRC resolutions concerning the HRS, and environmental, water and (sustainable) development instruments and political declarations (see 5.2 and Table 5.1). The recent UNGA Resolution 70/169 further reiterates that the HRS "entitles everyone, without discrimination, to have physical and affordable access to sanitation, in all spheres of life…"[89] Hence, accessibility has both a physical and an economic element (which I discuss under affordability below). Physical accessibility requires that

---

[88]   General Comment No. 3 indicates that the rights provided for in the ICESCR have a core content which guarantees that the minimum essential level of each right is achieved. States are expected to fulfil the core content except this becomes impossible due to resource constraints, for instance; see Riedel, Giacca and Golay (2014). Nonetheless, non-discrimination is an immediate obligation which does not require additional resources *per se* but requires available resources to be used to address the needs of the marginalised and vulnerable groups. I have relied on the normative content of the HRW, specified by the CESCR in paragraphs 10-12 of the General Comment No. 15, to determine the core content of the HRS.

[89]   Paragraph 2.

sanitation facilities and services are constructed, operated and maintained hygienically and without putting users at any risk during the day and night); negative socio-cultural practices (see 3.4.2) that prevent vulnerable groups like women and girls from accessing sanitation facilities are addressed; and sanitation facilities sited outside the house and in public places, to be within a reasonable distance from the vicinity of users (COHRE et al., 2008; de Albuquerque, 2009; Winkler, 2016).

Acceptability

Acceptability features in the various UNGA and HRC resolutions concerning the HRS (see 5.2) and the UNGA Resolution 70/169 also provides for universal access, that is: "socially and culturally acceptable and that provides privacy and ensures dignity."[90] Dignity is a shared basis for all human rights (McCrudden, 2008), though it lacks a definite meaning under HR (see 4.4.3), and while the notions of privacy and dignity also vary among different societies, the HRS is violated by: (a) "[d]egrading living conditions and deprivations of basic needs" (Schachter, 1983, p.852); (b) lack of access to sanitation facilities (de Albuquerque 2009); and (c) lack of privacy when using sanitation facilities (Lee & George, 2008).[91] Women and girls are more at risk of physical danger in addition to a violation of their dignity when they are forced to defecate in the open or use facilities that violate their privacy (Amnesty International, 2010). Acceptability therefore requires taking into account the local perspectives, especially from vulnerable groups like women and girls, when designing and distributing facilities, in order to counter the socio-cultural drivers that cause non-acceptance or violate dignity and privacy (Obani & Gupta, 2016). Nonetheless, some cultural practices are ill-informed (for instance, banishing women from dwellings during their menstrual cycle or prohibiting the siting of facilities within the living area) and require sanitation and hygiene awareness to promote good practices (Akpabio, 2012).

Affordability

Affordability features in the various UNGA and HRC resolutions (see 5.2) concerning the HRS and is an aspect of accessibility because the direct and indirect costs of accessing sanitation services can prevent access for the poor. Affordability therefore requires that the cost of sanitation does not affect access to sanitation, reduce the users' quality of life or in other ways compromise the realisation of their rights (COHRE et al., 2008; Fonseca, 2014).

---

[90] Paragraph 2.
[91] *Beja and Others v Premier of the Western Cape and Others* [2011] High Court (Western Cape) 21332/10, [2011] ZAWCHC 97 (South Africa); Municipal Council, Ratlam v. Shri Vardhichand and Others, (1981) 1 S.C.R. 97 (India).

This is best determined at local level, taking into account the local circumstances as what may constitute an affordable tariff to one set of individuals may be unaffordable to others. It requires a broad system of funding through a combination of tariffs, taxes, transfers and cross-subsidies for the poor (de Albuquerque, 2014; Heller, 2017; Obani & Gupta, 2016). The threshold figures for affordability specified in the literature often include both water and sanitation services and range from 5% (Fonseca, 2014; Deutsche Gesellschaft für Technische Zusammenarbeit [GTZ], 2009; UN-Water Decade Programme on Advocacy and Communication & Water Supply and Sanitation Collaborative Council, 2010), to 3 – 6% (Smets, 2008) of household income; whereas in the United Kingdom, for instance, 3% already indicates hardship on the users (United Nations Development Programme [UNDP], 2006). However, expressing affordability as a percentage of household income does not account for non-monetised economies (COHRE et al., 2008), labour exchange or other forms of non-monetary contributions (Obani & Gupta, 2015), nor households without regular incomes (hence, the recommendation to measure affordability using household expenditure instead of income) (COHRE et al., 2007). Further, practices like indexing tariffs to the USD or servicing contract loans in a foreign currency may also hamper affordability for poor countries (Barlow, 2009).

Availability

Availability is contained in the various UNGA and HRC resolutions concerning the HRS, and environmental, water and (sustainable) development instruments and political declarations (see 5.2 and Table 5.1). Availability means a sufficient quantity of sanitation facilities for domestic and personal uses in all spheres of life (de Albuquerque, 2009; Obani & Gupta, 2016; Winkler, 2016).[92] This requires establishing clear targets and timelines for progressively achieving universal access to the highest attainable sanitation standards (see 3.5.5) (de Albuquerque, 2014).[93] Unlike drinking water requirements that are clearly defined in quantitative terms (Howard & Bartram, 2003), the number of facilities necessary to fulfil the HRS is best determined on a case by case basis with due consideration of the special needs of vulnerable users like women and people living with disability and the ratio of users to sanitation facilities set (taking into account the special needs of users) in order to avoid

---

[92] Bangladesh, National Policy for Safe Water Supply and Sanitation 1998; South Africa - *Beja and Others v Premier of the Western Cape and Others* [2011] High Court (Western Cape) 21332/10, [2011] ZAWCHC 97, case summary in Waterlex, 2014.

[93] Some countries include sanitation targets and service standards in laws and policies as a benchmark service provision. See, Kenya, *The National Water Services Strategy (NWSS)*, 2007-2015, paragraph 3.3; South Africa, *Regulations relating to compulsory national standards and measures to conserve water* 2001.

long waiting times (see 3.5.5 and 3.6.1). Further, the question of whether or not shared facilities meet human rights standards is best addressed at scale and it depends on whether or not the facility ensures the dignity and wellbeing of users, as shared facilities may be necessary where space constraints prevent the building of personal facilities, for instance (see 3.5.5).

Safety

Safety is contained in the various UNGA and HRC resolutions concerning the HRS, and environmental, water and (sustainable) development instruments and political declarations (see 5.2 and Table 5.1). It requires the integration of environmental considerations in the design and implementation of sanitation services, in order to protect both humans and the environment from contamination by human excreta or wastewater (Feris, 2014; Zimmer et al., 2014); broad definition of the HRS to include waste collection, treatment and safe disposal or reuse services (see 5.2.3); universal access to public (sanitation) goods like sewer networks (see 3.3); development of alternatives to manual emptying of on-site sanitation facilities and regulation and monitoring to ensure safe handling and treatment of the waste collected from sanitation facilities (see 3.5); hygiene promotion and access to facilities for the hygienic disposal of menstrual products and water and soap for related hygiene needs (see 3.5.5); and the segregation of shared toilets by gender, to minimise assault on vulnerable users.

Accountability

Accountability features in the various UNGA and HRC resolutions concerning the HRS. It requires effective remedies to redress violations of the HRS; adequate reparation to the victims of such violations with measures to prevent the repeat of such violations, and mechanisms for monitoring the operations of duty bearers service providers (Baer, 2015; de Albuquerque, 2014; Obani & Gupta 2016, 2017). Accountability may either be achieved through substantive HR like access to information, or global administrative law principles like legality and due process (Harlow, 2006) that have been further legitimised with the adoption of the universal SDGs (Obani & Gupta, 2016). Accountability applies to States as the primary duty bearers (Martin, 2011), international organisations and NGOs (UNGA, 2010), private corporations that are directly involved in delivering sanitation services or interfering with the realisation of the right (Cavallo, 2013), and users who are required to cross-subsidise the poor through paying affordable tariffs (Obani & Gupta, 2014a) and hygienically maintain their facilities to minimise negative externalities (de Albuquerque,

2012). Accountability is further strengthened through the rule of law and pressure from the civil society (Abeysuriya et al., 2007; de Albuquerque, 2014; Giles, 2012; Kühl, 2009). The rule of law requires States to: (a) adopt a strong effective regulatory framework for the full realisation of the HRS; and (b) ensure that its agencies and non-State actors comply with the HRS (de Albuquerque, 2009; 2012; 2014). In addition, where local or national remedies are either lacking or inadequate, aggrieved individuals or groups may seek redress for violations through the international HR enforcement mechanisms like international courts or the Committee for Economic, Social and Cultural Rights (see 4.2.1), provided their country of origin is a State Party to the Convention and has accepted the competence of the Committee or the jurisdiction of the relevant court to consider individual complaints.

Equality and non-discrimination

Discrimination features in the various UNGA and HRC resolutions concerning the HRS, and one of the main objectives of the right is to address inequalities in access for the benefit of vulnerable and marginalised groups (see 5.2). Non-discrimination is also a central principle in the ICESCR (see 5.2.3) and discrimination can assume both formal and covert forms, where embedded in apparently neutral laws and practices that exacerbate inequities in access (de Albuquerque 2014).[94] Non-discrimination and equality require immediate steps to end discriminatory practices where they exist, targeted affirmative action to reduce inequalities in access with special attention to vulnerable and marginalised individuals or groups with special needs that are likely to suffer exclusion or discrimination in accessing sanitation, and an understanding of the disparities in access among various individuals and groups rather than relying on averages which might obscure inequalities in access (de Albuquerque, 2009; Obani & Gupta, 2016; WaterLex, 2014).

Participation

Participation is recognised in the various UNGA and HRC resolutions concerning the HRS and declarations like Rio, the Dublin Statement and Bonn (see 5.2) as an important principle for the protection of vulnerable and marginalised groups. The HR framework particularly emphasizes the need for participation to be meaningful and effective (de Albuquerque, 2014). It is through meaningful and effective participatory mechanisms that the special needs and

---

[94]   UN Committee on Economic, Social and Cultural Rights (CESCR), General comment No. 20: Non-discrimination in economic, social and cultural rights (art. 2, para. 2, of the International Covenant on Economic, Social and Cultural Rights), 2 July 2009, E/C.12/GC/20, available at: http://www.refworld.org/docid/4a60961f2.html [accessed 28 September 2017].

interests of vulnerable and marginalised groups can be reflected in sanitation governance. Participation requires States to provide the affected individuals and communities with full and equal access to information concerning sanitation and hygiene services and the implications for their health and immediate environment through accessible media, and the opportunity to effectively partake in all the processes for the planning, construction, operation, maintenance, and monitoring of services (de Albuquerque, 2014).

Transparency and access to information

The need for transparency in the governance process and access to information is recognised in the various UNGA and HRC resolutions concerning the HRS (see 5.2).[95] Transparency and access to information entail (including the right to seek, receive and share) information about sanitation laws, policies and programmes, service design, tariffs, the financial information of service providers, and the quality of services (de Albuquerque 2014; Obani and Gupta 2016). The information would need to be disseminated through accessible media and language and be freely available to the public without illegal restrictions like prohibitive access fees (de Albuquerque 2009; 2014).[96] Transparency and access to information foster democratic engagement, effective participation, accountability and informed decision making by rightsholders (de Albuquerque 2014). Further, access to information mitigates the power imbalance that exists between the poor and marginalised individuals and groups, like residents of informal settlements, and the State or service providers (de Albuquerque 2014).

Sustainability

The imperative of sustainable access is recognised in the various UNGA and HRC resolutions concerning the HRS (see 5.2). Although the meaning of sustainability is not clearly articulated in the resolutions, the literature shows that sustainability requires the progressive realisation of the HRS for everyone, without any retrogression. This entails a holistic consideration and balancing of the social, economic, and environmental aspects of sanitation services delivery (Feris, 2015; Obani & Gupta, 2016).[97] Sustainability further requires the operation, maintenance and repairs of sanitation facilities (Darrow, 2012), prioritisation of personal and domestic sanitation and hygiene needs over competing demands

---

[95] The right to information is enshrined in the Universal Declaration of Human Rights 1948, article 19 and the International Covenant on Cultural and Political Rights 1966, article 19(2).

[96] See *Federation for Sustainable Environment and Others v Minister of Water Affairs and Others* [2012] High Court (North Gauteng, Pretoria) 35672/12, [2012] ZAGPPHC 128 (South Africa); *Hernan Galeano Diaz c/ Empresas Publicas de Medellin ESP,y Marco Gomez Otero y Otros c/Hidropacifico SA ESP y Otros* [2010] Corte Constitucional T-616/10 (Colombia), case summary in Water Lex, 2014.

[97] General Comment No. 15 (n. 63), paragraphs 11 and 28.

for available resources (see 5.4.1), and inter-generational equity (de Albuquerque, 2014). Sustainability is also reflected in SDGs 6 (see 3.2) which incorporates safe access, and sustainable management of water resources.

Extra-territorial obligation

The CESCR obliges States to "take steps to prevent human rights contraventions abroad by corporations which have their main seat under their jurisdiction" (CESCR, 2011, paragraph 5) and "prevent their own citizens and companies from violating the right to water of individuals and communities in other countries" (CESCR, 2002, paragraph 33). Extra-territorial obligation thereby generally restrains States from using their resources either directly or indirectly to cause harm abroad (de Albuquerque, 2014).[98] The need for international cooperation is recognised in the various UNGA and HRC resolutions concerning the HRS, and the ICESCR itself (see 5.2). In relation to sanitation, the richer States therefore ought to respect, protect, and assist with the realisation of the HRS in poorer States, for instance through refraining from imposing embargoes that affect the supply of sanitation services or using sanitation as an instrument of political or economic pressure, and preventing their citizens from violating the HRS in other countries (de Albuquerque, 2014; Obani & Gupta, 2016).[99] Further, the Maastricht Principles on Extraterritorial Obligation of States in the area of Economic, Social and Cultural Rights 1997, developed by international human rights law experts, obliges Sates to *inter alia* ensure the availability of effective accountability mechanisms in dispensing their extraterritorial obligations, as well as "prompt, accessible and effective remedy before an independent authority, including, where necessary, recourse to a judicial authority, for violations of ESC rights" (De Schutter et al., 2012).[100]

5.4   HUMAN RIGHT TO SANITATION INSTRUMENTS AND INDICATORS

This section analyses the key regulatory (see 5.4.1), economic (see 5.4.2), and suasive (see 5.4.3) instruments through which the HRS principles (see 5.3) are operationalized.

5.4.1 Regulatory

---

[98]   See General Comment No. 15 (n. 63), paragraphs 23, 24, 33, 34, 35 36, 44 (c) (vii), 49, 50, 60; Sub-Commission Guidelines, paragraph 10; *Trail Smelter (United States v. Canada) Arbitration*, [1938/1941] 13 RIAA 1905; ICJ 25 September 1997, *Gabc'ikovo-Nagymaros Project (Hungary v. Slovakia)*, [1997] ICJ Rep 7; ICJ 20 April 2010, *Pulp Mills on the River Uruguay (Argentina v. Uruguay)*, [2010] ICJ Rep. 14. See also UN Convention on the Law of the Non-Navigational Uses of International Watercourses, article 32.

[99]   General Comment No. 15 (n. 63), paragraph 32.

[100]   See Principles 36 and 37.

This section discusses seven regulatory instruments.

Guaranteed free access to basic services

Guaranteed free access to basic sanitation services protects human dignity and assures access for vulnerable individuals and groups, like the residents of informal settlements which are often excluded from national development plans and in particular the formal sanitation networks (UN-Habitat, 2016). An example of statutory protection of guaranteed free access to basic services is in the Flemish region of Belgium which guarantees a minimum of 15 m^3 of water supply annually for domestic uses, including sanitation, progressive pricing for surplus consumption (see 5.5.2) and exempts the poorest citizens living under the Minimum Subsistence Level from the regional tax regime for sanitation services *inter alia*.[101] In the Brussels region of Belgium, there is also a Social Fund for Water that is financed by a tax of 0.01 euro/m^3 of water consumed, and the poorest people in the Brussels region receive a sanitation tax refund but the amount varies among the different municipalities. Nonetheless, the statutory protection of free basic water supply was challenged in the case of *Commune de Wemmel*,[102] as a violation of the competence of the municipal water companies to determine the water supply tariff. In that case, the Court of Arbitration upheld the municipality's right to legislate on the water tariff and stated that the provision advanced human dignity, and reduced water wastage using the progressive pricing mechanism. While the argument may be advanced that progressive pricing may increase the tariff for larger consumption users and thereby encourage water conservation and moderate consumption, the counter-argument can also be made that the increased prices at higher levels of consumption could negatively affect effective access for the vulnerable, poor, or marginalised people who need more than the free minimum quantity to meet their needs, especially in the absence of social safety nets (Obani & Gupta, 2014a).

Prohibiting arbitrary disconnections

Laws prohibiting arbitrary disconnections for non-payment of tariffs protect human dignity and prevent discrimination against poor users for instance by offering the option to pay in arrears, or subsidised or free basic services for users who cannot afford to pay for their domestic sanitation and hygiene services (de Albuquerque, 2014).[103] Such prohibitions are

---

[101] Décret de la Communauté Flamande concernant diverses mesures dáccompagnement du budget 1997, 20 December 1996, *Moniteur belge*, 31 December 1996, 3rd edition (unofficial translation), articles 3.1 and 34.

[102] 8 Commune de Wemmel, Cour d'arbitrage, Arrêt N°36/1998, 1 Avril 1998, Moniteur belge, 24 April 1998.

[103] In Brazil, defaulting customers are also protected from ridicule or embarrassment in the process of debt recovery through provisions like the *Consumers Defence Code, Law 8078* of 11 September 1990, as last amended by Law 12.039 of 2009 (unofficial translation), article 42.

incidental to the enjoyment of the HRWS.[104]   It is further important for service disconnections to be guided by procedures that are clearly stated in the legal framework. For instance, in South Africa, the Water Services Act requires that the procedures for limiting or discontinuing services must: "(a) be fair and equitable; (b) provide for reasonable notice of intention to limit or discontinue water services and for an opportunity to make representations, unless – (i) other consumers would be prejudiced: (ii) there is an emergency situation; or (iii) the consumer has interfered with a limited or discontinued service."[105] The prohibition of arbitrary disconnections has however been challenged in courts. In Belgium, the 1994 Belgian regional ordinance which prevents water distribution companies from disconnecting tenants who do not pay their water bill was challenged before the Court of Arbitration of Belgium.[106] Dismissing the petition, the Court held that access to a minimum quantity of drinking water and sanitation is a fundamental right and that a tenant's access must be guaranteed despite the inability to pay. Also, where the tenant is insolvent, the service providers are entitled to account the bill to the landlord on the basis of the property's connection to the distribution network. Similarly, in France, in the case of *Compagnie de dervices dénvironment v. Association des consommateurs de la Fontauliére*,[107] the Tribunal de Grande Instance of Privas upheld a similar prohibition stating that water is essential to life and the partial non-payment of water bills could not justify disconnection of services.

Direct provision of access by the State

The responsibility for sanitation provision is often assigned to a tier of government, a government organization, a line department or an agency. In India, the 12th Schedule to the 74th Constitutional Amendment Act, 1992 vests the urban local bodies with the responsibility for public health, sanitation, conservancy, and solid waste management; water supply for domestic, industrial, and commercial purposes; slum improvement and upgrading; protection of the environment and promotion of ecological aspects, among others. Similarly, the Constitution of South Africa, 1996, lists sanitation services as a local government responsibility, while in Kenya it is the responsibility of the county government.   This

---

[104]   City of Cape Town v Strümpher [2012] Supreme Court of Appeal 104/2011, [2012] ZASCA 54 (South Africa).

[105]   Section 4 (3).

[106]   Belgian Court of Arbitration, *A.s.b.l. Syndicat national des propriétaires et autre*, Case No.9/1996, 8 February 1996, in: *Moniteur belge*, 1996(02)00035, section I, paragraph III, 2.

[107]   Case No. 9800223, 5 March 1998. See also Tribunal de Grande Instance (District Court) of Meaux, xxx.v.xxx., 28 February 2001, in: *Droit Monde*, No. 37-28, 2004, p. 77, quoted in COHRE (2008), p. 295.

responsibility could either be exercised directly or indirectly through various forms of public-private partnerships involving private service providers (see 7.3.2).

Universal access & mandatory connections

National laws that require universal access and mandatory connection of users to sanitation service networks prevent discriminatory practices against the poor, marginalised and vulnerable individuals and groups like residents of informal settlements (de Albuquerque 2014). Some countries require mandatory provision of sanitation facilities in homes,[108] places controlled by organisations and people other than the users, like schools,[109] work places,[110] detention centres,[111] rented residences,[112] care homes and assisted living spaces,[113] and urban areas.[114] Some countries like Mali also mandate connection to the sewerage network where this exists.[115] For instance, in Uruguay,[116] it is mandatory for house owners and potential house buyers to connect their properties to the sewerage system within one-year or be liable to pay fines until they comply.[117]

Prioritisation of human needs

The depletion of resources and destruction of infrastructure that occur in humanitarian situations (see Chapter 6) may necessitate the prioritisation of human needs in the allocation of access to the remaining scarce resources. As a result, some national laws prohibit all other water intensive activities apart from human consumption during emergencies and humanitarian situations, depending on the nature of the emergency. In Niger for instance, during droughts the local authority may prohibit the watering of gardens or other water intensive activities except for direct human consumption.[118] Similarly, in Burkina Faso, the use of water for drinking and basic needs and human dignity (presumably including

---

[108] Republic of Benin, *Public Hygiene Code, Law No. 87-015* of 21 September 1987 (unofficial translation), article 20.

[109] England and Wales, the Education (School Premises) Regulations (1999 No. 2); Kenya, National School Health Policy 2009.

[110] England and Wales, Workplace (Health, Safety and Welfare) Regulations (1992 No. 3004); Benin, *Public Hygiene Code, Law No. 87-015* of 21 September 1987 (unofficial translation), article 93.

[111] Panama, Executive Decree 393 of 2005, article 12.

[112] Republic of Benin, Law No. 87-015, Public Hygiene Code, article 20.

[113] United States of America, State of Georgia, Rules and Regulations for Assisted Living Communities

[114] Mauritania, *Water Code, Law No. 2005-030* (unofficial translation), article 37.

[148] Mali, *Decree No. 01-395/P-RM* of 06 September 2001 determining the modalities for the management of wastewater and silt (Unofficial translation), article 8; Guinea Bissau, *Water Code, Law No. 5-A/92* (Unofficial translation), article 29(1)(2)(5); Australia, *Australian Utilities Act 2000 No. 65*, 2002, last amended by A2010-54 of 16 December 2010, sections 83, 84, 85 & 86; Guatemala, *New Health Code, Decreto 90-97* (unofficial translation), article 89.

[116] Law No. 18.840 of 2011, articles 6 and 7.

[117] Special Rapporteur on the human rights to water and sanitation, Mission to Uruguay, 2012 (A/HRC/21/42/Add.2), paragraph 18.

[118] Niger, *Water Law, Law No. 93-014* of 2 March 1993 (unofficial translation), article 9.

sanitation) is prioritised during emergencies and the hierarchy of other water uses is to be determined according to the local context and the principles of equity, subsidiarity, and where possible, participation.[119]

Litigation

Courts monitor and enforce compliance with the HRS principles (see 5.4) expressly contained in laws and policies, and in the absence of express recognition of the HRS they progressive interpret existing human rights provisions to imply the HRS at multiple levels of governance, which underscores the indivisibility and interdependence of human rights (Obani, 2015). For this purpose, the three standards generally applied by courts for the determination of the violation of economic, social, and cultural rights are also relevant for the HRS: reasonableness, minimum core, and proportionality.

Reasonableness is a measure for progressive realisation and prevents discrimination against segments of the society (WaterLex, 2014). In the case of *Lindiwe Mazibuko and Others v. City of Johannesburg and Others*,[120] the Constitutional Court stated that the socio-economic rights contained in the Constitution only entail an obligation to take reasonable legislative and other measures to ensure the progressive realization; it was inappropriate for a court to quantify what amounts to sufficient water; and the introduction of pre-paid meters in the poorer sections of the municipality was legal and non-discriminatory against the affected residents. This tersely illustrates the need to clearly define the HRS in national laws (see 5.2.3) and the limits of judicial activism in interpreting ambiguous provisions.

Defining the minimum core of HR ensures predictability of court decisions by specifying how a right is to be successfully implemented,[121] but this could be seen as undermining democratic values by being too prescriptive (Bernal, 2015). However, as HRS provisions do not generally state the minimum core in quantifiable terms, this is subject to contestation.[122] Establishing the minimum core of the HRS is especially difficult because of the contextual nature of access to sanitation and the slow development of its normative content, compared to water which has a minimum of 50 litres per person per day water (for drinking and domestic

---

[119]   Burkina Faso, *Decree No. 2005-191/PRES/PM/MAHRH* 4 April 2005 regarding priority uses and authority of government to control and allocate water in case of water shortage (unofficial translation), articles 2 and 3.

[120]   Case No. CCT 39/09, Judgement of 8 October 2009 (Constitutional Court of South Africa).

[121]   See *Hernan Galeano Diaz c/ Empresas Publicas de Medellin ESP,y Marco Gomez Otero y Otros c/Hidropacifico SA ESP y Otros* [2010] Corte Constitucional T-616/10 <http://www.corteconstitucional.gov.co/relatoria/2010/T-616-10.htm>.

[122]   Mazibuko (n. 120).

sanitation and hygiene uses) based on the World Health Organization standards (Howard & Bartram, 2003).

Proportionality is based on three sub-principles, namely, suitability, necessity, and balancing (Bernal, 2015). Suitability requires that any limitation of HR implementation contributes to a legitimate objective; necessity requires that such a limitation is the least restrictive means of achieving the desired objective; balancing requires that the limitation achieves the desired objective to an extent justifying the extent of limitation of the right in question.[123] Proportionality could be seen as interfering with legislative and executive functions but from a social and relational perspective, it is useful for prioritising HR over competing interests (Bernal, 2015).

Monitoring and accountability mechanisms

Monitoring and accountability mechanisms enhance the participation of users in sanitation governance and transparency (see 4.2.1 and 5.4.1). Some of the relevant monitoring and accountability mechanisms include supervising and reviewing the operations of service providers against the minimum service standards;[124] taking corrective action against providers who contravene the service standards;[125] establishing procedures for handling consumer complaints against the service providers;[126] and ensuring public access to information on consumer rights and the quality of services rendered by the providers (de Albuquerque 2014; WaterLex 2014).[127]

5.4.2 Economic

This section discusses three economic instruments.

Participatory budgeting for sanitation services

Participatory budgeting enables citizens to determine the priorities for public expenditure and budgetary allocations through negotiations with the government (Gonçalves, 2014). Public expenditure is essential for the provision of merit goods like sanitation (see 3.3.2) and participatory budgeting at the local level of governance has been linked to increased spending

---

[123] Venezuela - *Condominio del Conjunto Residencial Parque Choroní II c/ Compañía Anónima Hidrológica del Centro (Hidrocentro)* Corte Primera de lo Contencioso Administrativo (2005); Slovenia - *Ruling no Up-156/98* Ustavno Sodišče, *Ruling No Up-156/98* [1999] *Constitutional Court Official* Gazette RS, no 17/99; OdlUS VIII, 118, case summary reported in Waterlex, 2014.

[124] South Africa, *Water Services Act, Act 108* of 1997, amended 2004, sections 62 and 63.

[125] Namibia, *Water Resources Management Act, Act No. 24* of 2004, 26.

[126] Kenya, *The Water Act 2002, No 8 of 2002*, article 47(c); Ghana, *Public Utilities (Complaints Procedure) Regulation* 1999.

[127] See Brazil, *Law on basic sanitation,* 2007, Articles 26, 27 & 53; Colombia, *Law 142 establishing the regime for public household services* of 11 July 1994 (unofficial translation), article 9.

on basic sanitation and health services in some cases (Gonçalves, 2014; Wampler, 2017). However, some cases suggest a lack of systematic association between participatory budgeting and improvements in water and sanitation coverage as a result of inherent weaknesses in the process itself and weaknesses among different actors (Jaramillo & Alcázar, 2013). Some of the weaknesses include that the poor may bear high costs of participation as a result of participatory budgeting, the process requires political commitment to be effective, and the service providers and municipalities may lack the finances and technical resources necessary for service expansion to the poor (Boulding & Wampler, 2010; Gonçalves, 2014). The literature suggests ways of addressing the weaknesses of participatory budgeting: greater participation of women, improved access to information about the budgetary process, increased resources and independence from political influences for technical teams involved in the service delivery process, specific measures to enhance technical capacities and access to finances for the process, and the education and empowerment of social organisations to enhance auditing (Jaramillo & Alcázar, 2013).

Progressive pricing & cross-subsidies

Progressive pricing consists of charging more for higher blocks of consumption than for lower blocks; it may be funded through cross-subsidies from large users which also encourages conservative use (Government of Belgium, 2007). Whereas a flat rate subscription tariff structure imposes higher usage costs for individuals, which could hamper access for the poor (Biran et al., 2011), progressive pricing of water and wastewater services ensures affordability (see 5.4.1) for small users. In the Brussels region of Belgium, the water tariff is composed of a fixed connection cost, progressive pricing for water services, and sanitation tariff (Government of Belgium, 2007). Similarly, the Walloon region of Belgium[128] water supply system allows a minimum block of 30 m^3 of water per household per year at a lower price than the other three higher blocks of consumption, with cross-subsidies funded by the larger users (Government of Belgium, 2007). In France, water services also attract a reduced value added tax rate and there is financial aid to ensure affordable access for the poor (Thielbörger, 2014).

Targeted subsidies address poverty and related drivers of poor sanitation services that are linked to collective action problems (see 3.4.1) and enhance non-subsidy approaches like Community-Led Total Sanitation (see 7.3.3) (Guiteras, Levinsohn & Mobarak, 2015).

---

[128]   Décret relatif au Livre II du Code de l'Environnement constituant le Code de l'Eau, Moniteur belge, 23 July 2004, Article 1.1 (Walloon region of Belgium) (unofficial translation).

Critiques argue that subsidies hamper local innovation and lead to retrogression once the subsidy is exhausted (Harvey, 2011; Kar, 2003; Kar & Milward, 2011; Kar & Pasteur, 2005), while proponents advocate that subsidies are likely to increase ownership of sanitation facilities both among recipients of the subsidies and their neighbours, through a social multiplier effect, and the increased sanitation coverage/ownership is also likely to improve the use of facilities (Barnard et al., 2013; Guiteras et al., 2015).

Development finance

Development finance consists of official development finance (ODF), which includes official development assistance (ODA) and other official flows (OFF), and flows from non-official sources like NGOs and philanthropists (Cotton, 2013; Winpenny, Trémolet, Cardone, Kolker & Mountsford, 2016). Following the trend in official reports and the literature, this section mainly focuses on ODA.

From 1995 to 2009, annual ODA investments in basic and large-scale water and sanitation systems averaged 17% and 39%, respectively, with higher investments in water supply than sanitation (Winpenny et al., 2016). The GLAAS 2013/2014 external support agency survey also shows that an average of 60% of the ODA takes the form of loans rather than grants; only about 21% of the ODA for WASH is committed to basic systems and basic sanitation receives a meagre 27% of the disaggregated flows to WASH (UN-Water, 2014). This shows a recurrent pattern of underfunding sanitation services even at the international level, compared to water, and could be a reason for the poor state of sanitation services despite a general increase in ODA commitments to WASH as noted in the literature (Botting et al. 2010; Salami, Stampini & Kamara, 2012).

Unsurprisingly, nine out of the top ten recipients of ODA for WASH achieved their MDGs target for water, compared to only three who achieved their MDGs sanitation target; the tenth country (Kenya) was recorded as having made good progress (World Health Organization [WHO] & UNICEF, 2015). Further, although development assistance is increasingly targeting poor countries, with low sanitation and water access status, there are exceptions where some of the countries with the lowest sanitation and water access status are excluded from the list of priority countries for ODA suggesting a need to review the targeting of concessional flows (UN-Water, 2014); WaterAid, 2015).

In principle, donor countries committed to spending 0.7% of their gross national income (GNI) on ODA, and though this commitment has not been fully met by all the donors, there has been a total increase in ODA in general. Nonetheless, 10.8% of ODA in 2016 was spent on internal refugee costs within donor countries, an increase from 9.2% in 2015 and less than 5% in 2014; thus reducing the ODA available to poor countries by about 3.9% overall compared to 2015 (Organisation for Economic Cooperation & Development [OECD], 2017). Hence, it is imperative for States to explore new funding options that do not impose an unequal burden on the poor, while committing the available ODA towards basic sanitation services and progressive improvements for the poor.

5.4.3 Suasive

This section discusses two suasive instruments.

Human rights manuals and handbooks

Human rights manuals and handbooks expound legal provisions and practical ways of realising the HRS. A key example is the Handbook for realising the human rights to water and sanitation, authored by the former Special Rapporteur, with clear, practical guidelines and concrete examples to help States implement the right (de Albuquerque, 2014). It highlights issues like financing, monitoring, and translating the normative content of the rights into law, policy, budgets, and interventions. Another example is the Manual on the Right to Water and Sanitation written in non-legal language to provide practical implementation guidelines for policymakers and practitioners working on implementing the human right to water and sanitation (COHRE et al., 2007). This highlights the key components of the rights and States' obligations, the role of non-State actors, and practical policy recommendations for implementation.

Human rights reports

Human rights reports on sanitation services are important for tracking progressive realisation. An example is the biennial UN-Water, Global Analysis and Assessment of Sanitation and Drinking-Water (GLAAS), a multi-stakeholder process that assesses the inputs required to extend sustainable WASH systems and services in countries. GLAAS provides reliable, comprehensive, easily accessible and global analysis of the investments and enabling environment for WASH, for informed decisions in the WASH sector (Baquero, Jiménez & Foguet, 2015). In the post-MDGs framework, GLAAS monitors progress with the SDGs targets on international cooperation and capacity building support for developing countries in WASH related activities and programmes (Target 6a) and support and strengthen the participation of local communities in improving sanitation and water management (Target 6b), while the World Health Organization and UNICEF Joint Monitoring Programme (JMP) is monitoring the drinking water and sanitation targets (United Nations [UN], 2017).

5.5 HUMAN RIGHT TO SANITATION INDICATORS

This section analyses the existing indicators in the literature, because monitoring progress on the HRS has been historically hampered by a paucity of indicators and poor reporting practices by States (Riedel 2006). Indicators ensure comparison of States' compliance with HRS commitments, like the eThekwini Commitments on Sanitation, and are therefore critical for promoting accountability in sanitation governance (Obani & Gupta, 2016). I categorise the indicators for monitoring the realisation of the HRS into three: structural, process and outcome indicators (see 4.5) because realising the obligations imposed by the HRS depends as much on the policies established by the State (structures), and the actions which duty bearers take to advance progressive realisation (process), as on the result for the rightsholders (outcome), and link them to the HRS principles (see Tables 5.4, 5.5 and 5.6).

Structural

The literature presents two main indicators which I classify as structural indicators because they directly monitor the law and policy framework for the HRS, namely: (a) the inclusion of the HRS in law (Baquero et al., 2015), and (b) the prioritisation of universal access for vulnerable groups (UN Water & World Health Organisation [WHO], 2014) (see Table 5.4).

Table 5.4 Structural indicators for monitoring the human right to sanitation

Structural Indicators	Human Right to Sanitation Principles (see 5.3)	Key References
HRS included in the legal framework [a]	Rule of Law	Baquero et al., 2015
HRS and access for vulnerable groups included in the legal framework [a]	Equality and Non-discrimination	UN Water & WHO, 2014

a - Can be measured under the GLAAS framework

Nonetheless, beyond recognition of the principle in relevant laws, equality and non-discrimination would additionally require: a) an understanding of the factors that engender inequality and discrimination and identifying the resulting vulnerable groups; (b) disaggregating data on access to sanitation services based on the vulnerable groups identified; c) measuring the rate of progress required for both the worse-off and the better-off groups, in order to meet the universal access target; and d) calculating the disparity in the use of services between the worse-off and the better-off within each group (Satterthwaite, 2014; Winkler et al., 2014). In order to be considered on-track to realising the HRS and particularly promoting equality and non-discrimination, the progress of both the better-off and the worst-off groups would either follow or exceed the pre-determined rate of progress, in addition to a reduction in the disparity between the progress of the two groups (Satterthwaite, 2014; Winkler, Satterthwaite & de Albuquerque, 2014). Hence, in addition to structural indicators, it is also necessary to develop process indicators that monitor policy implementation, and outcome indicators that monitor the impact on rightsholders.

Process

The literature offers sixteen indicators which I classify as process indicators because they directly reflect on the milestones towards achieving the HRS (see 4.5), and I link these to eight HRS principles: (a) affordability (UN, 2012; UN Water & WHO, 2014), (b) rule of law (Baquero et al., 2015), (c) participation (Baquero et al., 2015), (d) equality and non-discrimination (Baquero et al., 2015; UN, 2012), (e) transparency and access to information (Baquero et al., 2015), (f) accountability (Baquero et al., 2015; UN Water & WHO, 2014), (g) sustainability (UN, 2012), and (h) extra-territorial obligations (Baquero et al., 2015) (see Table 5.5).

However, beyond establishing processes for the realisation of the HRS, the processes also need to be quantifiable, through milestones linked to the law and policy framework, to enhance measurement across temporal and spatial scales (Meier et al. 2017), and accessible to the most vulnerable rightsholders in order to improve the HRS outcomes (de Albuquerque, 2014). Further, the human rights literature does not specify process indicators for availability, accessibility, safety, acceptability and dignity.

Outcome

The literature offers nineteen indicators which I classify as outcome indicators because they directly measure the impact of HRS policies and processes on the quantity and quality of sanitation services used by rightsholders (see 4.5), and I link these to eight HRS principles (see 5.4): (a) availability (Baquero et al., 2015; Meier et al., 2017; UN, 2012), (b) accessibility (Baquero et al., 2015; UN, 2012), (c) safety (Baquero et al., 2015), (d) acceptability (Baquero et al., 2015; UN, 2012), (e) affordability (UN, 2012), (f) dignity

(Baquero et al., 2015), (h) equality and non-discrimination (Meier et al., 2017), and (i) sustainability (UN, 2012) (see Table 5.6).

Nonetheless, the outcome indicators which do not specify measurable targets for determining whether progress towards the HRS is on or off track paint a limited picture of HRS progress at best. For instance, household expenditures on sanitation need to be tied to a benchmark for assessing affordability expressed as a percentage of individual or household income or expenditure on the lifecycle cost of sanitation facilities (including operations, maintenance, and repairs) and other costs associated with accessing sanitation services or the number of maintenance hours per person per year for non-monetary contributions to HRS costs (see 5.3.1). Further, the human rights literature does not specify outcome indicators for the remaining five HRS principles (rule of law, participation, transparency and access to information, accountability and extra-territorial obligations).

Table 5.5 Process indicators for monitoring the human right to sanitation

Process Indicators	Human Right to Sanitation Principles (see 5.3)							
	Affordability	Rule of Law	Participation	Equality and Non-discrimination	Transparency/Access to Information	Accountability	Sustainability	Extra-territorial obligation
HRWS justiciability*		■						
International assistance by industrialised States* [a]								■
Private sector participation*								
Monitoring mechanisms [a]			■		■	■		
Civil society inclusion in monitoring process			■		■	■		
Complaints mechanisms in place [a]			■		■	■		
Financial flows to the vulnerable [a]				■				
Inequities reduction [b]				■				
Budgetary strategies for the vulnerable [a]						■		
Monitoring access among the vulnerable [a]						■		
Time-frame national strategy and plan of action for WASH provision* [a]								
Service users and communities participation in WASH supply decision making [a]	■							
Level of sector financing and human resources development [a]	■							
Direct or indirect discrimination by public and private actors [c]				■				
Share of public expenditure on provision and maintenance of sanitation in homes [d]								
Proportion of targeted population that was extended sustainable access [d]							■	

* - Presented as cross cutting and general indicators; a - GLAAS Report indicators; b - Can be measured under the JMP-post 2015 framework through the SDGs; c - Illustrative indicator on the sanitation aspect of the right to non-discrimination & equality; d - Illustrative indicator on the sanitation aspect of the right to adequate housing

Source: Compiled by the author, based on Baquero et al., 2015, UN, 2012, and UN Water & WHO, 2014

Table 5.6 Outcome indicators for monitoring the human right to sanitation

Outcome Indicators	Human Right to Sanitation Principles (see 5.3)							
	Availability	Accessibility	Safety	Acceptability	Affordability	Dignity	Equality & Non-discrimination	Sustainability
% access to improved/basic sanitation [a,b]	▪							
All members at any time [b]		▪						
Security at facilities and paths		▪						
Education/Health facilities [a,b]		▪						
Open defecation [b]			▪					
Safe management of excreta [b]			▪					
Sanitary conditions (facilities)			▪					
Wastewater treatment [a]			▪					
Hygiene awareness [a]			▪					
Hand washing device & soap [a,b]			▪					
Menstrual Hygiene Management (MHM) [b]				▪				
Household expenditure on sanitation					▪			
Assistance to low income groups [b]				▪	▪			
Privacy, comfort, dignity [a]	▪					▪		
% change in access to improved facility								
% change between rural and urban access							▪	
Proportion of population using improved sanitation facility [a,c,e] and waste disposal [c]	▪				▪			
Proportion of household budget spent on sanitation and waste disposal [d]					▪			
Proportion of targeted populations that was extended sustainable access to services [d]		▪						▪

a - GLAAS Report indicators; b - Can be measured under the JMP-post 2015 framework through the SDGs; c - Illustrative indicator on the sanitation aspect of the right to non-discrimination & equality; d - Illustrative indicator on the sanitation aspect of the right to adequate housing; e - Illustrative indicator on the sanitation aspect of the rights to life & health

Source: Compiled by the author, based on Baquero et al., 2015; Meier et al., 2017 and UN, 2012

5.6   HUMAN RIGHT TO SANITATION, DRIVERS & INCLUSIVE DEVELOPMENT

This section analyses the relevance of the human right to sanitation principles and instruments for addressing the drivers of poor sanitation services (see 5.6.1), and the wider implications for inclusive development (see 5.6.2); based on literature review and inductive analysis of the human right to sanitation institution.

5.6.1 Human Right to Sanitation and the Drivers of Poor Sanitation Services

The HRS principles address ten direct and six indirect drivers of poor sanitation services, as illustrated in Table 5.7 and elaborated below.

Principles for addressing the direct drivers

Affordability addresses household poverty and unaffordable tariffs where progressive pricing and cross-subsidies (see 5.5.2) are targeted at the poor and these and other pro-poor instruments are expanded to cover both the monetary and non-monetary sanitation costs borne by rightsholders (COHRE et al., 2008; Obani & Gupta, 2015). Equality and non-discrimination also address household poverty, where the poor are prioritised in budgetary allocations and the distribution of resources for sanitation (see 5.6.1). Further, transparency, access to information about the tariff structure, and accountability to users can improve the efficiency of the tariff collection system and affordability addresses prohibitive sanitation costs (see 5.3.1). Accessibility addresses the distance to facility driver where a maximum reasonable distance for siting sanitation facilities away from the vicinity of users is specified in the legal framework (see 5.3.1). Participation, equality and non-discrimination address the exclusion of minorities and other similarly negative social practices (see 5.3.1). Mechanisms for acceptability and participation of users in sanitation governance, including the choice, operation and maintenance of facilities, address non-acceptance and participation particularly improves knowledge about the proper use of facilities (see 5.3.1). Further, safety imposes an obligation on service providers and users to respectively ensure the physical safety of users and the hygienic maintenance of facilities (see 5.3.1). Equality and non-discrimination address tenure insecurity where the HRS law prioritises service provision to residents of informal settlements, in spite of the legal status of their dwellings (see 4.3 and 5.4.1). Sustainability addresses pollution and environmental degradation from the location and use of sanitation facilities, provided the HRS obligations are broadly interpreted (see 3.2 and 5.3.2) to include an environmental sustainability imperative (Feris, 2015). I further elaborate on addressing the remaining seven direct drivers which are not covered by the HRS principles (including discounting the future, preference distortion, risk aversion, space constraints, challenging or inaccessible topography, natural hazards and high temperatures/turbidity in source water) in Chapter 9.

Table 5.7 Human right to sanitation principles for addressing the drivers of poor domestic sanitation services*

Category of Drivers (see 3.4)	Details of Drivers (see 3.4)	Human Rights Principles for Addressing the Drivers (see 5.4)
Direct Drivers		
Economic	Household poverty	Affordability Equality and Non-discrimination
	Inefficient tariff collection system	Transparency and Access to Information Accountability
	Unaffordable tariffs & connection fees	Affordability
Social	Distance to the facility	Accessibility
	Exclusion of minorities	Participation Equality and Non-discrimination
	Negative cultural practices	Participation Transparency and Access to Information
	Non-acceptance of sanitation facility	Acceptability Participation
	Poor maintenance culture/ improper use of facilities	Participation Safety
	Tenure insecurity	Equality and Non-discrimination
Environmental	Pollution/water scarcity	Sustainability
Indirect Drivers		
Economic	Huge foreign debts	Extra-territorial Obligations
	Insufficient/poorly targeted funds	Affordability Equality and Non-discrimination Transparency and Access to Information Accountability
	National poverty	Extra-territorial Obligations
	Sanctions	Extra-territorial Obligations
Social	Low awareness about sanitation	Participation Transparency and Access to Information
	Population density	Availability

*Table 5.7 contains only direct and indirect drivers that are addressed by the HRS principles.

Source: Compiled by the author, based on COHRE et al., 2008; de Albuquerque, 2014; Feris, 2015; Obani & Gupta, 2016; Winkler, 2016

Principles for addressing the indirect drivers

Extra-territorial obligations require States to desist from negatively interfering with the realisation of the HRS outside their jurisdictions, for instance through foreign debts or sanctions, and generally encourage international assistance to enable poor States meet their human rights obligations and development goals (see 5.3.2). Affordability, and equality and non-discrimination can improve public expenditure on service provision to the poor, marginalised and vulnerable while transparency and access to information and accountability offer mechanisms for tracking the equitable distribution of public funds (see 5.3.1 and 5.4.2). Nonetheless, the HRS cannot directly tackle insufficient funds. Finally, participation and access to information improve the level of awareness about good sanitation practices (see 5.3.1). Availability addresses poor access to sanitation services due to population density where the HRS law specifies a minimum ratio of users to facilities as part of the service levels, while taking cognisance of the local conditions (see 3.5.5 and 5.3.1). I further elaborate on addressing the remaining three indirect drivers which are not covered by the HRS principles (including insecurity, conflicts and poor social cohesion, mass migration/urbanisation and climate variability and change) in Chapter 9.

5.6.2 Human Right to Sanitation and Inclusive Development

Recognising the HRS, and adopting HRS principles and instruments in sanitation governance does not guarantee ID, rather the way the HRS is operationalized will determine whether ID is achieved or not. As the HRS principles can be implemented differently in different contexts, which results in different outcomes for ID, it is further necessary to specifically analyse the impact of the HRS instruments discussed in Section 5.4, as they affect social and relational, and ecological inclusion. For example, if you have the principle of affordable access to basic sanitation services (see 5.3.1), this can be implemented through a variety of technologies and instruments. Figure 5.1 illustrates how instruments that are designed to operationalize the affordability principle can lead to different outcomes for ID. Further, although the same indicator of affordability is applied in the four quadrants, the outcome may also be different; the exclusion of the poor in Q1 may make sanitation services relatively less expensive for the rich living in formal settlements because there is no additional cost for wastewater treatment or cross-subsidies for poor users. However, the negative externalities in Q1 are also higher due to the exclusion of the poor from accessing sanitation, which is a public good (see 3.3). In Q4, the inclusion of wastewater treatment and poor users may increase the cost of sanitation services for the rich but this also reduces the negative externalities which would have resulted from excluding poor users.

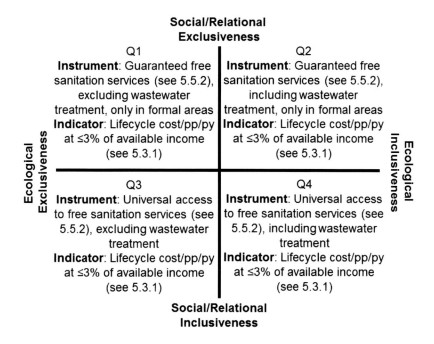

Figure 5.1 Assessing the affordability principle for inclusive development

Social and relational inclusion

Regulatory instruments (see 5.4.1) like guaranteed free access to basic sanitation services (that are designed with input from all the stakeholders), compulsory connections and prohibiting arbitrary disconnections (for instance, due to the inability to pay tariffs for basic services), promote social and relational inclusion by ensuring affordability, preserving access rights and eliminating discrimination against the poor, vulnerable and marginalised users. The instruments could fall within either of the bottom half quadrants (Q3 or Q4 of Figure 5.1), depending on their level of integration of environmental sustainability. Direct provision of services to the vulnerable may also fall within either Q3 or Q4 of Figure 5.1), depending on the level of integration of environmental sustainability. However, the exclusion of informal settlements from the process of implementing free access or the safeguards against arbitrary disconnections results in social and relational exclusion in Q1 or Q2 of Figure 5.1. Litigation provides an opportunity for preventing HRS violations or seeking redress, but may fall within either Q4 or Q3 in Figure 5.1 if the processes are accessible for the most vulnerable and/or do not integrate environmental sustainability, or the top quadrants (Q1 or Q2 in Figure 5.1) where the processes are inaccessible and/or do not integrate environmental sustainability. Monitoring and accountability mechanisms may similarly promote social and relational inclusion where they are freely accessible and facilitate participation by the vulnerable. Participatory budgeting (see 5.4.2) improves access to information and accountability in public expenditure and thereby advances social and relational inclusion (Q3 or Q4 of Figure 5.1), provided the process does not impose high costs of participation on the poor otherwise it leads to social and relational exclusion (Q1 or Q2 in Figure 5.1). Participation can engender a sense of ownership and responsibility for sanitation infrastructure, which is useful for addressing lack of social cohesion (McGranahan, 2013). Progressive pricing and cross-subsidies (see 5.4.2) also promote social and relational inclusion where the instruments are primarily designed to ensure access to basic sanitation services for the poor[129] or to allow in-kind contributions for services in densely populated areas where users would otherwise be unable to afford the monetary cost of accessing sanitation services inclusion (Q3 or Q4 of Figure 5.1).[130]

Ecological inclusion

Ecological inclusion requires a broad definition of basic sanitation services to include services like wastewater treatment that minimise the environmental impacts of sanitation services (see 3.2, 3.5.5, 5.2.3 and Q2 and Q4 in Figure 5.1). Ecological inclusion also requires instruments that encourage conservation such as progressive pricing of water and sanitation servicers beyond the basic consumption level (see 5.4.2).

5.6.3 Human Right to Sanitation and Legal Pluralism

The realisation of the HRS partly depends on the relationship between the HRS norm and other norms and principles for the governance of natural resources like water and land, required for the progressive realisation of the right. This section specifically analyses the interactions between the HRS and other human rights. The resulting pluralism can either manifest as rules coherence, where the HRS and the other rights and rules are mutually supportive, or incoherence and even competition, where there is tension between the rights

---

[129] Chile, *Law 18778 Establishing Services for the Payment of Drinking Water Consumption and Sanitation Services 1989/1994*, article 10; Nicaragua, *General Law on Drinking Water and Sanitation Services*, article 40.

[130] Rwanda, *Policy and Strategy for Water Supply and Sanitation Services 2010*, section 4.6.3.

and rules. Table 5.8 illustrates indifference, accommodation, competition and mutual support between the HRS and other rights and water governance rules.

Table 5.8   Typology of relationships between the human right to sanitation and other human rights

Type of Relationship	Description of the Relationship
Competition	Where the ICESCR requires States to apply maximum available resources for HR implementation but it does not clarify how much of the resources can be allocated to the HRS as a basic necessity
Indifference	Where the HRS is recognised in international law instruments, like the CEDAW and CRC, and captured without being implemented in the national legal system
Accommodation	Where the right to participation, included in the ICESCR, is used to try to encourage the rightsholders to participate in the policy process and contribute their unique perspectives on the prospects and challenges for realising their HRS
Mutual Support	Where the decisions of international courts expand on the meaning and principles of the HRS, in consonance with the ICESCR and other international law instruments recognising the right

Source: Compiled by the author, based on Bavinck & Gupta, 2014; Misiedjan & Gupta, 2014; Obani & Gupta, 2014b; 2016; WaterLex, 2014

Competition

The ICESCR which forms the main legal bases for the HRS, and the HRW, under HR treaty law does not expressly mention sanitation or water (see 5.2.1 and 5.2.2). However, the HRS, including the HRW, emerged in close connection with other rights that are expressly guaranteed in the ICESCR, like the rights to health, adequate housing, food and education, as illustrated in Table 5.2 (Obani & Gupta, 2015; 2016). The ICESCR requires States to use the maximum available resources to progressively realise the rights which it guarantees but does not clarify how the available resources are to be allocated to the various express and implied rights. This can create tensions between the HRS and other economic, social and cultural rights. The HRS as an individual right, may also be in competition with collective rights such as the indigenous' people's rights (Misiedjan & Gupta, 2014).

Indifference

As mentioned under the discussion on the emergence of the HRS (see 5.2), the HRWS emerged from local advocacy to being recognised within the UN HR system, and as a distinct legal right in various international law instruments and political declarations (see 5.2). However, the realisation of the HRS requires action from both States and non-State actors, all the way from the international to the lowest levels of governance (Kamga, 2013; de Albuquerque, 2014; Obani & Gupta, 2016; Winkler, 2016). Indifference results where the HRS norm that exists in international law is captured but not implemented at the national and sub-national levels and therefore fails to improve access to sanitation for the rightsholders.

Accommodation

The human rights framework offers procedural mechanisms for promoting HR, and obtaining redress for violations. These include the right to participation, guaranteed under the ICESCR (see 4.3 and 5.3.1). Accommodation results where mechanisms are established on the basis of procedural rights, like participation, in an attempt to encourage the rightsholders to engage with the policy process and express their unique perspectives on the HRS and the challenges which they face in accessing the right. Another instance is where efforts are made to enforce the HRS, using existing human rights mechanisms like the HR committees and Special Procedures (Obani & Gupta, 2016).

Mutual support

Mutual support exists where the decisions of national courts expand on the meaning and principles of the HRS, in consonance with the ICESCR and other international law instruments (see 5.2.2 and 5.2.3). For instance, the European Court of Human Rights has severally decided that lack of adequate sanitation in detention centres amounts to degrading and inhuman treatment of the detainees that are held in such centres, guaranteed under regional and international HR laws (see Table 5.3). Such decisions advance the normative content of the HRS and improve enforcement at various levels (Obani & Gupta, 2016; WaterLex, 2014).

5.7   INFERENCES

The chapter yields six main inferences. First, the HRS has evolved through implied recognition on the basis of the ICESCR provisions and has recently emerged as a distinct right in international law, but has not yet attained the status of CIL. In the absence of express recognition of the HRS at the national and sub-national levels, HRS advocates can nonetheless rely on the fundamental importance of HRS for the realisation of other guaranteed rights, some of which have attained the status of custom, like the right to life and freedom from inhuman and degrading treatment (see 4.2.1 and 5.2).

Second, the HRS can either be defined narrowly with limited consideration of environmental impacts or broadly to ensure human wellbeing and environmental sustainability (see 3.2, 5.3.1). At the international level, the definition advanced by the former Special Rapporteur extends beyond the provision of facilities for the collection, transport, treatment and disposal or reuse of human excreta and wastewater and associated hygiene in households to other spheres of existence where people spend a considerable amount of their time. However, the definition does not sufficiently capture environmental sustainability concerns, and some countries like the United Kingdom further restrict the scope of the HRS in their national legal framework (see 5.2.3).

Third, the definition of the HRS and the scope of HRS principles indicate that in addition to States as the primary duty bearers for the HRS, non-State actors also bear some of the responsibility for realising the right. For instance, the affordability principle imposes a corresponding duty on individuals who are able to contribute to the sanitation systems either in cash or in kind and a duty on non-State actors like private service providers to ensure equitable pricing of their services and not to disconnect users from accessing basic services due to inability to pay.

Fourth, the HRS principles, similar to the HR principles (see 4.3), mostly address the social pillar of sustainable development. The principles address ten direct and sixteen indirect drivers of poor sanitation services. This suggests a need to integrate complementary non-HRS principles in the sanitation governance framework in order to address the remaining seven direct and three indirect drivers, which may further contribute to plurality (characterised by rules incoherence, contradictions or support between the HRS and prevailing sanitation governance principles (Obani & Gupta, 2014b). The resulting rules incoherence may lead to the violation of the HRS, while coherence (mutual support) between the HRS and other dominant principles in the sanitation governance framework is the most desirable outcome capable of advancing ID.

Fifth, there is a paucity of quantitative indicators specifically developed for the HRS which further compounds the issue of evaluating the performance of HRS instruments. Nonetheless in this chapter, I derive two structural indicators, sixteen process indicators, and nineteen outcome indicators (n=37 indicators) from the literature which offer a good starting point for developing HRS indicators in national legal frameworks (see 5.5). However, the indicators are unevenly spread across the thirteen principles of the HRS and this could mean that some principles get overlooked in the monitoring process. Further, there are few structural indicators but considerably more process indicators and the majority are outcome indicators. This could mean that the law and policy framework for the HRS becomes underreported, and the milestones for measuring progressive realisation of the HRS are insufficiently linked with the final outcomes for rightsholders.

Sixth, an inclusive development assessment (see 2.4.3) shows that the true impact of the HRS principles are more closely linked in practice and dependent on the local context. For instance, although safety is focused on social and relational inclusion when equated with safe use of sanitation services it could also address ecological inclusion if it is broadly applied to the safe management of wastewater and conservation of sanitation resources (like water for wet sanitation systems) to ensure environmental sustainability. In addition, the effect of HRS instruments on ID largely depends on how the implementation at scale (see 5.6). Taking the example of affordability, determining an affordable tariff for sanitation services requires empirical support based on local circumstance because what constitutes an affordable tariff for sanitation in one context may prove unaffordable under different circumstances.

Chapter 6. Human Right to Sanitation in Humanitarian Situations

6.1 INTRODUCTION

Humanitarian situations occur where technical issues, human negligence, weak infrastructure in disaster prone areas, civil war, or a rare event, with retrospective predictability, results in system failures and the affected population have to rely on external assistance for their recovery (Labib & Read, 2015). Although international humanitarian law focuses on protecting vulnerable groups during armed conflicts, and limiting the methods and effects of warfare, natural hazards are also increasingly creating humanitarian situations and vulnerable groups that require some forms of humanitarian assistance and protections (Internal Displacement Monitoring Centre [IDMC] & Norwegian Refugee Council [NRC], 2017). Further, low and middle income countries account for significant new internal displacement resulting from conflicts and natural hazards and displacement will likely continue to hamper human rights realisation, social development and economic growth except the drivers of poverty, environmental change and state fragility are tackled (Bennett et al., 2017).

This chapter therefore broadly analyses the human right to sanitation (HRS) in humanitarian situations resulting from both armed conflicts and natural hazards, though it does not specifically distinguish between specific groups of vulnerable persons in humanitarian situations, like poor migrants, refugees and internally displaced persons. The preoccupation with humanitarian situations in this chapter is because humanitarian situations often obscure or escalate lack of access to basic infrastructure like sanitation and the inequities within a population as a result of destruction of the existing infrastructure or high influx of displaced persons creating additional stress on the infrastructure and the environment (ACF-France, 2009; Nicole, 2015). Further, humanitarian situations involve a variety of humanitarian actors, State actors, communities and their representatives, and the operation of multiple normative orders including human rights, humanitarian principles and local customs and practices. The plurality of rules inherent in humanitarian situations offers an avenue for investigating the fit between the HRS and other principles for sanitation governance, in this case humanitarian principles (see 2.4.2).

Humanitarian law principles are mostly discussed in connection with regulating the use of force in armed conflict but rarely in connection with the promotion of human rights like the HRS. This chapter therefore also relies on the human rights literature and content analysis to address the subsidiary research questions: (a) how is sanitation defined in the context of humanitarian situations? (b) How is the human right to sanitation reflected in international humanitarian law and how can progress towards realising the right be monitored? (c) How do the humanitarian principles for sanitation governance promote inclusive development? (d) How does an understanding of the incoherence between the HRS and humanitarian principles, using legal pluralism theory (see 2.4.2), affect the design of sanitation governance frameworks in humanitarian situations?

The chapter analyses the meaning of sanitation as provided for in international humanitarian law (IHL) and the sources of humanitarian law (see 6.2), and principles (see 6.3) that support the HRS; the humanitarian instruments for sanitation governance (see 6.4); and the existing indicators for monitoring progress on the HRS in humanitarian situations (see 6.5). Section 6.6.1 analyses the humanitarian principles for addressing the drivers of poor domestic sanitation services, while Section 6.6.2 assesses humanitarian principles for sanitation governance for inclusive development and analyses the incoherence between humanitarian principles and the HRS principles. Finally, Section 6.7 highlights the key inferences from the chapter.

6.2 Legal Bases and Meaning of the Human Right to Sanitation in International Humanitarian Law

International humanitarian law (IHL) is the body of international law that protects specific vulnerable groups in situations of armed conflicts, and limits the methods and effects of warfare, especially as it applies to non-combatants. IHL also protects civilian objects from military attacks or use as weapons of war during conflicts, and mainly comprises of customs; treaties (like the Hague Conventions of 1899 and 1907 and their regulations, and the Four Geneva Conventions with the First and Second Additional Protocols); and soft law (see for instance, Baker, 2009). Though IHL is the *lexis specialis* for the protection of persons during armed conflicts, human rights (HR) law still operates as a complementary law for the protection of persons. Specifically, the International Covenant on Economic, Social and Cultural Rights 1966 (ICESCR) from which the HRS is mainly derived in international law (see 5.2) does not permit derogation from its provisions, while the International Covenant on Civil and Political Rights 1966 (ICCPR) permits derogations under exceptional circumstances like war or public emergencies.[131] Nonetheless, contrary to the perception that

---

[131] ICCPR, article 4.

HR derogations follow humanitarian situations or emergencies (Brookings-Bern Project on Internal Displacement, 2011), the literature suggests that modern democracies are strongly committed to observing HR obligations even during humanitarian situations (Hafner-Burton, Helfer & Fariss, 2011). In addition, soft law instruments like the Guiding Principles on Internal Displacement offer principles for the protection of internally displaced persons, which may be as a result of humanitarian situations, and humanitarian actors have rules of practise like the Sphere Handbook (see 6.2.4) which imposes moral obligations on them in delivering humanitarian assistance. Hence, this section explores IHL provisions that promote the HRS (see 6.2.1), before exploring the meaning of sanitation under the humanitarian law framework (see 6.2.2).

6.2.1 Treaties

Article 3 common to the Four Geneva Conventions provides that persons protected by the Conventions should "in all circumstances be treated humanely, without any adverse distinction founded on race, colour, religion or faith, sex, birth or wealth, or any other similar criteria". The Geneva Conventions III and IV further protect the rights of prisoners of war[132] and civilian populations in war times[133] to access sufficient water and soap for their daily toilet needs and doing personal laundry. Victims of international and non-international armed conflicts are also entitled to the protection of their drinking water installations and supplies and irrigation works as civilian objects, while the latter are entitled to additional health and hygiene safeguards.[134] These protections are remarkable especially as the Geneva Conventions have been ratified by at least 192 States and most of the provisions are considered part of customary international law.[135]

At the regional level, the 2009 African Union Convention for the Protection and Assistance of Internally Displaced Persons in Africa (Kampala Convention),[136] which came into force in 2012, prohibits combatants from depriving internally displaced persons (IDPs) of their right to live in satisfactory conditions of dignity, with access to basic needs like sanitation.[137] It also obliges States to take necessary measures to ensure that IDPs are received in host

---

[132] Geneva Convention III – Treatment of Prisoners of War, article 29.

[133] Geneva Convention IV – Protection of Civilian Persons in Times of War, article 85

[134] Additional Protocol I – protection of Victims of International Armed Conflicts, articles 54 and 55; Additional Protocol II – Protection of Victims of Non-International Armed Conflict, articles 5 and 14.

[135] Customary International Law, Vol 1.

[136] The Kampala Convention is the first legally binding treaty for the protection of the rights and wellbeing of persons displaced by various forms of conflict, disasters, human rights abuses and violence.

[137] Article 7(5)(c). The Kampala Convention is yet to enter into force, subject to ratification.

communities without any discrimination, and provided adequate humanitarian assistance including sanitation services, to the fullest extent possible and with minimal delay.[138]

6.2.2 Customary International Humanitarian Law

There are 161 rules of customary international humanitarian law; at least 136 of these are based on the rules of the Additional Protocol 1 to the Geneva Conventions. These rules however also apply to non-international armed conflicts and include the rules relating to the principle of distinction, specifically protected persons and objects, specific methods of warfare, weapons, and the treatment of civilians and persons *hors de combat*, *inter alia* (Henckaerts & Doswald-Beck, 2005). The rules relevant for the HRS discussed in sections 6.2, 6.4.1 and 6.4.4 reveal two main limitations of customs as a source of international humanitarian law: (a) custom is constantly evolving and therefore provides a weak basis for the uniform application of law in humanitarian situations, and (b) custom also includes the actual practise of belligerents which would otherwise be considered unacceptable (Sassòli, Bouvier & Quintin, 2011).

6.2.3 Soft Law

The Guiding Principles on Internal Displacement obliges authorities undertaking displacements to ensure that IDPs are afforded satisfactory hygiene and other similar basic services.[139] All IDPs also have the right to an adequate standard of living, including safe access to sanitation.[140] Further, soft law instruments on the HRS, like the General Comment No 15 of the Committee on Economic, Social and Cultural Rights (CESCR) integrates the humanitarian law obligations of States during armed conflicts[141] to protect the rights of vulnerable groups like refugees and IDPs;[142] articulate the duty of States to adopt comprehensive and integrated strategies for emergency response;[143] prohibit the use of water (presumably also including sanitation and hygiene) infrastructure as a weapon of war or punishment;[144] promote the HRS and other ICESCR rights in disaster relief and emergency

---

[138]   Article 9(2)(a)(b).
[139]   Guiding Principles on Internal Displacement 1998, principle 7(2).
[140]   Ibid, Principle 18.
[141]   General Comment No. 15, paragraph 22.
[142]   General Comment No. 15, paragraph 16.
[143]   General Comment No. 15, paragraph 28(h).
[144]   General Comment No. 15, paragraph 21.

assistance efforts[145] and advance the central role of humanitarian actors in HR implementation. [146]

6.2.4 Defining the Human Right to Sanitation under a Humanitarian Law Framework

Although neither sanitation nor the HRS is expressly defined under IHL, sanitation, water and hygiene services are addressed in connection with the protection of civilian populations and objects during armed conflicts (see 6.2.1 and 6.2.2) and general HR protections for IDPs (see 6.2.1 and 6.2.3) (Maunganidze, 2016). Additionally, the Sphere Handbook prepared by the Sphere Project,[147] which is considered the most widely recognised and accepted set of principles and minimum standards for humanitarian assistance by the international humanitarian community, has provisions on water, sanitation and hygiene services (WASH) and contains seven related themes that are broadly relevant for sanitation in humanitarian situations: WASH, hygiene promotion, water supply, excreta disposal, vector control, solid waste management, and drainage (Obani, 2017). The sanitation theme in particular has five components: excreta disposal, vector control, solid waste management, and drainage (Obani, 2017). Humanitarian sanitation interventions also generally include direct provision, support for capacity building and preparedness, and response coordination at the international and domestic levels, and during recovery to support the development of the local water, sanitation and hygiene infrastructure (Ramesh, Blanchet, Ensink & Roberts, 2015; UNICEF, 2017). At the national level, some laws also permit the abstraction of water from any source for human consumption and fire fighting during emergencies without any license (see 5.4.1). Presumably, such provisions could be interpreted in order to extend the application of the human rights protection of access to sanitation (and water) to emergencies and humanitarian situations generally, depending on the express wording of the relevant statutory provision (see 2.3.5).

Further, humanitarian situations present additional nuances for operationalizing the HRS principles (see 5.3). In this regard, the Sphere Handbook provisions expatiate on the HRS principles of acceptability, non-discrimination, safety, accountability, participation and access

---

[145] General Comment No. 15, paragraph 34.
[146] General Comment No. 15, paragraph 60.
[147] The Sphere Project was established in 1997, by the representatives of global networks of humanitarian agencies, to improve accountability and the quality of humanitarian response. See further, http://www.spherehandbook.org/en/what-is-sphere/ accessed on 27 September 2017. The Sphere Handbook is however not legally binding but imposes moral obligations on humanitarian actors at various levels of governance, as an epistemic community.

to information (Sphere Project, 2011). To ensure the acceptability of humanitarian sanitation services, it is important for the humanitarian actors to understand the varied needs of the population based on disaggregated assessment data, embark on hygiene promotion to ensure proper use of the facilities provided, use appropriate technology options for different phases of emergency response, and integrate user preferences, the local culture, existing infrastructure, local conditions, and special needs of vulnerable groups in the decision support systems (see 6.4.2). Accountability requires: (a) coordinating and implementing effective response with other agencies and governmental authorities engaged in impartial humanitarian action; (b) conducting systematic assessments to understand the nature of the disaster, identify the impacts and the affected population, and assess people's vulnerability and capacities; (c) examining the effectiveness, quality and appropriateness of humanitarian responses; (d) recognising the obligation of humanitarian agencies to employ suitably qualified aid workers to deliver an effective humanitarian response and support the workers with effective management and support for their emotional and physical well-being, and (e) assisting the affected population with rights claims, access to remedies and recovery from abuse (Sphere Project, 2011).

Effective sanitation interventions require the active participation of the affected population, including through: (a) establishing explicit links between project strategies and community-based capacities and initiatives, (b) facilitating the active participation of the affected population in regular meetings on how to organise and implement humanitarian response, (c) increasing the number of self-help initiatives by the affected community and local authorities during the response period, and (d) establishing mechanisms for investigating complaints against humanitarian actors and taking appropriate action, including (e) a feedback mechanism through which the affected population can influence the planning and implementation of humanitarian interventions using tools like a balanced representative participation (Humanitarian Accountability Partnership [HAP], 2008; Sphere Project, 2011).

The affected population are entitled to accurate and updated information with special consideration of the local language, including adapted media for children and adults who cannot read for instance or people with visual impairments (HAP, 2008; Sphere Project, 2011), and the right to formal mechanisms for complaints and redress (Sphere Project, 2011). It is also necessary to disaggregate the data during assessment of humanitarian interventions, in order to capture the impact of humanitarian assistance on vulnerable groups (Sphere Project, 2011).

6.3 PRINCIPLES OF HUMANITARIAN ASSISTANCE AND PROTECTION OF PEOPLE IN HUMANITARIAN SITUATIONS

The discussion here mainly focuses on the seven principles of humanitarian assistance and the four principles for the protection of people in humanitarian situations (n=11) that are most relevant for sanitation governance in humanitarian situations, based on literature review, content analysis and an inductive analysis of the framework for humanitarian assistance. However, this section does not focus on the general principles of international humanitarian law that govern armed conflict, including: (a) prohibition on causing unnecessary suffering, (b) necessity, (c) independence of *jus in bello* (humanitarian rules to be respected in warfare) from *jus ad bellum* (legality of the use of force), (d) distinction between civilians and combatants, civilian objects and military objectives, (e) humanity (Meron 2000; Slim 1998),[148] and (f) proportionality of military action.[149] The principles of humanitarian assistance and protection principles discussed in this section are predominantly social principles, even though in practise they may also affect the economic and environmental pillars of sustainable development.

6.3.1 Principles of Humanitarian Assistance

Humanity

Humanity is established in the Additional Protocol (1) to the Geneva Conventions and soft law IHL instruments like the General Assembly Resolution 46/182 (see Table 6.1). It limits the suffering and destruction incident to warfare by generally prohibits the use of force, or infliction of any form of damage, suffering or injury that is not essential to realising the legitimate purpose of a conflict (Arai, 2011; International Committee of the Red Cross, 2014). It also requires humanitarian actors to take all steps possible to address all forms of human suffering (Sphere Project, 2011; United Nations Office for the Coordination of Humanitarian Affairs [OCHA], 2012). In relation to the HRS, humanity therefore prohibits damage to sanitation infrastructure that does not advance the legitimate purpose of conflicts and obliges humanitarian actors to take all possible steps to improve access to sanitation for persons living in humanitarian situations.

---

[148] See Case No. 153, ICJ, Nicaragua v. United States, Para. 242.
[149] See Case No. 123, ICJ/Israel, Separation Wall/Security Fence in the Occupied Palestinian Territory [Part B., paras 36-85]; Case No. 124, Israel, Operation Cast Lead [Part I, paras 120-126; Part II, paras 230-232]; Case No. 257, Afghanistan, Goatherd Saved from Attack; Case No. 290, Georgia/Russia, Human Rights Watch's Report on the Conflict in South Ossetia [Paras 28-30, 41-47].

Neutrality

Neutrality is enshrined in soft law instruments like the UNGA Resolution 46/182 (see Table 6.1). Neutrality imposes three duties on a State, to: abstain from providing military assistance to belligerents; prevent belligerents from using its territory for purposes that contravene neutrality; maintain impartiality in relating with the belligerents (Schindler, 1991). It also demands that a humanitarian actor desists from taking sides in hostilities or engaging in political, racial, religious or ideological controversies, in order to retain the trust of the parties (Harroff-Tavel, 1989; International Committee of the Red Cross, 2002; Schindler, 1991). There is a wide spectrum of approaches to neutrality by humanitarian actors, ranging from the classicists who believe that it is both possible and desirable to divest humanitarian assistance of all forms of political influences, to solidarists that abandon neutrality to pick sides and reject consent as a requirement for humanitarian interventions, and the minimalists and maximalists across the spectrum between classicists and solidarists (Minear, 1999; Weiss, 1999, cf. Tanguy & Terry, 1999). In relation to the HRS, some humanitarian actors may be inclined to provide sanitation services based on the needs of the victims, without regard for political considerations (classicists), take sides with the victims and on that basis provide them with the sanitation services they need (solidarists), only intervene to relieve the suffering of victims provided such intervention is sustainable and does not aggravate the conflict (minimalists), or offer sanitation services as part of a political strategy to tackle the causes of the conflict (maximalists) (Weiss, 1999).

Independence

Independence is contained in soft law instruments like the 1991 General Assembly Resolution 46/182 (see Table 6.1). It similarly entails expunging extraneous influences, such as political, economic or military objectives from the humanitarian system (OCHA, 2012). Independence is easily applied to non-State humanitarian actors like international and local NGOs, but is also relevant to the intricacies of how the foreign policy of Member States affects the operations of inter-governmental organisations (like the United Nations Children's Fund and the United Nations High Commissioner for Refugees) who also play a significant role in humanitarian interventions (OCHA, 2012). In relation to the HRS, impartiality requires that sanitation services are delivered as part of humanitarian assistance and without any constraints imposed by extraneous political, economic or military objectives, for instance.

Impartiality

Impartiality is contained in soft law instruments like the 1991 General Assembly Resolution 46/182 (see Table 6.1). It requires non-discrimination (see 4.3 and 5.3.1) and proportionality in humanitarian assistance, such that the most urgent and greater needs are to be prioritised and rendered the greatest assistance (Harroff-Tavel, 1989; Sphere Project, 2011). Impartiality enhances distributive justice in humanitarian assistance. Similar to neutrality, impartiality can be interpreted in different ways in practise: classicists deliver humanitarian assistance on the basis of non-discrimination and proportionality, while solidarists allocate humanitarian assistance on the basis of their support for the main victims of a conflict, and there are minimalists and maximalists across the spectrum between classicists and solidarists (Weiss, 1999). In relation to the HRS, impartiality obliges humanitarian actors to prioritise the most vulnerable members of the population for assistance, and provide sanitation services, in a non-discriminatory manner.

Proportionality

Proportionality, which is the second element of impartiality, is contained in the Additional Protocol (1) to the Geneva Conventions (see Table 6.1). It qualifies impartiality and non-discrimination on the basis of human needs, with the effect that: (a) the more urgent needs are prioritised in humanitarian assistance, (b) the greater need receives the greater amount of assistance, and (c) needs that are qualitatively different are nonetheless addressed in an equitable manner (Lensu, 2003; Sphere Project, 2011). In relation to the HRS, proportionality means that people who are likely to suffer the most injury or death as a result of lack of access to sanitation services are prioritised and receive the necessary amount of assistance to ensure their survival and those with greater needs also receive greater assistance that is equitable.

The right to give and receive humanitarian assistance

The right to give and receive humanitarian assistance is an emerging custom (see Table 6.1) that simultaneously obliges belligerent States to provide humanitarian assistance to citizens of enemy States within their jurisdiction, and grant access to humanitarian actors to operate (Lensu, 2003). The principle applies to both international and non-international armed conflicts,[150] but its relevance to other forms of humanitarian situations resulting for instance from natural hazards is unclear (Lensu 2003), while natural hazards significantly affects

---

[150] See Additional Protocol I, article 70 and Additional Protocol II, article 18.

internal displacement (Bennett et al. 2017). In relation to realising the HRS, the right to give and receive humanitarian assistance obliges States to grant humanitarian actors free access to provide sanitation services to the affected population, and obliges humanitarian actors to provide the necessary assistance.

Table 6.1 Legal status of the principle of humanitarian assistance

Humanitarian Principles	Source Documents	Legal Status
Humanity	1977 Additional Protocol (1) to the Geneva Conventions 1991 General Assembly Resolution 46/182 2004 UNOCHA Guiding Principles	Established in treaty and soft law
Neutrality	1991 General Assembly Resolution 46/182 2004 IFRCRCS and the ICRC Code of Conduct	Established in soft law
Impartiality	1991 General Assembly Resolution 46/182 2004 UNOCHA Guiding Principles	Established in soft law
Independence	2003 General Assembly Resolution 58/114 2004 IFRCRCS and the ICRC Code of Conduct	Established in soft law
Non-discrimination	1977 Additional Protocol (1) to the Geneva Conventions 2004 UNOCHA Guiding Principles	Established in treaty and soft law
Proportionality	1977 Additional Protocol (1) to the Geneva Conventions	Established in treaty and soft law
Right to give and receive humanitarian assistance	2004 IFRCRCS and the ICRC Code of Conduct	Emerging custom
Avoid causing harm	1977 Additional Protocol (1) to the Geneva Conventions	Established in treaty and soft law
Ensure people's access to impartial assistance	1977 Additional Protocol (1) to the Geneva Conventions 1991 General Assembly Resolution 46/182 2004 UNOCHA Guiding Principles	Established in treaty and soft law
Protect people from violence and harm	1977 Additional Protocol (1) to the Geneva Conventions 2011 Sphere Handbook	Established in treaty and soft law
Assist people claim their rights claims and remedies, and recover	1977 Additional Protocol (1) to the Geneva Conventions 1966 International Covenant on Civil and Political Rights 1966 International Covenant on Economic, Social and Cultural Rights	Established in treaty and soft law

Source: Compiled by the author, based on Harroff-Tavel, 1989; Hilhorst, 2005; Kälin, 2008; Lensu 2003; Sphere Project, 2011; Tsui, 2009; Weiss, 1999

6.3.2 Protection Principles

Avoid causing harm

To avoid causing harm is established in the Additional Protocol (1) to the Geneva Conventions (see Table 6.1). Also, the first protection principle adopted by Sphere is to: "avoid exposing people to further harm as a result of your actions" (Sphere Project, 2011). This requires that: (a) the nature of the humanitarian assistance and the environment where it is provided do not violate the rights of the recipients or otherwise expose them to physical hazards or violence; (b) humanitarian assistance and protection efforts do not compromise the capacity of the recipients for self-protection; and (c) humanitarian actors manage sensitive information to protect the security and anonymity of informants (Sphere Project, 2011). In relation to the HRS, this principle therefore requires humanitarian actors to offer sanitation services in a form and within an environment that does not expose the users to further harm, for instance by ensuring the safe management of wastewater generated from sanitation and hygiene uses.

Ensure people's access to impartial assistance – in proportion to need and without discrimination

The need to ensure access to impartial assistance is enshrined in treaty and soft law instruments (see Table 6.1). Additionally, the second Sphere protection principle requires that people be able to access humanitarian assistance based on need and without unlawful discrimination and that humanitarian actors are allowed access necessary to discharge their obligations in accordance with the Sphere standards. This principle therefore combines the principles of impartiality, non-discrimination, proportion, and the right to give and receive humanitarian assistance, already discussed above (see 6.3.1).

Protect people from physical and psychological harm arising from violence and coercion

The imperative to protect people from harm is enshrined in the Additional Protocol (1) to the Geneva Conventions (see Table 6.1). The third Sphere protection principle specifically requires that "people are protected from violence, from being forced or induced to act against their will and from fear of such abuse" (Sphere Project, 2011). This principle has three elements, including: (a) taking all reasonable steps to protect the affected population from violent attack and from coercion to act in ways that may bring them to harm or violate their rights; and (b) supporting local initiatives to ensure safety, security and human dignity (Sphere Project, 2011). In relation to realising the HRS, this protection principle requires the design and management of sanitation facilities in a way that ensures the safety and dignity of users from physical and psychological harm arising from either violence or coercion.

Assist people claim their rights, access available remedies and recover from the effects of abuse

To assist people to get redress for human rights violation which they suffer is an established principle in treaties and soft law instruments (see Table 6.1). It is also the fourth protection principle under Sphere and requires that: "the affected population is helped to claim their rights through information, documentation and assistance in seeking remedies. People are supported appropriately in recovering from the physical, psychological and social effects of violence and other abuses" (Sphere Project, 2011). Though the principle *prima facie* reinforces the HRS principles of accountability and transparency and access to information (see 4.1 and 5.3.1), the effects may be limited in practice due to incoherence in the meaning of shared principles like participation and accountability as employed among humanitarian actors and human rights practitioners (Klasing et al., 2011). Nonetheless, in relation to realising the HRS in humanitarian situations, this principle obliges the humanitarian actors to assist the affected population with enforcing their HRS, and accessing available remedies and recovery mechanisms.

6.4 HUMANITARIAN INSTRUMENTS FOR SANITATION

Humanitarian WASH is delivered through the cluster approach. Box 6.1 describes how the WASH cluster works at the global and country levels, while the remainder of this section analyses economic (see 6.5.2), and suasive instruments (6.5.3) used by humanitarian actors. Nonetheless, the technologies for domestic sanitation (see 3.5) and the human right to sanitation instruments (see 5.5) earlier considered may also apply in humanitarian situations, depending on the context (ACF-France, 2009; Harvey, 2007; Johannessen et al., 2012).

Box 6.1 Global and country-level WASH cluster

Global WASH Cluster

Where the UN agencies are contacted to intervene in the aftermath of a disaster, the UN first decides internally whether it is a Level 1, Level 2, or Level 3 emergency and the necessary response based on an analysis of the scale, urgency and complexity of the response required, as well as the capacity and potential reputational risk to the UN and/or its agencies. The Under-Secretary General and the Emergency Relief Coordinator   (USG/ERC) act as the central point for the coordination of governmental, intergovernmental, and non-governmental humanitarian response. The USG/ERC may appoint a Humanitarian Coordinator to coordinate efforts at the level of the affected country. A major challenge to humanitarian agencies' response in emergencies used to be lack of proper coordination, until the adoption of the cluster approach in 2005 as a way of promoting smooth coordination and multi-sectoral participation of various international humanitarian actors (Heeger, 2011).

There are currently 11 Clusters identified by the Inter Agency Standing Committee, the inter-agency forum of the UN and non-UN humanitarian partners; each of these clusters is headed by a Cluster Lead Agency (CLA) responsible to the USG/ERC at the global level or the HRC at the country level. The global CLA, UNICEF in the case of the WASH Cluster, usually acts as the default country CLA except when it either lacks the necessary capacity or presence within the affected country or if another agency, NGO or humanitarian partner has comparative advantage, in which case the head of cluster (HC), the global CLA, and the humanitarian country team (HCT) can assign the role to another competent agency.

WASH Cluster partners share a strategic operational framework (SOF) which provides details about operational ways of working but allows sufficient flexibility for respective partners to develop their individual approaches in line with their respective mandates, capabilities, capacities, and comparative advantages. The global WASH Cluster generally adopts Sphere as its guidance for standards and indicators for humanitarian WASH interventions, including the principles of consultation, participation, and non-discrimination.

Country- level WASH cluster

The country level CLA appoints a WASH Cluster Coordinator (WCC) to establish the WASH cluster and improve coordination and equal partnership between the WASH Cluster Partners. Where there is no agency responsible to provide WASH services, the CLA automatically assumes the role as the 'Provider of Last Resort' (POLR). The in-country WASH Cluster collaborates with field staff from other clusters such as Education, Emergency Shelter, Health, Nutrition, and Camp Coordination, Camp Management (CCCM), in establishing, organising, and managing WASH in humanitarian situations. The collaboration clarifies responsibilities, improves the accountability of clusters, and prevents any operational overlaps among them. For instance, the 'WASH in Nut' strategy was employed in Niger in 2014 to reach over 118,000 malnourished child/mother pairs with WASH facilities and nutrition, simultaneously.

The in-country WASH Cluster is also responsible for integrating Early Recovery in WASH at the earliest possible stage, through the restoration of basic services, livelihoods, shelter, security and rule of law, governance, sustainability, capacity development for local and national ownership, participatory planning and implementation of programmes, disaster risk reduction, and cohesive multi-sectoral activities.

Source: Compiled by the author, based on expert interviews and literature review

6.4.1 Economic

Budgeting

The budgeting process benefits from financial standards like the International Public Sector Accounting Standards, and specialized Finance and Administration Dashboard which improve the accuracy and reliability of financial records for donor reporting and grants management, and also strengthen corporate financial risk management through closer monitoring of key financial and administrative performance areas (UNICEF, 2017). The budget heads for emergency WASH response include dedicated needs assessment teams, support of counterparts, coordination and the cluster management, adequate information management for the WASH Cluster, adequate technical and management support to WASH, personnel costs like daily subsistence allowance and rest and recuperation leave pay where appropriate, third party monitoring in areas with limited access, headquarter retention costs (5% for thematic funding), baseline surveys, evaluations, hygiene promotion, Non-Food Items and other supplies, logistics and transport, water quality monitoring and appropriate corrective action/alternative supplies, decommissioning of latrines and other temporary facilities, security, transport, and utilities, and waste disposal; at the minimum a realistic budget proposal would include a USD 5 to USD 10 investment per capita depending on the local context and type of emergency (United Nations Children's Fund [UNICEF] & United Nations High Commissioner for Refugees [UNHCR], 2014).

Flash Appeal

A Flash Appeal can be used to structure coordinated humanitarian response within the first three to six months of an emergency and incudes an overview of urgent lifesaving needs and early recovery (ER) projects (see for instance, United Nations Office for the Coordination of Humanitarian Affairs [OCHA], 2016; 2017). A Revised Flash Appeal is launched after 30 days of the initial appeal based on a more detailed assessment of the emergency situation and new or revised response plans.[151] The Consolidated Appeal Process (CAP) is coordinated by the OCHA in close collaboration with the UN agencies, NGOs, donors, government, and other relevant stakeholders to ensure a strategic approach to humanitarian response where an emergency lasts beyond the timeframe of a Flash Appeal and is declared to be 'complex' or

---

[151] Interviewees 20 and 21.

'major' (OCHA, 2016; 2017). The CAP is launched annually, and depending on the scale of the emergency, UNICEF could also develop a proposal for funding from donors.[152]

Regular resources and Emergency Programme Fund

The operations of the WASH cluster (see Box 6.1) and UNICEF emergency response are generally largely funded by governments and intergovernmental organizations, at 80%, while the private sector and other inter-organisational arrangements provide 10%, respectively (UNICEF, 2017). The funds include regular resources, and earmarked or other resources like pooled funding modalities such as thematic funding for UNICEF Strategic Plan outcome areas, humanitarian action and gender. Regular resources are flexible and unrestricted contributions that are allocated to country programmes, based on under-5 mortality rates, child population, and gross national income per capita, for investment in the development priorities of the relevant country. The Emergency Programme Fund (EPF), on the other hand, is the fastest and most flexible source of immediate funding to enable the scaling up of humanitarian action within the first 24 to 48 hours of intervention. The EPF is currently an annual revolving fund of USD 75 million after the original biennial limit of USD 75 million was depleted in 2014 alone, with the declaration of six major emergencies mostly occurring simultaneously (UNICEF, 2015). Despite the importance, regular resources have continued to decline from 50% to just over 25% in the 2000s, and there was an 8% decrease in EPF from 2015 to USD 1.600 billion in 2016 (UNICEF, 2015; 2017).

Thematic funding

Thematic funding which has less restrictions than other resources and complements regular resources was just under 9% in 2014, compared to 42% in 2005; global thematic humanitarian funding which is the most flexible type of thematic funding because it can be invested where needs are greatest was less than 1% of total thematic humanitarian funding (UNICEF, 2015). WASH only received 6% of the thematic contributions to strategic plan outcome areas and humanitarian action in 2014, for instance despite its central importance to survival and recovery in emergencies, and 29% of the Other Resources Emergency contribution followed by Health and Nutrition (21% and 20%, respectively) (UNICEF, 2015). Similarly, thematic funding declined in 2016 and this threatens the shared commitment by UNICEF partners for more flexible and pooled funding (UNICEF, 2017).

---

[152]   Interviewees 20 and 21.

6.4.2 Suasive

Operational guidelines for humanitarian actors

Humanitarian actors work closely with relevant national government agencies and are guided by operational guidelines, such as the influential Sphere Standards (2011), the Operational Guidelines on Human Rights and Natural Disasters (2006), and the IASC Gender Marker launched by the Sub-working Groups on Gender and on the Consolidated Appeal Process to improve the impact of humanitarian programming on gender equality; these guidelines operationalize the principles of humanitarian assistance (see 6.3.1).

Decision support systems and other assessment tools

Decision support systems and other assessment tools used in humanitarian situations improve disaster preparedness, coordinated response by humanitarian actors, and early detection of the immediate needs of the affected population (Jahre, Ergun & Goentzel, 2015), and related technological options (Zakaria, Garcia, Hooijmans & Brdjanovic, 2015), provided they incorporate the perspectives of the affected population. Assessment tools generally enhance risk management and operational continuity (Veeramany et al., 2016), promote learning and theory development across different disciplines and multiple levels of governance (Zommer, 2014), and advance the evaluation of vulnerabilities and the impacts of emergencies (Zachos, Swann, Altinakar, McGrath & Thomas, 2016).

Disaster risk management

Disaster risk management (DRM) is "[T]he systematic process of using administrative directives, organizations, and operational skills and capacities to implement strategies, policies and improved coping capacities in order to lessen the adverse impacts of hazards and the possibility of disaster" (Luber & Lemery, 2015, p.67). The Sendai Framework for Disaster Risk Reduction 2015-2030,[153] paragraph 19(c), also emphasises that DRM "...is aimed at protecting persons and their property, health, livelihoods and productive assets, as well as cultural and environmental assets, while promoting and protecting all human rights, including the right to development." (United Nations 2015b:13). DRM assumes that (even natural) disasters result from a failure of human systems and aims to reduce vulnerability and exposure through addressing the drivers of risk and is therefore focused on minimising the causes of risks and enhancing the adaptive capacity of institutions and structures (Thomalla,

---

[153] This succeeds the Hyogo Framework for Action (HFA) 2005-2015: Building the Resilience of Nations and Communities to Disasters as a non-binding agreement recognising the primary role of States to reduce disaster in collaboration with other stakeholders including the private sector.

Downing, Spanger-Siegfried, Han & Rockström, 2006; United Nations 2015a), including through re-distributive mechanisms like insurance. Its seven elements include: (a) threat recognition and identifying risk and vulnerability, (b) risk analysis and assessment, (c) risk control options based on feasibility, effectiveness and cost/benefit analysis, (d) strategic planning with economic, political and institutional support considerations, (e) response, recovery, reconstruction and rehabilitation, (f) knowledge management and sustainable development, and (g) resilience building and community participation (Carreño, Cardona & Barbat, 2007; Nirupama 2013). DRM requires the participation of local communities in building their sanitation infrastructure to comply with the HRS principles (see 5.3.1), and ensure resilience to the impacts of hazards.

6.5   MONITORING PROGRESS ON THE HUMAN RIGHT TO SANITATION IN HUMANITARIAN SITUATIONS

The literature on access to water in humanitarian situations lists sanitation as a medium term use of water; falling behind other medium term water uses like subsistence agriculture, home cleaning, washing clothes, personal washing, and short term survival uses like cooking and drinking (Reed 2005). Based on literature review, content analysis and inductive analysis, I categorise the available indicators for sanitation in humanitarian situations into outcome and structural indicators for three HRS principles (accessibility, availability and safety). Table 6.2 illustrates outcome indicators for the availability of hygiene promotion services and the safety of toilets and soak away facilities, while Table 6.3 illustrates outcome indicators for the accessibility, availability and sustainability of sanitation facilities, including toilets, water for sanitation and hygiene and waste collection and disposal. Table 6.3 also distinguishes between indicators for the short and long terms.

Table 6.2 Outcome indicators for the availability of hygiene promotion and safety of toilets and soak away facilities

Principle	Sanitation Components	Indicators
Availability	Hygiene Promotion	2 hygiene promotors: 1000 community members
Safety	Toilets and soak away facilities	Site at least 30 metres away from any groundwater source
		Maintain a distance of at least 1.5 metres between the bottom of the facility and a water source

Source: Compiled by the author, based on Sphere Project, 2011

Table 6.3 Outcome indicators for measuring accessibility, availability and sustainability of sanitation facilities in the short and long terms

Sanitation Services	Locations	Minimum Standards	
		Short term	**Long term**
Access to toilets	Locations		
	Markets	1:50 stalls	1:20 stalls
	Hospitals	1:20 beds or 50 outpatients	1:10 beds or 20 outpatients
	Food Points	1:50 adults; 1:20 children	1:20 adults; 1:10 children
	Transit Points	1:50 individuals; 3:1 female to male	
	Schools	1:30 girls; 1:60 boys	1:30 girls; 1:60 boys
	Households	1:50 people, lowering the number to 20 users as soon as possible	1:20 people
	Offices		1:20 staff
Access to water for sanitation and hygiene purposes*	Public toilets	1-2 litres per user daily for hand washing 2-8 litres per cubicle daily for toilet cleaning	1-2 litres per user daily for hand washing 2-8 litres per cubicle daily for toilet cleaning
	All flushing toilets	2-5 litres per user daily for pour-flush toilets 20-40 litres per user daily for conventional flushing toilets connected to sewer	2-5 litres per user daily for pour-flush toilets 20-40 litres per user daily for conventional flushing toilets connected to sewer
	Anal washing	1-2 litres per user daily	1-2 litres per user daily
Solid waste management – refuse collection and disposal*	Households	Daily removal of waste from the immediate living environment and at least twice weekly from the settlement	Daily removal of waste from the immediate living environment and at least twice weekly from the settlement
	Shared refuse collection & disposal	Minimum of 100-litre:10 households (where domestic refuse is not buried on site and private refuse containers are not available); containers are ≤ 100 metres from a communal refuse pit and emptied at least twice a week	Minimum of 100-litre:10 households (where domestic refuse is not buried on site and private refuse containers are not available); containers are ≤ 100 metres from a communal refuse pit and emptied at least twice a week

*The Sphere Handbook does not distinguish between the minimum standards for short term and long term in relation to access to water for sanitation and hygiene purposes, and solid waste management. The distinction drawn in this table is for illustration only and mirrors the same standards for both long and short term scenarios for access to water and solid waste management services.

Source: Compiled by the author, based on Sphere Project, 2011

6.6 HUMANITARIAN FRAMEWORK, DRIVERS, INCLUSIVE DEVELOPMENT AND LEGAL PLURALISM

This section analyses whether humanitarian principles and instruments address the drivers of poor sanitation services (see 6.6.1). In doing so, I do not distinguish between the legally binding and non-legally binding principles as I already addressed the legal status of the principles in Section 6.3 and Table 6.1). Further, I discuss the wider implications for inclusive development (see 6.6.2) and legal pluralism (see 6.6.3), based on the literature and an inductive analysis of the humanitarian framework.

6.6.1 Humanitarian Framework and the Drivers of Poor Sanitation Services

The humanitarian assistance and protection principles address six direct and five indirect drivers of poor sanitation services, as illustrated in Table 6.4 and elaborated below.

Principles for addressing the direct drivers

The principle of avoiding causing harm has implications for the design of sanitation facilities and the distance from the users, as well as the safe management of human waste in order to prevent pollution (see 6.3.1 and 6.5). Nonetheless, the right to receive assistance, impartiality and non-discrimination address   are also relevant where people living far from the offices of the relevant government agencies or humanitarian actors are denied access to sanitation services (see 6.3.1). Humanity, impartiality, non-discrimination and the right to receive humanitarian assistance also protect against the exclusion of minorities (see 6.3.1). The principle of assisting people with their rights' claims further addresses negative social practices like discrimination and poor maintenance and improper use of facilities that affect access (see 6.3.1 and 6.3.2). This could also include the use of suasive instruments to improve the level of awareness about the HRS within the affected population, thereby addressing non-acceptance of the sanitation facility based on culture (see 6.3.2). Protection from violence also addresses negative social practices that lead to violence against minorities and other forms of insecurity affecting access to sanitation services (see 6.3.2). The remaining two direct drivers that are relevant for humanitarian situations (space constraints and challenging or inaccessible topography) are least affected by the principles for humanitarian assistance and require technological instruments suited to the local context (see 3.5). I further elaborate on addressing the remaining four direct drivers which are not covered by humanitarian assistance and protection principles (including space constraints, challenging or inaccessible topography, natural disasters and high temperatures/turbidity in source water), *vis-à-vis* HRS principles, in Chapter 9 (see 9.2).

Principles for addressing the indirect drivers

Impartiality, non-discrimination and proportionality address poor targeting of sanitation funding by obliging humanitarian actors to allocate all available resources for sanitation services to the affected population based on their individual needs, and in a manner that ensures equity (see 6.3.1). Further, assisting people to claim their rights, including sanitation and hygiene education, tackles low awareness about sanitation and protects the target population from discrimination in the allocation of available resources for sanitation; it also enables them to seek redress for HRS violations (see 6.3.2). Humanity obliges States and non-state humanitarian actors to assist people living in humanitarian situations with basic services for their survival and recovery (see 6.3.1). Avoiding causing harm and protecting people from violence addresses poor sanitation services due to insecurity and conflicts (see 6.3.2). The outcome indicators for availability and accessibility address population density and growth where there are available resources to meet the minimum ratio of facilities required (see 6.5). I further elaborate on addressing the remaining four indirect drivers which are not covered by humanitarian assistance and protection principles (including huge foreign debts, sanctions, mass migration, and climate variability and change), *vis-à-vis* HRS principles, in Chapter 9 (see 9.2).

Table 6.4 Humanitarian principles for addressing the direct and indirect drivers of poor domestic sanitation services

Category of Drivers (see 3.4)	Details of Drivers (see 3.4)	Humanitarian Principles/Instruments (see 6.3)
Direct Driver		
Environmental	Distance to the facility	Avoid causing harm Impartiality Non-discrimination Right to give and receive assistance
	Pollution/water scarcity	Avoid causing harm
Social	Exclusion of minorities	Humanity Impartiality Non-discrimination Right to give and receive assistance
	Negative social practices	Assistance with rights claims Protection from violence
	Non-acceptance of sanitation facility	Assistance with rights claims
	Poor maintenance culture/ improper use of facilities	Assistance with rights claims
Indirect Drivers		
Economic	Insufficient/poorly targeted funds	Assistance with rights claims Impartiality Non-discrimination Proportionality
	National poverty	Humanity
Social		
	Insecurity, conflicts and poor social cohesion	Avoid causing harm Protection from violence
	Low awareness about Sanitation	Assistance with rights claims Right to get and receive assistance
	Population density/growth	Outcome indicators

Source: Compiled by the author, based on Harroff-Tavel, 1989; Hilhorst, 2005; Kälin, 2008; Lensu 2003; Sphere Project, 2011; Tsui, 2009; Weiss, 1999

6.6.2 Humanitarian Framework and Inclusive Development

The principles for humanitarian assistance for sanitation can lead to different outcomes for ID, depending on how the principles are interpreted, in relation to the HRS norm, and the choice of instruments for sanitation service provision in any given context. Figure 6.1 illustrates how the participatory disaster risk management can lead to different outcomes on social and relational inclusion, depending on whether the implementation process integrates environmental sustainability. Further, Figure 6.1 shows that indicators are value laden because expanding the social, relational and ecological values will also increase the cost of providing sanitation services with the least cost possibly arising from Q1 and the highest cost in Q4. Ordinarily, humanitarian situations exacerbate human vulnerabilities and inequities in access to WASH (see 6.1), thereby hampering social and relational inclusion (Akhter et al., 2015; Nakhaei et al., 2015). Humanitarian situations could also pose stress on infrastructure and deplete the available natural resources for sanitation services, thereby further hampering ecological inclusion (Obani, 2017). This makes it necessary to further analyse the impact of the humanitarian framework on ID components (social and relational, and ecological inclusion).

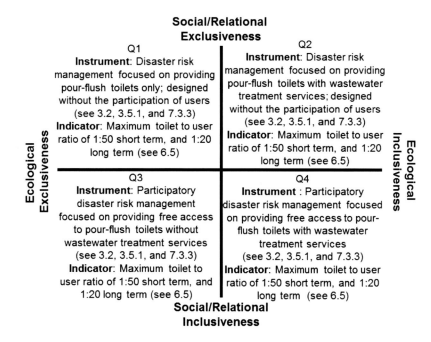

Figure 6.1 Assessing disaster risk management for inclusive development

Social and relational inclusion

Economic instruments (see 6.4.1) *prima facie* ensure social and relational inclusion (Q3 and Q4 of Figure 6.1), to the extent that the budgets and the application of emergency programme funds, regular resources, and thematic funding reflect the needs of the affected population and prioritise the most vulnerable people for assistance based on the right to receive humanitarian assistance, impartiality, non-discrimination and proportionality principles, for instance. Suasive instruments (see 6.4.2) can also enhance social and relational inclusion by highlighting vulnerabilities within the affected population and ensuring timely interventions to improve recovery. Specifically, assessment tools promote participation and enhance disaster preparedness and effective response (Obani, 2017). Participation can foster unity within the affected community during the recovery phase (Samaddar, Okada, Choi & Tatano, 2017). However, the overall effect on ID would also depend on the extent to which the WASH Cluster engages with the local population in the design of interventions, and how well the instruments promote environmental sustainability (see Figure 6.1).

Ecological inclusion

The protection of drinking water installations and supplies and irrigation work, as civilian objects, and during armed conflicts (see 6.2), though primarily designed to protect the affected population, also promotes ecological inclusion (Obani, 2017). Similarly, the integration of broader sanitation components beyond excreta management coupled with minimum indicators for sustainability under Sphere (see 6.5) and the principle of avoiding harm in humanitarian assistance (see 6.3.4) also enhances ecological inclusion by preventing the contamination of the environment (Obani, 2017). Nonetheless, environmental sustainability may also constrain the choice of technologies and modalities for service provision where necessary to avoid environmental pollution (Feris, 2015; Holden, Linnerud & Banister, 2016).

6.6.3 Humanitarian Framework and Legal Pluralism

The interactions between the HRS and the humanitarian framework may result in competition, indifference, accommodation, or mutual support (see Table 6.5).

Competition

Competition results where military action violates the HRS; for instance, where sanitation facilities are destroyed on the grounds of military necessity during armed conflicts. Tension could also arise where a traditional approach to humanitarian assistance situates sanitation as a competing priority, rather than a basic human need, against other basic needs like safety, shelter, food, water and healthcare (Breau & Samuel, 2016). For instance, although the WHO Technical Notes on Drinking-Water, Sanitation and Hygiene in Emergencies recommends 70 litres of water for sanitation and waste disposal during emergencies, this is listed as a medium term function for the maintenance of the affected population, whereas 10 litres of water for drinking and 20 litres of water for cooking are listed as short-term functions for the survival of the affected population (Reed & Reed, 2011). Another instance of competition is where DRM (see 6.4.2) is conducted without the participation of the affected local community. Conversely, the recognition of sanitation as a distinct human right underscores its critical importance for human dignity and survival, and the ICESCR rights also apply during emergencies (see 5.2 and 6.2).

Indifference

Indifference occurs where although the HRS principles are captured in the principles of humanitarian assistance, the principles are not implemented in humanitarian situations. Rather, humanitarian situations are governed as a general exception to the operation of HR norms and humanitarian assistance does not integrate HRS principles in practise.

Accommodation

Accommodation occurs where some efforts are made to adopt HRS principles in the humanitarian framework and encourage the participation of the affected population in humanitarian situations. However, there are marked differences in the application of principles like accountability and participation by HR practitioners and the humanitarian actors (Klasing et al., 2011). This hints at a mere accommodation of the HR principles in the operations of the humanitarian actors (Klasing et al., 2011).

Mutual support

Mutual support exists where IHL protects drinking water installations and supplies and irrigation work as civilian objects, and guarantees access to sanitation and hygiene facilities including access to water, toilets and soap for personal and domestic uses for protected persons during armed conflicts, and these provisions improve the realisation of the HRS in humanitarian situations. Mutual support also exists where there is a broad conception of sanitation that is supported by minimum indicators in the humanitarian framework, for instance under Sphere, enrich the development of the HRS norm.

Table 6.5 Types of legal pluralism relationship which arise between the human right to sanitation and humanitarian principles

Legal Pluralism Relationship	Description of the Relationship
Competition	Where sanitation facilities are destroyed as a military necessity
Indifference	Where the HRS principles like non-discrimination and accountability are captured in the humanitarian framework but humanitarian situations continue to be governed as an exception to the application of human rights principles
Accommodation	Where efforts are made to incorporate the HRS principles like participation in the domestic humanitarian framework
Mutual Support	When the humanitarian framework adopts a broad definition of sanitation which integrates environmental sustainability and thereby enhances the HRS

Source: Compiled by the author, based on Klasing et al., 2011; Obani, 2017; Obani & Gupta, 2014b

6.7 INFERENCES

There are eight inferences drawn from this chapter. First, IHL is the *lexi specialis* applicable to armed conflicts and related emergencies, and distinguishes between the rights and protections of combatants and non-combatants, though IHL customs also apply to humanitarian situations resulting from natural hazards. Human rights law on the other hand *stricto sensu* also applies to humanitarian situations and complements IHL in the general protection of vulnerable populations in humanitarian situations, without bias to their status as combatants or non-combatants. Further, while States may temporarily suspend civil and political human rights obligations, subject to some exceptions, during life threatening public emergencies, to the extent that is strictly required by the exigencies of the situation, the International Covenant on Economic, Social, and Cultural Rights 1966, which largely forms the legal basis for the HRS, does not provide for similar derogation. Rather, gross human rights violations which constitute war crimes, genocides, and crimes against humanity can be considered violations of customary international law and can be prosecuted either through the International Criminal Court or by national courts based on the principle of positive complementarity (see Chapter 4). Hence, HRS law can significantly complement IHL for the protection of vulnerable populations in humanitarian situations.

Second, within the humanitarian framework, Sphere, which is widely adopted in practice by humanitarian actors though not legally binding, offers a broad approach to sanitation (see 6.3) and minimum indicators (see 6.5) that can enrich the definition and implementation of the HRS even outside humanitarian situations. This is a paradox because although the definitions of the HRS in some national laws are more extensive than the definition under the international HRS framework (see 5.3.2), none goes as far as Sphere in addressing WASH components and stipulating minimum standards.

Third, there are ten humanitarian assistance and protection principles; six of these principles arise from the international humanitarian law framework and therefore focus mainly on armed conflicts and emergencies like political tensions, but do not extend to non-conflict related humanitarian situations. This is a limitation given the increasing numbers of non-conflict related humanitarian situations, caused by hurricanes for instance, that are nonetheless capable of disrupting sanitation systems both in the affected locations and host communities for the resulting internally displaced and refugee populations. Whereas, the four protection principles under Sphere offer much broader protection in humanitarian situations resulting from both armed conflicts and natural hazards, they lack the binding legal status of hard law as such and are therefore of limited legal effect in advancing the HRS.

Fourth, the HRS may not be prioritised in the design of regulatory, economic and suasive instruments for the governance of humanitarian situations depending on whether sanitation is classified as a survival need to be provided immediately after a humanitarian situation arises or whether sanitation is classified as a medium term need to be satisfied after the provision of basic necessities like water for drinking and cooking. Failure to recognise sanitation as a survival need however undermines humanitarian assistance and jeopardises the chances of effectively addressing other basic survival needs like drinking water, human security, nutrition and healthcare (see 1.2).

Fifth, the humanitarian framework offers seventeen relevant outcome and process indicators that extend beyond the existing three HR indicators (see 5.5, 3.2?), which can strengthen the monitoring of progress on toilet availability especially for vulnerable groups like women, girls, children and the sick (see 6.5).

Sixth, humanitarian assistance and protection principles address eleven drivers of poor sanitation services. However, they may be ineffective because: (a) the crises situation may overwhelm the coping capacity of the responding institutions; (b) humanitarian situations are regarded as exceptional circumstances in which compliance with HR may not be feasible; (c) there is lack of shared understanding among humanitarian and HR practitioners of participation and accountability; (d) humanitarian actors adopt a predominantly technocratic approach to emergency response; or (e) sanitation is not prioritised as a basic human need in humanitarian situations. Further, the remaining ten drivers that are not addressed by humanitarian principles require complementary principles such as the existing HRS principles which can address sixteen drivers (see 5.6.1)

Seventh, although the humanitarian principles promote predominantly social and relational inclusiveness, they also address environmental sustainability since the principle of avoiding harm in delivering humanitarian assistance requires a safe environment (free of contamination) for the wellbeing of the affected population. The actual impact of humanitarian principles on the components of ID depends on the extent to which the affected population are involved in the design and implementation of interventions (which may be more feasible in the medium to long term than in the immediate aftermath of an emergency, for instance), and the integration of environmental sustainability in the design of implementation instruments.

Eight, the apparent complementarity between HRS and humanitarian and protection principles and instruments, as illustrated throughout this chapter, is marred by strong forms of rules incoherence leading to HRS violations in humanitarian situations. This occurs where for instance, there is a lack of shared understanding of common principles like participation that are nonetheless applied both within the human rights and humanitarian frameworks. Rules incoherence in humanitarian situations hampers social and relational inclusion, where it compromises the HRS for the vulnerable or excludes the affected population from participating in sanitation governance, and ecological inclusion, where humanitarian assistance or military action compromises environmental sustainability.

Chapter 7. Non-human Rights Principles for Sanitation Governance

7.1 INTRODUCTION

Besides human rights and humanitarian principles, there are other principles for sanitation governance. This chapter focuses on these other principles since: (a) the HRS framework does not sufficiently address all the drivers of poor sanitation services; (b) other frameworks influence national sanitation governance; and (c) there is need to investigate the fit between the HRS and other frameworks which generally stem from other discourses. It examines: (a) how and under what circumstances can non-human rights frameworks for sanitation governance promote the HRS and ID outcomes? (b) How does an understanding of the incoherence between the HRS and non-human rights frameworks for sanitation governance, using legal pluralism theory (see 2.4.2), affect the design of sanitation governance frameworks? The chapter first discusses the other principles (see 7.2), analyses the instruments through which the principles are operationalised (see 7.3), assesses how these principles address the drivers of poor domestic sanitation services (see 7.4.1) and the consequent impact on inclusive development (ID) (see 7.4.2) and legal pluralism (see 7.4.3), before presenting the inferences (see 7.5).

7.2 NON-HUMAN RIGHTS PRINCIPLES FOR SANITATION GOVERNANCE

This section assesses other principles of sanitation governance (see Table 7.1). I group the based on my conceptual framework, and they mainly fall under two categories: (a) social (see 7.2.1), and (c) environmental principles (see 7.2.2).

7.2.1 Social

There are three social principles for sanitation governance: (a) capacity building, (b) poverty eradication and equality, and (c) subsidiarity.

Capacity building

The capacity building principle is contained in the Rio Declaration (1992: principle 9) and the SDGs.[154] Capacity building is the process through which communities, groups and organisations acquire the technical and administrative skills which they need to enable them to participate maximally in governance processes to whatever level they may desire (National

---

[154] See Goals 6, 13 and 17.

Research Council [NRC], 2006). The origins of capacity building is linked to the neo-liberal aversion of dependence on external aid and support for individual agency as a means of promoting social development; the attraction is the potential to empower marginalised people through organised trainings to effectively participate in governance and adapt to social change while reducing the pressure on the government and international donors to continue to administer local development programmes (Kenny & Clarke, 2010). This paradigm however ignores the fact that some communities fail to develop not due to lack of capacity but as a result of structural, political and resource drivers that impede their development (Kenny & Clarke, 2010). Conversely, an asset-based approach to capacity building starts from the premise that the target population have existing capacities which can be harnessed for social development (Eade, 1997).

Applied to the HRS, capacity building requires: (a) an understanding of the local capacities and how these can be harnessed for the realisation of the HRS; (b) knowledge exchange between users and other stakeholders in the sanitation sector, to improve coproduction and local solutions for sanitation; and (c) addressing gender inequality and equipping the vulnerable and marginalised with opportunities for meaningful participate in sanitation governance.

Poverty eradication and equity

Poverty eradication and equity is enshrined in the Rio Declaration (1992: principles 5, 6) and the SDGs.[155] Poverty (see 3.4) is a multi-dimensional concept that encompasses a lack of basic necessities, low health and academic outcomes, and various forms of vulnerability, exposure to risks, and marginalization (Daemane 2014). Thus poverty is not just material deprivation measured through income or consumption levels but also in terms of productive employment (sustainable livelihoods approach), participation in social life (social and relational inclusion), and the agency of individuals in promoting social development (capabilities approach). Poverty implies inequities in access to assets, resources, and opportunities for participation in markets and governance processes, and often has a gender perspective (Reddy & Moletsane, 2009; UN Chronicle, 1996). Consequently, poverty eradication and equity require deliberate efforts to reduce the inequities in human development especially by reducing the disparities in access to assets and resources for human wellbeing and enhancing the opportunities for public participation among the impoverished. Poverty eradication in developing countries is mainly funded through domestic

---

[155] See Goals 1 and 7.

resources (like taxes and export-earnings) but with stagnating or declining tax-to-GDP ratios in many developing countries and aid fatigue in industrialised countries (Gupta & van der Grijp, 2010), developing countries need to (be supported to) develop their tax capacities (Chowdhury, 2016; World Bank Group, 2014). Applied to the HRS, this principle requires well-targeted pro-poor instruments like cross-subsidies (see 5.5.2 and 5.7.2), access to assets and mechanisms for the meaningful participation of the vulnerable and marginalised in sanitation governance, within ecological limits, in the interest of ID (see dos Santos & Gupta, 2017 for details).

Subsidiarity

The principle of subsidiarity (Rio Declaration 1992: principle 10; Treaty on European Union, 2012, article 5) requires decisions to be taken at the lowest possible level.[156] It further requires the higher-level entity to: (a) directly maintain the public good; (b) provide the necessary support (monitoring, regulating and coordinating) for individuals and associations (subsidium) to contribute to the common good through their free initiative; and (c) refrain from interfering with the activities of the lower-level entities which do not undermine the common good (Martini & Spataro, 2016; Murphy, 1999). Subsidiarity presumes that it is inefficient to burden the State with local concerns, and human development "depends crucially on freedom for individual self-direction and for the self-government of voluntary associations and that human beings flourish best through their own personal and cooperative initiatives rather than as the passive consumers or beneficiaries of the initiatives of others" (Murphy 1999, p.887). Subsidiarity is reflected in cultural relativism (under HR law), which enables national bodies to determine how to implement a right based on their unique local circumstances though this does not permit deviations from a peremptory norm (see 4.2.1), and the devolution of State powers (among different levels of government) (Arden, 2015; Barber & Ekins, 2016). Subsidiarity requires that while the public good aspects of sanitation is maintained and supported by the state, relevant decisions are made at the lowest level of governance to promote free initiative and innovation in addressing the problems in a contextually relevant manner (Obani & Gupta, 2014b). Like other principles, it is influenced by political considerations which may undermine its regulatory efficacy (Craig, 2012).

---

[156] The Bonn Recommendations 2001 also advocates water management at the lowest possible level and the decentralisation of responsibilities for water and other services to local governments.

7.2.2 Environmental

This section discusses three relevant environmental principles.

Polluter-pays principle

The polluter-pays principle (Rio Declaration 1992: principles 13 and 16; UNECE Convention 1992, article 5) requires the polluter to internalize the external costs arising from the pollution (de Sadeleer, 2002).[157] The polluter-pays principle precludes the repetition of environmental degradation (de Sadeleer, 2002). In relation to the HRS, the polluter-pays principle can promote equitable access and the funding of sanitation services by ensuring that the external costs of pollution (including the cost of waste management) are fully internalized by those who produce waste or in other ways pollute the environment. In theory, this could mean that businesses or individuals who pollute the environment have to pay for the clean-up costs, thereby minimising environmental degradation. Nonetheless, the polluter-pays principle may impose a huge burden on the poor, where it is coupled with full cost-recovery, except they are guaranteed basic services at an affordable rate through regulatory and economic HRS instruments (see 5.4.1 and 5.4.2).

Precautionary principle

The precautionary principle (Rio Declaration, 1992: principle 15) requires that even where the causality between A and B is uncertain, if B happens and is irreversible, this is a reason to take precautionary action (Aven 2011). "If there is (1) a threat, which is (2) uncertain, then (3) some kind of action (4) is mandatory" (Sandin 1999:891; Sandin et al. 2002). The principle may be applied based on: (a) considerations of irreversible damage and cost effectiveness, (b) any threat to health or the environment irrespective of cost-benefit considerations, and (c) additional considerations like personal experiences in the evaluation of risks (de Sadeleer 2002; Fisher 2002; Sheng et al. 2015). While the principle may stifle development (Sunstein, 2003), it can enable the adoption of safer techniques and procedures.[158]

Applied to the HRS, it calls for improving the safety of sanitation infrastructure and technologies to reduce its harmful impacts on humans and the environment; This principle reinforces the pubic goods characteristic of sanitation (see 3.3) where the threats of poor

---

[157] See generally OECD Recommendation on the Implementation of the Polluter Pays Principle 1974, C(74)223.
[158] See European Commission White Paper on Strategy for a New Chemicals Policy, COM (2001).

sanitation services to human health or the environment spur universal service provision irrespective of cost-benefit considerations, and make the State the primary duty bearer.

Table 7.1 Legal status of the non-human rights principles that are relevant for sanitation governance

Non-HR Principles	Source Documents	Legal Status
Capacity building	1992 Rio Declaration	Emerging
Equity	1992 Rio Declaration	Established in treaties and soft law
Polluter-pays Principle	1972 European Program of Action on the Environment 1972 OECD Council Recommendation on Guiding Principles Concerning the International Economic Aspects of Environmental Policies 1992 UNECE Water Convention 1992 Rio Declaration	Established in treaties and soft law
Poverty Eradication	1992 Rio Declaration	Emerging
Precautionary Principle	1992 Rio Declaration	Established in treaties and soft law
Prevention Principle	1972 Stockholm Declaration 1979 LRATP Convention 1985 Vienna Convention for the Protection of the Ozone Layer 1992 CBD 1992 Rio Declaration Preambles of the UNFCCC1997 UN Watercourses	Established custom, also in treaties and soft law
Subsidiarity	EU law generally	Established in treaties and soft law

Source: Compiled by the author, based on Arden, 2015; Craig, 2012; Daemane, 2014; de Sadeleer, 2002; Gupta & van der Grijp, 2010; Kenny & Clarke, 2010; Sheng et al., 2015

Prevention principle

The prevention principle[159] requires States to apply due diligence to prevent harm but does not make every transboundary harm automatically unlawful (de Sadeleer, 2002; Sheng et al., 2015). It may also require States to: "… co-operate in a spirit of global partnership to

---

[159] The prevention principle has been upheld in a number of decided cases concerning a mediated type of environmental damage like the *Trail Smelter Case (United States v. Canada)*, Arbitration, 1938 and 1941 Decisions, UNRIAA, vol. III, 1941, pp. 1905–1982, and the Pulp Mills on the River Uruguay (Argentina v. Uruguay), Judgment, ICJ Reports 2010, pp. 14–107. A mediated type of environmental damage is "damage from an 'agent' delivered from the territory of one state to that of another through a shared physical medium (air, water) abutting the two states, whether or not in conjunction with an alteration in the condition of that medium (such as the poisoning, or change in the natural course, of a river)" (Zahar , 2014, p.225). In other cases, like the *Gabčíkovo-Nagymaros Project (Hungary v. Slovakia)*, Judgment, ICJ Reports 1997, pp. 7–84, and *In the Arbitration Regarding the Iron Rhine ('Ijzeren Rijn') Railway (Belgium v. Netherlands)*, Award of 2005, Permanent Court of Arbitration, UNRIAA, vol. XXVII, 1941, pp. 35–125, the prevention principle was applied to stop development activities that were expected to cause immediate and geographically confined damage to the environment of a neighbouring State. The principle has also been proposed for wider application to cumulative environmental damage, like climate change (Mayer 2015; cf. Zahar 2014).

conserve, protect and restore the health and integrity of the Earth's ecosystem" as stated in the Rio Declaration 1992, Principle 7 (see Hayat & Gupta (2016) for a general discussion of the kinds of freshwater and the related ecosystem services for human wellbeing).

Applying the prevention principle to the HRS would reinforce the need to use the Best Available Techniques (BAT) for sanitation. The BAT is the "most effective and advanced stage in the development activities and their methods of operation."[160]

7.3   NON-HUMAN RIGHTS INSTRUMENTS FOR SANITATION GOVERNANCE

This section classifies the existing most commonly used non-HR instruments for sanitation governance as regulatory (see 7.3.1), economic (see 7.3.2), management (see 7.3.3), and suasive (see 7.3.4). Technologies are also an important instrument for sanitation governance (see 3.5) but are not discussed here.

7.3.1 Regulatory

Prioritisation of sanitation and hygiene in licensing exemptions and permissible use of water

The prioritisation of sanitation and hygiene in licensing exemptions and permissible uses of water improves access to water resources for sanitation and hygiene purposes. Many of the licensing provisions in national laws relates to the permissible uses of water. While some national laws generally permit the abstraction of water from public sources as a common pool resource for the purpose of meeting individual and domestic needs, without need for a license,[161] others exempt additional domestic uses like water for livestock[162] and irrigation of domestic gardens;[163] and water for fire fighting during emergencies.[164] Some laws further rank the permissible socio-economic uses of water, with human consumption (drinking water and presumably sanitation and hygiene needs) as the top priority. For instance, in Peru, the Water Resources Act 2009 prioritises human need as primary uses of water, over other uses.[165]

---

[160]   European Union Directive 2008/1/C Official Journal of the EU, 2008, p. 4. See also Stockholm Convention on Persistent Organic Pollutants, 2001.

[161]   Cape Verde, *Water Code, Law No.41/II/84*, as amended by Decree No. 5/99 (unofficial translation), article 61; Tanzania, *The Water Utilization (Control and Regulations) Act*, 1974, as revised 1993, section 10.

[162]   Zambia, *Water Act*, 1949, sections 2 & 8; Zimbabwe, *Water Act 1998*, Preliminary.

[163]   Cambodia, Law on Water Resources Management of the Kingdom of Cambodia 2007, Article 11; Guinea Bissau, *Water Code, Law No. 5-A/92* (Unofficial translation), Article 7; Swaziland, *Water Act, Act No. 7* of 2003, Interpretation; Angola, *Water Act*, 21 June 2002 (unofficial translation), articles 10(2), 22, 23.

[164]   South Africa, *National Water Act, Act 36* of 1998, sections 4(1) and 22(1)(a)(i), Schedule 1 and Item 2 of Schedule 3. See also, Uganda, *The Water Statute, Statute No. 9* of 1995, section 7(1)(2).

[165]   Peru, *Water Resources Act*, June 2009 (Unofficial translation), articles 35 and 36. See also, Central African Republic, *Water Code 2006; LAW No 06.001 of 12 April 2006* (unofficial translation), article 44; Chad,

7.3.2 Economic

This section discusses two economic instruments.

Microfinance

Microfinance could provide loans for sanitation to people where beneficiaries save small amounts of money to finance their own sanitation investments or access micro credit (Mader, 2012). It is hinged in the need for the State to restrict its functions to regulation, national defence, and international policy while leaving other functions (including the provision of public goods) to "a social-consciousness-driven private sector" (Yunus, 2003, p.204). Microfinance is a private business model which further requires users to fully appreciate and internalise the costs of sanitation to be effective (Mader, 2012; Mehta, 2008); the extent of internalisation required of users for microfinance to be effective runs counter to the inherent multi-dimensional economic (public, private and merit goods) properties of sanitation (see 3.3). Microfinance is ill-suited to improving access to sanitation goods and services for the poor where: (a) poor households are excluded from the benefits of privatised sanitation services; (b) political elites interfere with the sanitation projects for their own benefits; (c) public funding is insufficient to provide the common good or network components of sanitation services, and (d) sanitation marketing mainly focuses on public health gains rather than capitalising on the extraneous incentives for households' sanitation investments, such as social status (Mader, 2012).

Public-private partnerships

Public-private partnership (PPP) includes a wide variety of relationships between the State or public sector entities and private actors, with the purpose of collaborating to implement an agreed policy objective on the basis of their individual agenda (Jamali, 2004), while the State retains regulatory control (Baud, 2004). The individual agenda for public actors may be the provision of public goods, and for the private actors it may be profit making. Whereas PPPs

---

Water Code, Law 016/PR of 18 August 1999 (unofficial translation), articles 149 & 150; Benin, *Water Code, Law No. 87-016* of 21 September 1987 (unofficial translation), article 54; Burundi, *Government Decree No. 1/41* of 26 November 1992 on the establishment and organisation of the public water domain (unofficial translation), articles 1 & 14; Cape Verde, *Water Code, Law No.41/II/84*, as amended by Decree No. 5/99 (unofficial translation), articles 58(a) & 59; Ethiopia, *Ethiopian Water Resources Management Proclamation, Proclamation No.197/2000*, article 7(1); Guinea, *Water Code, Law No. L/94/005/CTRN* of 14 February 1994 (unofficial translation), article 20; Lesotho, *Water Act, 2008*, articles 5(2) & 13(2)(a); Madagascar, *Decree No. 2003-941 concerning monitoring of water resources, control of water destined for human consumption and priorities of access to water resources* (unofficial translation), articles 1 & 2; Mauritania, *Water Code, Law No. 2005-030* (unofficial translation), article 5(1)(2); Senegal, *Water Code, Law No. 81-13*, 1981 (unofficial translation), articles 75&76. See also Guyana, *Water and Sewerage Act*, 2002, article 25(a)(d).

is sometimes used loosely to describe liberalization, private sector participation, and privatization (Felsinger, 2010; Sovacool, 2013), each of these types of market reforms has distinct targets and policy implications (Bakker, 2007). The attractions of PPP include improvements in access and quality of services, transfer of technology and skills from the private sector to the public sector, and increased operational efficiency of the public sector (Marin, 2009). The pro-poor public-private partnership (5P) model recently emerged to ensure service provision to the poor who are often considered a business risk under the traditional PPPs and excluded from service delivery (Sovacool, 2013). The model is built on inclusive partnerships with users, the private sector, and other partners like NGOs, CBOs, development banks, co-operative organisations and philanthropists (United Nations Development Program [UNDP], 2011); it thereby merges profit making with social objectives and ensures the participation of the affected population (Mukherjee, 2005).

7.3.3 Management

This section discusses management instruments grouped into two categories, based on the lead actors/stakeholders involved in the implementation: (a) child-led participatory approaches, and (b) community-led participatory approaches.

Child-led approaches

Child-led approaches harness the capabilities of children in the process of co-producing sanitation goods and services for children and their communities. One example is the child-to-child (CtC) which facilitates an understanding of hygiene among children and enables them to identify health and development priorities through active learning, by gathering information on sanitation, taking necessary action, and sharing their knowledge with the wider community (WaterAid, 2013). A second example is the School-Led Total Sanitation (SLTS) which is similar to CtC but specifically focused on educating school children about sanitation and hygiene and encouraging them to spread the knowledge in their homes and communities (Wicken, Verhagen, Sijbesma, Da Silva & Ryan, 2008). The advantages of SLTS include that schools may have better access to funding from the government or other donors, which can be used for SLTS programmes, the rate of progress is fast and the outcomes are sustainable because of children as the lead change agents (WaterAid, 2013; Wicken et al., 2008). There are three main phases involved in child-led management approaches: (a) ignition, which involves training children on sanitation and hygiene, mapping the community and exploring the sanitation cycle; (b) triggering, during which children work in groups to trigger their schools or communities to take action on sanitation and end open defecation; and (c) post-triggering, when the children are expected to monitor and follow-up the triggered schools and communities, to ensure continuity. Child-led approaches raise concerns for the protection of children from overexploitation or overworking as sanitation change (WaterAid, 2013).

Community-led approaches

Community-led approaches harness the capabilities of communities in the process of co-producing the sanitation goods and services. One example is community management which is the collective ownership and management of infrastructure or facilities for sanitation service delivery by community-based organisations or neighbourhood groups. This is appropriate for labour-intensive operations, and fosters skills acquisition, capacity building, income generation, and a strong sense of community ownership of facilities, requires close-knit community structure and coordination to be successful, and may be unsuitable for sanitation infrastructure or services which require specialised skills without special training of the community members in charge (WaterAid, 2013).

A second example is the community health club (CHC), a voluntary and free community-based membership organisation aimed at improving community health through encouraging the members to practise what they have learned using home assignments and monitored home visits (Waterkeyn & Cairncross, 2005).

A third approach is the participatory hygiene and sanitation transformation (PHAST) which involves using the local languages, situations, and perceptions to promote awareness of the sanitation situation within a community, and empowering the community to develop and implement their own plans to improve services (Malebo et al., 2012; Wicken et al., 2008).

A fourth approach that has gained a lot of ground as an alternative to subsidies (see 5.4.2) is the community-led total sanitation (CLTS) which enables communities (especially in rural areas) to recognise the problem of open defecation and take positive steps to become open defecation free (ODF), through community mapping, walks, and the use of the equivalent of the word 'shit' in the local language to create disgust for open defecation (Harvey, 2011; Kar, 2003; Kar & Milward, 2011; Kar & Pasteur, 2005). This requires skilled facilitators to conduct triggering exercises in communities, strong local institutional capacity, and moving beyond the preoccupation with "open defecators" to also tackle complex issues of power and politics in sanitation governance (Bardosh, 2015; Galvin, 2015). Although CLTS has been successful in Asia and parts of Africa (Kar & Milward, 2011), communities with prior subsidies have been less responsive to it (WaterAid, 2013).

7.3.4 Suasive

This section discusses four suasive instruments.

Sustainable development goals and other voluntary assessment and reporting mechanisms

The Sustainable Development Goals (SDGs) offers an international framework of a (water and) sanitation goal, targets and indicators, and the Joint Monitoring Programme process for monitoring and reporting progressive realisation of universal access to safely managed sanitation services (see 3.2). The recently released SDGs assessment report, coming two years after the SDGs implementation process commenced, does not report on the SDGs sanitation indicator, that is, the proportion of population using safely managed sanitation services, including a hand-washing facility with soap and water. Rather, it falls back to the MDGs terminology and uses a threshold of 30% to produce global estimates of safely managed services, which is lower than 50% that was used to report on improved facilities in the previous JMP reports; this report also uses 50% as the threshold for global estimates of basic services and states that "[I]n 2015, 2.9 billion people (39% of the global population) used a 'safely managed' sanitation service - a basic facility that safely disposed of human waste" (United Nations [UN], 2017, p.6). The lower estimate for safely managed services means that the estimates will be less robust than under the previous JMP reports for improved facilities, under the Millennium Development Goals framework. The 2017 report also does not cover all the indicators, for instance affordability which is an important synergy between the SDGs and the HRS (see 5.3.1).

Three other voluntary assessment mechanisms are: (a) the public expenditure review from the perspective of the water and sanitation sector guidance notes which offers guidelines for analysing the allocation of public funds to the water and sanitation sector (WaterAid, 2013; (b) the Methodology for Participatory Assessments which assesses the link between sustainability, demand-responsiveness, and gender and poverty sensitivity (Dayal, van Wijk, & Mukherkee, 2000); and (c) the Tracking Financing to Drinking Water, Sanitation and Hygiene, which aims to develop and implement a global methodology for tracking WASH financing at the national level, improve national and global systems for collecting and analysing financial flows for the WASH sector (World Health Organization [WHO] & UN-Water, 2015). These three reflect fewer HRS principles than the SDGs.

Education, training and advocacy

Education, training and advocacy materials, like posters, stickers, and other memorabilia, are designed to improve sanitation and hygiene awareness and elevate the global sanitation crisis in the political and development agenda. An example is the 'Sanitation Resources for Media', offering infographics, facts, figures, press releases, feature stories, blogs and other practical resources, for use by media professionals in sanitation training and advocacy (The World Bank, 2014; WaterAid, 2013).

Other examples include: (a) Saniya, a hygiene communication campaign focused on improving good practices like hand washing after contact with faeces through home visits, theatre groups, radio programmes, and similar media (WaterAid, 2013); (b) the Menstrual Hygiene Management Knowledge Space (MeHMKS), offering resources for menstrual hygiene training, advocacy and enhanced research strategies (WaterAid, 2013); (c) SaniFOAM, a framework for analysing the determinants of sanitation behaviours classified into three categories: Opportunity, Ability and Motivation, while the F stands for Focus, in order to transform poor sanitation practices (Devine 2010); (d) Self-esteem Associative Strength Resourcefulness Action Planning Responsibility (SARAR), built on strengthening self-esteem, associative strength, resourcefulness, action planning, and a sense of responsibility among local people, to enable them think critically about their sanitation challenges and employ their creative abilities in planning, problem-solving, and evaluation (WaterAid, 2013); (e) Technical Notes for WASH in Emergencies, which consists of illustrated notes providing practical, evidence-based recommendations on 15 key issues relevant for immediate and medium-term WASH interventions in emergency situations, whether caused by natural or human-induced factors (Water Engineering Development Centre [WEDC], 2013).

Sanitation networks

Sanitation networks exist that provide a platform for collective action by private actors. Examples are: (a) WASH Journalist Networks which promotes policy changes and accountability of key stakeholders through media advocacy; (b) Women Leaders for WASH which consists of prominent women from different fields of endeavours with the aim of advocating for improved WASH services, both as individual women and as representatives of the broader constituency they represent; and (c) World Business Council for Sustainable Development (WBCSD) whose members (as well as non-members who have signed up to the

WBCSD WASH Pledge) are committed to providing all employees in all the premises that they directly control with access to safe WASH services (The World Bank, 2014).

Sanitation safety planning manual and technical guidelines

Sanitation safety planning and technical guidelines promote the safe construction, operation, maintenance and repair of sanitation infrastructure. An example is the World Health Organization (WHO) Sanitation Safety Planning Manual which offers a step by step description of how to implement the guidelines for the safe use of excreta, greywater and wastewater, while the standards and guidance for WASH in healthcare facilities provides guidelines for safe, efficient, and environmentally sound methods for the handling and disposal of healthcare wastes in emergency situations and normal settings (World Health Organization [WHO], 2016). Another example is the WHO Guidelines on Sanitation and Health that is designed to raise awareness on sanitation and hygiene especially, which offers a normative basis for sanitation interventions, summarises effective sanitation interventions and provides recommendations for sanitation policy and programme implementation (Sanitation Guidelines, n.d.). At the national and sub-national levels, there are also norms for the safe planning, design, construction, operation and maintenance of sanitation facilities which may not be mandatory except where they are included in the terms of a service contract (see 3.5.5) or in regulatory instruments (see 8.4.2).

7.4 NON-HUMAN RIGHTS FRAMEWORKS, DRIVERS, INCLUSIVE DEVELOPMENT AND LEGAL PLURALISM

This section analyses whether non-HR principles and instruments can address the drivers of poor sanitation services (see 7.4.1) and the implications for inclusive development (see 7.4.2) and legal pluralism (see 7.4.3).

7.4.1 Non-human Rights Frameworks and Drivers

The non-HR principles address twelve direct and four indirect drivers of poor sanitation services, as illustrated in Tables 7.2 and 7.3, respectively.

Principles for addressing the direct drivers

Capacity building, especially when integrated in suasive instruments for education and training on the importance of good sanitation practices and combined with pro-poor instruments, addresses preference distortion, discounting the future, inefficiencies in the tariff system by improving the capacity of the collectors, the exclusion of minorities, non-acceptance, negative social practices affecting access to sanitation and poor maintenance as a result of inadequate local capacity (see 7.2.1). Subsidiarity also improves knowledge and participation in the sanitation governance process, and can be used to empower minorities and other marginalised groups (see 7.2.1). Poverty eradiation and equity address household poverty, unaffordable tariffs and connection fees, exclusion of poor users and distance to the facility by empowering the poor with the necessary resources to invest in household sanitation services and ensuring that the available resources for sanitation are distributed in an equitable manner; for instance, this may be achieved through the use of targeted subsidies for the construction of sanitation facilities (Guiteras et al., 2015). Poverty eradication and equity also address tenure insecurity especially where households are unable to regularise their land titles due to poverty or where informal settlements are excluded from accessing basic services due to their socio-legal status (Dagdeviren & Robertson, 2009; Murthy, 2012). Further, environmental principles like precaution, prevention and polluter-pays address pollution and water scarcity when environmental sustainability is prioritised in the sanitation governance process (de Sadeleer, 2002; Holden et al., 2016). I elaborate on addressing the remaining five direct drivers which are not covered by humanitarian assistance and protection principles (including risk aversion, space constraints, a challenging or inaccessible topography, high temperatures and turbidity in source waters and natural hazard), *vis-à-vis* HRS principles, in Chapter 9.

Table 7.2   Non-human rights principles and instruments for addressing the direct drivers of poor domestic sanitation services

Direct Drivers (see 3.4)	Details of Drivers (see 3.4)	Non-human Rights Principles for Addressing the Drivers (see 7.2)
Economic	Discounting the future	Capacity building
	Household poverty	Poverty eradication and equity
	Preference distortion	Capacity building
	Inefficient tariff collection system	Capacity building Poverty eradication and equity Subsidiarity
	Unaffordable tariffs & connection fees	Poverty eradication and equity
Social	Distance to the facility	Poverty eradication and equity
	Exclusion of minorities	Capacity building Subsidiarity Poverty eradication and equity
	Non-acceptance of sanitation facility	Capacity building Subsidiarity
	Negative cultural practices	Capacity building Subsidiarity
	Poor maintenance culture/ improper use of facilities	Capacity building Subsidiarity
	Tenure insecurity	Poverty eradication and equality
Environmental	Pollution/water scarcity	Precautionary principle Pollution prevention Polluter-pays principle

Source: Compiled by the author, based on Akpabio, 2012; Dagdeviren & Robertson, 2009; de Sadeleer, 2002; Ersel, 2015; Evans et al., 2009; Mader, 2012; Mathew et al., 2009; Poulos & Whittington, 2000; USAID Egypt, 2015

Principles for addressing the indirect drivers

Capacity building, poverty eradication and equity and subsidiarity can address economic drivers like insufficient or poorly targeted funds and inefficient tariff collection system, by harnessing the capacities of the poor, improving their access to assets and resources for sanitation and other basic needs, and improving their participation in the governance process, respectively (COHRE et al., 2008; Kenny & Clarke, 2010; USAID Egypt, 2015). Capacity building can also address national poverty, where local resources, including knowledge and skills, are adapted to meet the sanitation needs of the population rather than relying on predetermined technocratic solutions that may be unaffordable given the national GDP (Zorn & Shamseldin, 2015).

Further, capacity building and subsidiarity can be integrated in suasive instruments for education and training to improve good sanitation practices and enhance the participation of vulnerable and marginalised members (see 7.2.1). DRM can address adaptation to environmental drivers like weather variability, natural hazards, and high temperatures/high turbidity in source water particularly when it is pursued through participatory approaches that involve the affected communities in the disaster risk planning and implementation processes (Vietnam Red Cross, 2011; Samaddar et al., 2017). Further, the precautionary principle obliges States to promote local capacity for climate change adaptation and mitigation measures, even in the face of scientific uncertainties over some aspects of climate change impacts (de Sadeleer, 2002; Holden et al., 2016; Kenny & Clarke, 2010). I further elaborate on addressing the remaining five indirect drivers which are not covered by humanitarian assistance and protection principles (including huge foreign debts; sanctions; insecurity, conflicts and poor social cohesion; mass migration/urbanisation; and population density/growth), *vis-à-vis* HRS principles, in Chapter 9 (see 9.2).

Table 7.3   Non-human rights principles and instruments for addressing the indirect drivers of poor domestic sanitation services

Indirect Drivers (see 3.4)	Details of Drivers (see 3.4)	Non-human Rights Principles for Addressing the Drivers (see 7.2)
Economic	Insufficient/poorly targeted funds	Capacity building Poverty eradication and equity Subsidiarity
	National poverty	Capacity building
Social	Low awareness about sanitation	Capacity building Subsidiarity
Environmental	Climate variability and change	Capacity building Precautionary principle

Source: Akpabio, 2012; de Sadeleer, 2002; Vietnam Red Cross, 2011; Kenny & Clarke, 2010; Munamati et al., 2016; Zorn & Shamseldin, 2015

7.4.2 Non-human Rights Frameworks and Inclusive Development

The implementation of the non-HR principles (see 7.2) and instruments (see 7.3) for sanitation governance in different country contexts can lead to different outcomes for IDID (i.e. see different results in different quadrants of Figure 7.1). Figure 7.1 illustrates how microfinance loans for sanitation services can either promote ID (Q4 in Figure 7.1) or exacerbate one or more forms of exclusion (Q1-Q3 in Figure 7.1); the outcome depends on whether or not the loans are offered to the poor, marginalised and vulnerable, the borrowers can pay back, and the funded services integrate environmental sustainability. Figure 7.1 also illustrates that indicators which disaggregate data on access to resources and sanitation services based on vulnerability will enhance inclusion better than indicators which only portray the average level of access.

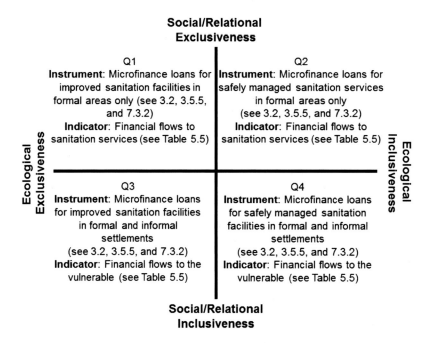

Figure 7.1 Assessing microfinance loans for sanitation services for inclusive development

Social and relational inclusion

Regulatory instruments like the prioritisation of sanitation and hygiene in licensing exemptions and permissible uses of water (see 7.3.1) improve access to water resources for sanitation and related human development benefits, thereby advancing social and relational equity. Economic instruments and market-based approaches provide options for sanitation cost recovery and financial sustainability of service provision to the poor (Jenkins and Scott 2007), and may also be used to promote positive sanitation practices and sanitation services (Evans et al., 2014). At the individual and community level, economic instruments are mainly applicable in stable conditions (for instance within the context of formal and informal settlements) but there is a growing interest in providing people living in humanitarian situations with non-food items like sanitation and hygiene materials through economic instruments like cash transfers and market based programming that can stimulate the recovery of the local markets (Albu, 2010; Martin-Simpson, Parkinson & Katsou, 2017). However, where economic instruments are poorly targeted (WaterAid, 2015) or solely focused on full cost recovery, this may exclude people who are unable to afford the cost of sanitation services from accessing them, leading to negative externalities for the wider population and the environment (Barlow, 2009; Mader, 2012; Sovacool, 2013). Practices like promoting conspicuous consumption and assaulting the dignity of non-users which commonly occur in sanitation marketing also affect human wellbeing and exacerbate inequalities (Barrington et al., 2017).

Management instruments provide users with an opportunity to participate in the sanitation governance process; through active participation, the users can communicate their unique preferences and challenges to the relevant stakeholders, in order to inform better sanitation policy and interventions and lead to sustainable outcomes (Malebo et al., 2012; Samaddar et al., 2017; Wicken et al., 2008). The participation of the local population can uncover local traditions and practices which may be adapted at low cost to address poor sanitation (McGranahan, 2013). However, participatory approaches without human rights safeguards can be subject to the power structures in communities and still exclude the marginalised and vulnerable who do not participate in the political/policy making process, like residents in informal settlements.

Suasive instruments are useful for creating a learning effect and influencing policy direction and character of actors from the international to the local levels of governance, but they generally lack legal force and require the voluntary compliance of actors. Hence, suasive instruments may need to be supported with regulatory instruments (for instance, mandatory technical guidelines for sanitation safety) to enhance compliance (Majoor & Schwartz, 2015). Suasive instruments also require affordable technology to promote upscaling (Gross & Günther, 2014), and continual monitoring for sustained behavioural changes (Mathew et al., 2009).

Ecological inclusion

Ecological inclusion requires the integration of environmental principles like disaster risk management, precaution, pollution prevention, polluter-pays and sustainable development, in order to operationalize a broad definition of sanitation (see 3.2 and 3.5.5) and meet human sanitation and hygiene needs within ecological limits (Feris, 2015; Holden et al., 2016). Though the SDGs mostly promote social and relational inclusion (Arts, 2017; Gupta & Vegelin, 2016), the SDGs 6 on water and sanitation particularly aims to ensure availability and sustainable management of sanitation services for everyone, safe drinking water, and safely managed water resources; this promotes ecological inclusion.

7.4.3 Non-human Rights Frameworks and Legal Pluralism

The interactions between the HRS and non-human rights frameworks may result in accommodation, competition, indifference, or mutual support (see Table 7.5).

Competition

Competition occurs where tensions between the HRS and non-human rights frameworks lead to HRS violations. For instance, where an insistence on full cost recovery results in a strict requirement for rightsholders to pay the full cost of their sanitation services, including waste treatment services, this may lead to the exclusion of the poorest who cannot afford the cost. Competition also results where the 'no priority' of use in Article 10 of the UN Watercourses Convention competes with the priority of human sanitation and drinking needs among water uses (Obani & Gupta, 2014b).

Indifference

Indifference occurs where although the HRS is captured in the non-human rights framework for sanitation governance, the framework does not prioritise access to sanitation for the poor, vulnerable and marginalised rightsholders. Indifference also results where the operationalization of environmental principles like precaution excludes low cost sanitation technologies and shared facilities which may be the more affordable and practical solution in densely populated informal settlements (see 3.5).

Accommodation

Accommodation occurs where efforts are made to adopt HRS principles (see 5.3) in non-human rights frameworks for sanitation governance. An example is where non-human rights principles promote the participation of the poor, vulnerable and marginalized individuals and groups in sanitation governance. For instance, the sanitation ladder can be used to try to create an opportunity for local stakeholders to participate in the policy process and selection of sanitation technologies (see 3.5.5).

Mutual support

Mutual support exists where: (a) the precaution and pollution prevention principles promote a broad definition of sanitation, and the use of the best available technology for sanitation, irrespective of cost considerations; (b) the polluter-pays principle generates funds for cross-subsidies to enable poor people access sanitation services; and (c) protecting and preserving the marine environment and ecosystems (Part IV of the UN Watercourses Convention of 1997) promotes pollution prevention and the elimination of open defecation (Obani & Gupta, 2014b).

Table 7.4 Types of legal pluralism relationship which arise between the human right to sanitation and non-human rights principles

Legal Pluralism Relationship	Description of the Relationship
Competition	Where the polluter-pays principle results in the exclusion of the poor from accessing sanitation services due to their inability to pay; the 'no priority' of use in article 10 of the UN Watercourses Convention competes with the priority of human sanitation and drinking needs among water uses
Indifference	Where a non-human rights framework captures HRS principles but excludes low cost and shared sanitation facilities needed by vulnerable and marginalised rightsholders, for instance under the MDGs where shared facilities were strictly considered unimproved
Accommodation	Where the sanitation ladder is used to try to create an avenue for local stakeholders to participate in the policy process and selection of sanitation technology
Mutual Support	Where the polluter-pays principle generates funds for cross-subsidies to enable poor people access sanitation services; non-human rights instruments like the UNECE Water Protocol contain obligations for water supply and sanitation services; protecting and preserving the marine environment and ecosystems (Part IV of the UN Watercourses Convention) promotes pollution prevention and the elimination of open defecation

Source: Compiled by the author, based on Bavinck & Gupta, 2014; de Sadeleer, 2002; Heijnen et al., 2015a, 2015b; Klasing et al., 2011; Obani & Gupta, 2016; UN, 2017

7.5 INFERENCES

This chapter yields five inferences. First, the non-HR framework for sanitation governance offers seven (three social, four environmental) principles that mainly stem from environmental law (see 7.2). The environmental principles are increasingly recognized in international customary law *inter alia* (see Table 7.1). The social and relational principles are still emerging in international law but have gained wide acceptance in international development practice. Also, the non-HR framework significantly offers environmental principles which may be supportive of the predominantly social and relational principles of the HRS in practice.

Second, non-HR principles address eighteen drivers of poor sanitation services, which is more extensive than the HRS and the humanitarian frameworks. This potential mainly stems from the wide array of environmental principles that is lacking in the previously considered frameworks (see Chapters 5 and 6). Nonetheless, the non-HR principles that are devoid of a relational component (with the exception of poverty eradication and equity) may exacerbate inequities in access to sanitation and therefore need to be complemented with the HRS to reduce contradiction.

Third, because the non-HR principles stem from different normative foundations and have varying legal status in international law, there is a potential for trade-offs in the absence of rules to address incoherence or contradictions in the implementation process. For instance, the improvement of living standards based on the principles of capacity building and poverty eradication and equity could improve access to sanitation where household poverty is a driver, but undermine sustainable development in the absence of measures (like progressive pricing) to curb unsustainable human consumption patterns and environmental degradation. This makes it important to design sanitation governance instruments based on a good understanding of the drivers of poor sanitation services in each context, the interplay between the non-human rights principles, and the suitability of the principles for addressing each driver without undermining other principles.

Fourth, in addition to the internal rules incoherence in the implementation of non-HRS frameworks for sanitation governance, non-HR principles can result in improved access to sanitation, yet compound the drivers of poor sanitation services and even mask HRS violations. For instance, applying the requirement for sewerage treatment (which supports the HRS principle of safety) to the JMP definition of access to improved sanitation facilities drastically reduces the estimate of people who had access to improved sanitation in 2010 from 4.3 billion as estimated by the JMP to 2.8 billion only (Baum et al., 2013).

Fifth, the effect of the non-HR framework on ID ultimately depends on the context and how they are used. Instruments focusing only on increasing coverage to sanitation services for the poor, for instance, without integrating environmental sustainability would fall within the top-left quadrant if they simultaneously increase equity (see Q1 in Figure 7.1) or the bottom-left quadrant if they simultaneously increase inequity due to failure to prioritise the poor (see Q2 in Figure 7.1). Similarly, economic instruments that are administered without consideration of social and relational inclusion may also be counterproductive for inclusive development (see Q3 in Figure 7.1). Management instruments that promote the participation of vulnerable people without the integration of environmental sustainability result in ecological exclusion, while those that integrate environmental sustainability promote ID (see Q4 in Figure 8.1).

Chapter 8.   Architecture of Sanitation Governance in Nigeria

8.1   INTRODUCTION

I now turn to analyse the multi-level architecture of sanitation governance in Nigeria, an emerging economy. The case study investigates the normative framework for sanitation governance and its implementation to understand: How does the human right to sanitation (HRS) influence the normative framework for sanitation governance towards inclusive development (ID) outcomes across different levels of governance in Nigeria? The chapter starts with contextualising the sanitation problem in Nigeria (see 8.2), the legal basis for the HRS including the other principles for sanitation governance (see 8.3), and the HRS and non-HRS instruments used for operationalizing sanitation governance principles (see 8.5.2). Section 8.6 analyses how the sanitation governance framework in Nigeria addresses the drivers of poor domestic sanitation services (see 8.6.1); the impact of the governance framework on ID (see 8.6.2); and the incoherence between sanitation governance framework in Nigeria and the international HRS (see 8.6.3).   Section 8.7 highlights the key recommendations for redesign. The methods which I used for my data collection and analysis have been previously discussed in Section 2.3.

8.2   BACKGROUND TO THE CASE STUDY AND THE STATUS OF THE HUMAN RIGHT TO SANITATION

This section sets the background for the rest of the chapter through providing a general overview of Nigeria's geographic, socio-economic and political context (see 8.2.1), the national development challenges (see 8.2.2), the status of the HRS in terms of actual realisation of the right (see 8.2.3), and the key stakeholders and organisational roles in the sanitation sector across the three levels of government in Nigeria (see 8.2.4).

8.2.1 Geographic, Socio-economic and Political Context

Nigeria is between latitudes 4°16' and 13°53' north and longitudes 2°40' and 14°41' east along the west coast of Africa. It occupies approximately 923,768 square kilometres of land stretching from the Gulf of Guinea on the Atlantic coast in the south to the fringes of the Sahara Desert in the north. Nigeria shares the country borders Niger in the north, Chad and Cameroon in the east, Lake Chad to the northeast, and the Republic of Benin in the west, while its coast lies on the Gulf of Guinea in the south.

Nigeria is the tenth most populous country in the world, and one of the fastest growing countries in Africa with a population growth rate of around 3.2% per annum (National Population Commission (NPC) [Nigeria] and ICF International 2013). In 2015, Nigeria had an estimated population of 183 million inhabitants, calculated based on the projected population estimates from the published figures of the 2006 Population and Housing Census which stood at 140,431,790 (National Bureau of Statistics, 2016). Slightly less than half of the population (49.5%) are female and over half of the population (50.5%) are male (National Bureau of Statistics [NBS], 2016). Nigeria has around 123,240 communities (Federal Ministry of Water Resources [FMWR], European Union, UKAid & UNICEF, 2016).

Nigeria has abundant natural resources that are currently exploited at statistically lower levels than the available deposits, except for oil which is the main foreign exchange earner. Nigeria's economy has suffered from the fall in international oil prices since the middle of 2014 and the inflation rate was as high as 17.26% (17.24% year-on-year) in March 2017 (National Bureau of Statistics [NBS], 2017), triggering a renewed focus on economic diversification beyond oil exploitation, revenue generation from internal sources through taxation, and the deregulation of the economy. Although there is a high rate of poverty which affects home ownership and access to assets generally (Nwakanma & Nnamdi, 2013), there are no official figures on which areas constitute informal settlements or the population living in such areas. As a result of the high cost of perfecting land ownership titles, people tend to discover the illegal status of their properties on an ad hoc basis when the government either issues eviction notices or embarks on forced evictions and demolitions, mostly in urban areas.[166]

Nigeria practices a federal system of government and is divided into the Federal Capital Territory, Abuja, and 36 states[167] for political administrative purposes. Each state is further divided into local government areas (LGAs) and there are 774 LGAs across the country within which around 250 different ethnic nationalities can be found. I conducted my household surveys in Benin, the capital city of Edo State in Nigeria. The justifications for my choice of Benin City are specified in Section 2.5.7. Benin City has extended to include four out of the eighteen local governments in Edo. The four local governments are Egor, Ikpoba-Okhan, Oredo and Ovia South-West.[168] During the 2006 national population census, the

---

[166] Interviewees 10, 17, and 18.
[167] The use of "states" in this chapter is not to be confused with the contemporary use "States" in reference to countries in my analysis of international law within this thesis.
[168] Interviewee 7.

combined population of the four local governments was 1,224,954 people out of which 49.8% were female and 50.2% were male (National Population Commission [NPC], 2010).

8.2.2 Development Challenges

The current federal government assumed office in May 2015 and made a commitment to address corruption, security challenges, unemployment, environmental degradation and the low living standards resulting from inequities in access to basic amenities, among its policy priorities. The current President, a former Military Head of State, was especially popular for his commitment to fighting corruption (see Box 8.1), maintaining national security and improving the nation's socio-economic performance. Two years into the four-year tenure of the government, Nigeria faces the same development challenges and the humanitarian situations resulting from the Boko Haram insurgency in the North-East, militancy within the oil-producing communities in the Niger-Delta, and inter-communal conflicts across the Middle-belt region and various parts of the country involving suspected Fulani herdsmen and host communities where they take their cattle to graze persist. The instability in the North-East especially is creating high numbers of internally displaced persons (IDPs) who have to flee to camps for IDPs all across the country. An important policy challenge that arises is how to improve human wellbeing and realisation of the human rights of the IDPs, host communities and the entire population of Nigeria while addressing the myriad development challenges in an inclusive manner. The IDP population is currently dispersed in camps and host communities both within the North-East and other parts of the country, like Benin City, the capital of Edo. This is one of the reasons why I chose Benin City for my household survey and field study of the HRS in humanitarian situations (see 2.3.2).

Box 8.1 Corruption and the human right to sanitation

Nigeria ranked 136 in the 2016 Transparency International (TI) corruption perception ranking, and has maintained the same position since 2014 after dropping from the 144th position in the TI ranking in 2013. Corruption is the abuse of public power for private gains and it persists at various levels of governance in all countries, but the manner in which it is handled has significantly marked the relationship between the global north and the south (Baillat, 2013). During the 1970s and 1980s, corruption was regarded as a cultural phenomenon, and accepted by States in the North as a 'necessary' cost of business in the South (Baillat, 2013). Multiple disciplines developed theories on corruption; economists proposed economic reforms to tackle corruption while the social scientists highlighted that the politicians in the South adopted a patron-client relationship model to explain corruption and saw politicians in the South as the main offenders (Baillat, 2013; Theobald, 1999). The United States of America took a bold step to pass the Foreign Corrupt Practices Act which prohibits American companies from bribing foreign officials and there are now several international and regional instruments initiatives that are aimed at tackling the corruption menace, including the African Union Convention on Preventing and Combating Corruption 2003, and the United Nations Convention Against Corruption 2005 (UNCAC).

Beyond the impact of corruption on the economy, there is also an emerging discourse on the relationship between corruption and human rights (HR) (Baillat, 2013; Gathii, 2010; International Council on Human Rights Policy [ICHRP], 2009b). Critics however argue that linking corruption and HR targets countries in the South, and hides neoliberal policies that are the real cause of development failures, and not corruption (Rose-Sender & Goodwin, 2010). Linking corruption and HR may oversimplify the complexities of corruption in practice, such as the complexity that may arise where a corrupt practice (like bribing a public official) ensures access to sanitation for people living in informal settlements or other vulnerable populations without access to sanitation services (Baillat, 2013). Nonetheless, a HR framework offers instruments, including principles, such as participation and accountability, to empower rightsholders to tackle corruption (Baillat, 2013).

Based on the understanding that corruption affects human rights, it is now a pressing concern for both UN Human Rights mechanisms like the special procedures, treaty bodies and the Universal Periodic Review and international organisations like TI, as well (Baillat, 2013; International Council on Human Rights Policy [ICHRP] & Transparency International [TI], 2010; Mbonu, 2009). Linking corruption with HRS gives a human face to the corruption menace and illustrates practically how corruption can reinforces exclusion and exacerbates HR violations. Nonetheless, corruption exacerbates discrimination, inequality and human rights violations (de Beco, 2011; ICHRP & TI, 2010); it violates the obligation of a State to apply the maximum available of its resources to progressively achieve the HRS where public resources that are meant for sanitation are mismanaged, misappropriated or diverted for private gains instead; abuse of functions by sanitation service providers or public officials (UNCAC, article 19) violates the obligation to respect; and failure to take appropriate legislative and other measures to prevent corruption that affects access to sanitation violates the obligation to fulfil the HRS (see generally, Baillat (2013)).

8.2.3 Status of the Human Right to Sanitation

National sanitation policies and laws relevant to sanitation governance contain HRS principles relating to sustainable, appropriate, acceptable and affordable sanitation services, as well as non-HR principles that may advance the realisation of the HRS. Table 8.1 illustrates the key national policies and laws relevant to sanitation, with the HRS principles they each contain, while indicating the extent to which the principles are respectively operationalized within the sanitation sector. This section provides facts and figures from desk research, augmented with the data from my field study, in an attempt to show the status of the HRS in Nigeria, and the instruments through which the principles are operationalized are analysed later in the chapter (see 8.5).

Table 8.1 Principles for sanitation governance in Nigeria

Principles		National Policies			
		NWSP	NESP	NWRP	NPE
Social	Acceptability	■			
	Access to information			■	■
	Accessibility	■	■	■	
	Accountability	■		■	■
	Affordable	■			
	Autonomy of the service providers			■	
	Availability	■		■	■
	Capacity building	■		■	■
	Cost sharing	■			
	Demand responsiveness	■			
	Economic good			■	
	Environmental offsetting				■
	Equitable access and poverty reduction	■		■	
	Extra-territorial obligations	■			
	Gender responsiveness	■			
	Participation	■	■	■	■
	Private sector participation	■	■	■	■
	Right to a healthy environment			■	■
	Safety	■			
	Social good		■	■	
	Subsidiarity			■	
	Sustainability	■		■	■
	Water rights for sustenance			■	■
Env.	Integration	■		■	■
	Polluter pays		■		■
	Precaution				■
	Prevention		■	■	■
Env. = Environmental; NWSP = National Water Sanitation Policy 2004 (draft); NESP = National Environmental Sanitation Policy 2005; NWRP = National Water Resources Policy 2016; NPE = National Policy on the Environment 2017					

National level

Nigeria loses over USD 147 million annually due to poor sanitation services and related impacts on human health and the economy (see Box 8.2) (Lawrence-Agbai, 2016; Oredola, 2016). It is however difficult to objectively assess the number of people who enjoy the HRS in Nigeria. This is because the existing indicators, assessments and data mainly focus on: (a) access to toilets and hand washing facilities; (b) access to solid waste management services;

and (c) the prevalence of malaria (see Figure 8.1). Beyond physical access, the HRS requires that sanitation services are made safe, acceptable, affordable, and available for users at all times during the day and night and the governance process empowers the marginalised and vulnerable populations through accountability, non-discrimination, participation, and access to information *inter alia* (see 5.3). There are discrepancies in the data from different official reports because they apply different measurement criteria and methods; this further compounds the complexity of ascertaining public spending on sanitation services (African Ministerial Conference on Water [AMCOW], 2011). For instance, the Water and Sanitation Monitoring Programme (WSMP) initially classified communal latrines as improved while the World Health Organisation and UNICEF Joint Monitoring Program (JMP) classified it as unimproved (USAID, 2015). The figures from various official datasets are now largely compatible because of the prevalence of the JMP definition of improved sanitation facilities, although the JMP now monitors a different set of criteria under the current Sustainable Development Goals (SDGs) framework (see 3.2 and 3.5.5).

Box 8.2 Statistics on the sanitation crisis in Nigeria

- Ranks 3rd globally for being behind on access to sanitation & having the longest toilet queues (WaterAid, 2015)
- Ranks 6th globally among the countries with the most OD per square km (WaterAid, 2015)
- Only about 40% of the total population of women that live in urban areas have access to sanitation services (Interviewee 3, June 5, 2013)
- Maternal mortality rate of 57.7 per 100,000 births due to sepsis, a harmful bacterial infection (WaterAid, 2015)
- 11 per 1,000 child deaths, under 5s due to diarrhoea (WaterAid, 2015)

The data is also mainly disaggregated according to rural and urban areas, without indicating the level of access for specific vulnerable and marginalized groups within each sub-national administrative unit and the challenges they face in realising the HRS. The most recent Nigeria Demographic and Health Survey of 2013 (NDHS), conducted from December 2012 to January 2013, uses the Enumeration Areas (EAs) prepared for the 2006 national population census as a sampling frame to select 40,680 households as a representative sample, with a minimum target of 943 completed interviews for each province (National Population Commission (Nigeria) [NPC] & ICF International, 2013). It shows that only 34% of the population used an improved sanitation facility, not shared; 28.7% practiced open defecation; while 59.6% used an improved drinking water source (NPC & ICF International, 2013). It also shows that 42.7% of the population in urban areas used an improved sanitation facility, not shared, compared to 28.2% in urban areas. The NDHS additionally shows the manner of disposing faeces for children under 5 years of age (see Figure 8.1). Edo accounts for 37.9% of the population with an improved sanitation facility and 48.7% of children whose faeces are safely managed; there is no further disaggregation of the data for Benin City or the individual EAs.

The final JMP update report shows that Nigeria did not meet the MDGs target for sanitation because only 25% of the population used an improved sanitation facility by 2015, as against 38% in the baseline line year, 1990; but Nigeria met the MDGs target for water with 69% of the population relying on an improved water source (World Health Organization [WHO] & UNICEF, 2015). The JMP update further shows that 33% of the population in urban areas use an improved sanitation facility compared to 25% in rural areas, but does not disaggregate the data for the various administrative units in Nigeria. The National Health Policy 2016 further indicates that 30% of households use an improved toilet facility, not shared with other households, and 61% use an improved water source; there is no further disaggregation of the data (Federal Ministry of Health [FMH], 2016).

The SDGs indicators baseline study however states that 60.3% use safely managed sanitation facilities, 69.6% of the population access safe drinking water, 48% access hand washing facilities with soap and water, and 0.04% of wastewater is safely treated (Office of the Senior Special Assistant to the President on SDGs [OSSAP-SDGs] & National Bureau of Statistics [NBS], 2017). The report attributes the drastic improvements to "the possible effect of awareness programmes implemented to educate Nigerians on managing sanitation services in their places of abode" (OSSAP-SDGs & NBS, 2017, p.28). However, awareness programmes are not enough to ensure improvements in sanitation coverage in a situation where people lack physical access to sanitation facilities, to start with. The level of progress that the study suggests is not supported by any other official record, or the data from the sanitation experts, stakeholders and households in the case study (see 2.3.2). Instead it appears to hide the real state of poor access to sanitation in Nigeria (see Figure 8.1). Hence, I do not include it in the compilation of official records on access to sanitation in Figure 8.1.

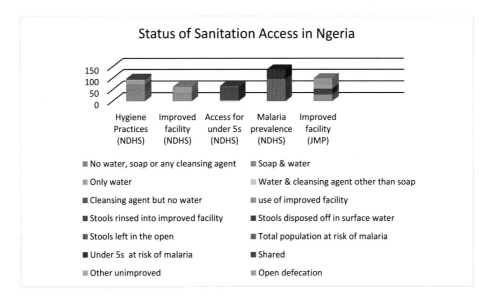

Figure 8.1 Percentage of the total population with access to sanitation in Nigeria (Based on the NPC & ICF International, 2013; WHO & UNICEF, 2015)

Datasets like the NDHS and the JMP are therefore indicative of the level of access to sanitation in Nigeria, at best, but do not reflect the status of the HRS because they only measure physical access and apply different methods. Nonetheless, the poor level of physical access to sanitation services which the datasets highlight already indicates a violation of the HRS for a majority of Nigerians at least. Only some parts of Lagos and Abuja have sewerage networks and some households resort to channelling untreated sewage into public drains.[169] Further, despite the broad definition of environmental sanitation in the NESP, there is no formal sanitary inspection of premises except for some private establishments which are visited by environmental health officers for legal formality; sanitation is very poor in most areas, especially in public markets, abattoirs and schools; there is indiscriminate disposal of dead animals especially though dead humans are buried according to cultural rites; no controls for reared and stray animals which leads to animal litter along the roads and clashes between herdsmen and indigenous people; weed and vegetation control is the responsibility of private landowners with little or no regulation except in mega cities like Lagos where the government recently passed a law requiring planting of flowers, and in Abuja parks and gardens are actively maintained through private public partnerships; there is no operational national strategy on hygiene education and promotion.

---

[169] National Policy on Environmental Sanitation and Implementation, 2005.

The household survey (see 2.3.2) provides valuable empirical insight into the status of the HRS in Nigeria, including the extent to which HRS principles like affordability, accessibility and participation have become localised.

Benin City

A majority of the households I surveyed (79.1%) used wet sanitation systems but only 46.5% had access to public water supply and the remaining relied on bore-holes, wells, water vendors or other private alternatives. Only 7.1% employed the services of a waste manager to evacuate their soakaways although a majority employed self-help mechanisms like adding chemicals to the waste (51.2%), transferring to another pit (7.9%), or even manual evacuation using buckets (4.3%). A majority of the households (54.7%) poured their wastewater into the public drainage, and 37% disposed their solid waste through unapproved dumpsites, burning or other unsanitary means. Further, 71.7% of the households surveyed were aware of good health and hygiene behaviours like hand washing and the use of hand sanitizers; this was as a result of public health campaigns following the 2014 Ebola Virus Disease (EVD) health emergency. A majority (98.4%) also practiced what they had learnt from the public health campaigns and this was not affected by household income distribution mainly because the hand sanitizers were often provided for free by the government, politicians, NGOs and some businesses as part of their corporate social responsibility programmes. A majority of the household respondents did not reuse their waste/wastewater and lacked knowledge of safe waste and wastewater management options. The population mostly relied on public waste management services for the collection of their non-degradable solid waste for a fee.

The NWSP recognises cost sharing (see Table 8.1), as a way of ensuring the financial sustainability of the sanitation sector. This requires that users contribute to the cost of operation and maintenance of sanitation infrastructure, including through paying for the services they use while the government also provides counterpart funding for capital investment in sewage, storm water, and on-site sanitation systems. This makes individuals and households directly responsible for the ownership, operation and maintenance of private sanitation facilities. The HRS does not obviate the need for users who have the financial capability to pay for their sanitation services, in order to ensure financial sustainability of the service delivery system, but it requires that the cost is affordable and the poor are not denied access as a result of their inability to pay (see 5.2.3 and 5.3.1).[170] The factors affecting WTP

---

[170] Interviewees 2, 10 and 11.

among the households surveyed are the cost, quality and regularity of service, and other personal considerations which they may not have disclosed. This corresponds with the evidence in the literature that WTP is influenced by the level of income, perceptions of the benefits derivable from the use of a good or service, the quality of the service, and the level of priority of sanitation. Another important factor affecting WTP which I discovered during my field visits and interactions with experts and stakeholders in Edo and Rivers is the proximity to waste management facilities and the state of the facilities. Residents of Ikhueniro, Uhunhunode L.G.A (Edo) decried the foul odour and contamination of groundwater and the environment by a government dumpsite in their vicinity and were outraged at being asked to pay for waste collection, in addition to the nuisance they were already suffering. Similarly, residents of Oyigbo L.G.A. (Rivers) decried the foul odour and health risks from a government-owned landfill maintained along the Oyigbo-axis of the Port Harcourt/Aba expressway, until the landfill was eventually closed sometime around 2014.

The average monthly cost of formal solid waste disposal services ranges from ₦200 (≈ USD0.56) to ₦300 (≈ USD0.83) for a one room apartment, irrespective of the actual income of the residents or the amount of waste generated; yet the principles of polluter pays, cost recovery and affordability form part of the sanitation policy framework (see Table 8.1). The figures do not even include the costs of construction, operation and maintenance of sanitation facilities like toilets which could be as high as ₦ 18,000 (≈ USD50) for digging a pit latrine; ₦30,000 (≈ USD83.3) for the evacuation of a soak away tank,[171] ₦145,450 (≈ USD404) for constructing a pour-flush toilet connected to a septic tank; and between ₦120,000 to 650,000 (≈ USD333 - 1805.6) for heavy duty borehole drilling or ₦160,000 – 300,000 (≈ USD444.44 - 833) for manual drilling of a borehole which service wet sanitation systems for about 10 years.[172] These costs are unaffordable for the poor, without subsidies or cross-subsidies for basic sanitation needs. Rather, a majority of the respondents were willing to use clean public toilets and pay between ₦10 and ₦20per use; and considered paying ₦1,000 monthly or around ₦30 daily to be affordable for household sanitation services. As much as 99% of the respondents were able to access their toilets at all times during the night and day, and 72.8% use a shared facility.

---

[171] This appears to be lower than the global average of USD 180.8 for basic improved sanitation without faecal sludge management and USD 304.7 per person for faecal sludge management (Hutton and Varughese, 2016).

[172] Interviewee 14. The factors affecting the cost include the geology of the environment, type of pump head used, number of casings, and water treatment and analysis.

Only about 40% of the respondents participated in the selection of their facilities and most were unaware of how to go about obtaining information about their sanitation services. Nonetheless, the Freedom of Information Act 2011 makes public records and information freely accessible and the participation principle is enshrined in the NWSP (see Table 8.1). However, the NWSP encourages participation from both rightsholders and the private sector operating sound business principles, as a way of strengthening the sanitation sector.

The households survey showed that the penetration of formal waste collection services is still very low[173] and many households have to either rely on the sparse services offered by public-private partnerships (mainly in the city centres and surrounding wealthy neighbourhoods) or on private facilities. But there is a limit to which individuals and households can invest in public goods like roads and sewer networks for centralised wastewater treatment (see 3.3). School sanitation services are often directly provided, operated and managed by the school authorities, and the most recent available national report shows that on the average, there is only one toilet for every 500 students in schools (Federal Ministry of Education [FMEdu], 2003).[174] Further, informal settlements are sometimes excluded by the government policy from accessing the formal waste collection services, and the informal service providers may be criminalised on the grounds of poor quality of services rendered and non-compliance with legal requirements for operations despite the immense relevance for augmenting formal service provision.[175] This finding corroborates the findings in the sanitation governance literature that informal services can be mainstreamed to improve sector efficiency in developing countries with weak formal sanitation service provision structures (Ezeah & Roberts, 2014; Nzeadibe & Anyadike, 2012).

Although there were no official figures of sanitation coverage in humanitarian situations at the time of this study, the experts and stakeholders interviewed generally indicated that: (a) IDPs are either sheltered in camps run by the National Emergency Management Agency (NEMA) and the States' Emergency Management Agencies (SEMA) or other humanitarian actors, in spontaneous settlements, or in local communities; (b) the sanitation facilities in the camps and spontaneous settlements are grossly inadequate; and (c) the influx of IDPs in host communities often creates stress on existing infrastructure. This is supported by the fact that the households in the IDP camps I visited mostly rely on a limited number of pour-flush toilets and pit latrines for their excreta management. For instance one of the camps has only 8

---

[173] Interviewees 10, 11, 12, 13, 14 and 15.
[174] *Nigeria Education MIS*, Federal Ministry of Education, 2003.
[175] Interviewees 10, 16, 17 and 19.

toilets used by over 2,000 registered inhabitants. Although a recent outbreak of scabies in the camp had triggered donations of additional sanitation and hygiene materials like medicated soaps, the efforts were mainly sporadic and sanitation was still ranked low on the list of priorities, often falling behind shelter, food, and drinking water.[176] This also corresponds with the findings of non-prioritisation of sanitation services in humanitarian situations from the literature review (see Chapter 6). People living in humanitarian situations are generally unable to invest in sanitation and are therefore heavily dependent on humanitarian assistance,[177] for dignity kits which cost between ₦3,000 – 4,000 (≈USD 8.3 - 11.1), depending on the content.[178]

The expert and stakeholder interviews revealed that informal settlements are not prioritised in the government's sanitation policies. Indeed, government policies only focus on rural, semi-urban and urban areas, while the provision of services to informal settlements is highly politicised. Although the regulators generally agreed on the need for universal coverage, they were divided on the need to provide services to informal settlements (see 8.5.1). One social actor cautioned against the strict distinction between formal and informal settlements in sanitation policy process, because: "[I]f you do that, then you are giving the government an excuse not to act. The people who live in the places where they should be (formal settlements) still do not have (access).... Just ask the straight question to the relevant government agencies: what are you doing? There are many agencies, just wasting money."

8.2.4 Key Stakeholders and Organisational Roles

Historically, Nigeria's water and sanitation policy focus has been on providing the population with access to water and sanitation services, but in most cases the government interventions have focused on providing water infrastructure, like boreholes, and monitoring water quality for public consumption (Akpabio, 2012). The government's sanitation interventions on the other hand were often reactionary and borne out of a public health emergency, like the outbreak of scabies in one of the IDP camp camps I visited for my field work (see 8.2.3). Further, the interventions were implemented independent of a specialised government ministry, department or agency, resulting in fragmentation and lack of progress on sanitation targets. Following the failure of the interventions and public utilities to ensure full service coverage for the population, the current policy reforms in the water and sanitation sector,

---

[176] Interviewee 18.
[177] Interviewees 6, 12 and 18.
[178] Interviewee 6.

starting from the 2000s, advance neo-liberal principles like demand responsiveness, cost sharing, full cost recovery, water as an economic good, and private sector participation (see Table 8.1), with timelines for achieving service improvements in urban and rural areas. This is reflected in policies like the National Water Supply and Sanitation Policy 2000 (NWSSP) that lapsed in 2011; the National Water Sanitation Policy 2004 (NWSP) and the National Water Resources Policy 2016 (NWRP) (see Table 8.1). This section specifically identifies the key stakeholders and organisational roles for sanitation across the three tiers of government (federal/national, states and LGAs). Table 8.1 further illustrates the main events in the evolution of sanitation governance in Nigeria.

Federal level: Policy formulation, coordination and advocacy

The National Water Supply and Sanitation Policy 2000 specifies policy making, the regulatory role of the government and appropriate legislation as some of the guiding principles for the coordination of the water and sanitation sector. The National Water Sanitation Policy also emphasises the importance of political will in this regard. Sanitation policy formulation, co-ordination and advocacy are conducted by the federal government.[179] The main objectives of sanitation governance at this level include: (a) protecting water quality; (b) preventing public health risks; (c) improving service coverage in line with international development goals and human rights obligations; (d) regulating activities in the sector; and (e) advancing the political agenda of the government. The NWSP highlights the need for an integrated approach to sanitation in order to ensure water safety, control of water pollution and increased awareness on water contaminants.[180] As a result, water and sanitation are still closely linked in the national policy process.

Sanitation governance at the federal level is steered by the Federal Ministry of Water Resources (FMWR), Federal Ministry of Health (FMH), and Federal Ministry of Environment (FMEnv). Within the FMWR, the Department of Water Quality Control and Sanitation conducts: advocacy visits to policy makers at state level; awareness creation/sensitization to stakeholders in the sanitation and hygiene sector; capacity development of States and LGA WASH personnel on the Community Led Total Sanitation approach implementation nationwide; policy development; verification exercise on Open Defecation Free Communities in States implementing Community Led Total Sanitation; production and dissemination of information, education, and communication materials in

---

[179]   Interviewees 2, 3, 5, and 6.
[180]   Interviewees 2 and 3.

different Nigerian languages and jingles; hand washing campaigns; and promotion of appropriate sanitation technologies.[181] Further, in 2002, the federal government established a National Task Group on Sanitation (NTGS) as an inter-ministerial/agency group to coordinate national efforts for sanitation and hygiene programming and advised that a household toilet may be shared by a maximum of 15 persons, while communal toilets may be shared by 45 persons, though it is not clear how this standard can be enforced ("Group insists 15 persons maximum", 2017), but this is at odds with the National Environmental (Sanitation and Wastes Control) Regulations of 2009, that recommends 2 toilets for 11 to 20 persons.[182]

Although the NTGS coordinates national efforts for sanitation and hygiene programming, there is no clear definition of the HRS in the policy process. Rather, the members of the NTGS[183] each tend to adopt a meaning of sanitation that aligns most closely with their respective core mandates and their respective targets for sanitation coverage show some level of incoherence (see 8.3.2). For instance, one of the targets of the FMWR authored National Water Sanitation Policy 2004 (NWSP) is: "achieve 100% sanitation coverage by 2025" and the NWSP emphasises improved sanitation technology for households (see 3.2 and 3.5.5). On the other hand, the targets of the FMEnv authored National Environmental Sanitation Policy 2005 (NESP) include to: "increase access to toilet facilities … in public places and … in households by … 75% and 100% respectively by 2010". Both policies therefore promote universal access to sanitation services but their targets and timelines are incoherent. Both policies also contain different definitions. The NWSP states that "[A]dequate sanitation means access to safe excreta disposal facilities, services to households, public facilities, and disposal of liquid and solid waste without contamination of water sources, health hazards to people or deterioration of the environment" and improved sanitation means "[U]pgrading traditional latrines to reduce flies and odour, and provide superstructures; provision of water flush system (septic tank/soakaway) and sewerage system; and provision of hand washing facilities after use." On the other hand, the NESP defines environmental sanitation comprehensively, as: "the principles and practice of effecting healthful and hygienic

---

[181] Interviewee 2.
[182] See Regulation 10(1)(b), Guidelines for recommended approved number of toilets.
[183] Other members of the NTGS include: (a) the Federal Ministries of Education, Environment, Health, Women's' Affairs, and Housing and Urban Development; (b) MDGs Office; (c) National Orientation Agency; (d) National Planning Commission; (e) National Agency for Food, Drug Administration and Control; (f) NGOs like WaterAid; (g) international organisations like UNICEF, World Bank and DfID; (h) the European Commission; and (i) civil society groups and sanitation networks like the Society for Water and Sanitation (NEWSAN).

conditions in the environment to promote public health and welfare, improve quality of life and ensure a sustainable environment" (NESP 2005, paragraph 1.3.1).

Other key stakeholders at the federal level are: (a) the 12 River Basin Development Authorities (RBDAs), responsible for developing, operating and managing the bulk water reservoirs for water supply for irrigation and domestic uses, hydroelectric power generation, navigation, recreation, and fisheries projects; and the (b) National Emergency Management Agency (NEMA) leads on sanitation response in humanitarian situations, in collaboration with the State Emergency Management Agencies (SEMAs) and other local and international humanitarian actors.[184]

States: Policy coordination, adaptation and implementation

Each state in Nigeria is responsible for policy coordination, adaptation and implementation of the sanitation policies, including federal policies, within their jurisdiction.[185] This responsibility is exercised through the relevant ministry or agency in charge of environment, waste management, or sanitation services, depending on the local political structure. The focus of the states mainly centres on waste management and environmental protection (see 8.3.2). Sometimes, solid waste is regulated by the sanitation ministry or agency while wastewater is regulated by the water resources ministry or agency, similar to the federal level. Newer laws like the Environmental Law of Lagos State of 2017, and the Environmental Management (Miscellaneous Provisions) Law of Ogun State of 2005 integrate both solid and liquid waste under a similar framework for environmental management. The state laws do not define the HRS but generally define waste to include any substance that is scrap material or needs to be disposed of as a result of being damaged, contaminated, or worn out; an unwanted surplus from any process; and anything that is discarded or treated as waste.[186] The states have recently started embarking on donor-driven sector reforms, largely aimed at addressing deficits in rural water and sanitation services and urban water supply (leaving out sanitation services in urban centres).[187] Examples of the reforms include the establishment of a regulatory framework for the sanitation sector and the commercialisation of the public sanitation and water utilities (USAID, 2015).[188] Further, there are task groups on sanitation at

---

[184] Interviewee 2.
[185] Interviewees 2, 3, 5, 6, 7 and 8.
[186] See Environmental Law of Lagos State 2017, Section 38(1)(b)(g)(i)-(iii); Abuja Environmental Protection Board Act 1997, Section 46
[187] Interviewees 7 and 8.
[188] Interviewees 8 and 13. See also the National Water Supply and Sanitation Policy 2000.

the state level, and the State Emergency Management Agency coordinates emergency sanitation response in partnership with other stakeholders.[189]

Local level: Policy implementation, programme management, co-ownership and sustainability

The local government is responsible for policy implementation, programme management, co-ownership and sustainability.[190] Some states have established a full-fledged Rural Water Supply and Sanitation Agency (RUWASSA), and WASH Departments or Units within the LGSs in their jurisdictions, to promote decentralisation, and efficient planning and implementation of WASH governance and advocacy.[191] This is especially common among the states with funding from donors (Federal Ministry of Environment et al. 2016; USAID 2015). Otherwise, the responsibility for sanitation is delivered by a department of environment or health, or a water and sanitation committee or department.[192] Other stakeholders at the local level include: (a) private sector service providers who deliver services like waste collection and treatment, supply of sanitation goods and other related services, and construction, drilling, operation and maintenance of facilities; (b) NGOs (like WaterAid) who support local communities to strengthen sanitation governance through initiatives like the community-led total sanitation (see 7.3.3); (c) donors who fund local sanitation programmes; and (d) other international organisations like UNICEF.

8.3 DRIVERS OF POOR SANITATION SERVICES

Nigeria has various sanitation targets, policies and programmes in place, towards the progressive realisation of universal coverage. There was a national target of constructing one million household latrines annually, as from the International Year of Sanitation (2008) until 2015, to meet the MDGs target of 65% sanitation coverage, but this failed abysmally by the end of 2015. The failure informed the current Partnership for Expanded Water Supply, Sanitation and Hygiene (PEWASH) strategy, to ensure that Nigeria is open defecation free (ODF) by 2025, and meets the SDGs Targets 6.1 and 6.2 by 2030. This section highlights the direct and indirect drivers of poor access to sanitation services   that emerged from the case study.

---

[189]   Interviewees 2 and 6.
[190]   Interviewee 3.
[191]   Interviewees 2 and 3.
[192]   Interviewees 2 and 3.

Direct

Economic Drivers

Many interviewees identified poverty and high tariffs as important drivers of poor sanitation services for domestic users, in informal settlements for instance.[193] Nonetheless, there was a marked difference between the importance assigned to poverty as a driver among regulators and social actors interviewed; the latter generally attached higher importance to poverty than the regulators. One social actor stated that, "now in renting accommodation, ensuite houses with toilets, and accommodation with toilets outside the house, or those without toilets attract different rents. The choice is all down to poverty! ... If you do not pay, then your waste is uncollected and it is a breeding ground for diseases."[194] Another similarly stated that "access is also linked to affordability. ... Some landlords do not believe in locating toilets in their houses due to trying to cut costs. As a result, there is a lot of open defecation."[195] Conversely, one of the regulators felt that although   poverty may be an issue, it is not the root cause for people who can afford other basic necessities like food for themselves.[196] The feedback from the regulators also suggests that the government may be unwilling to increase the cost of sanitation services due to political considerations; this point is also indicated in the literature (for instance, World Bank (1999)). The stakeholders also revealed that inefficient tariff collection reduces the available funds for sanitation;[197] preference distortion makes some users unwilling to pay for sanitation;[198] and risk aversion makes some private service providers unwilling to cover the informal settlements without any assurance of cost recovery through government-funded bridging or subsidy.[199] Some households complained about being charged for waste collection services even though the waste managers assigned to their areas did not actually collect their waste.

Social Drivers

Many interviewees identified poor design and the siting of facilities in inaccessible locations as some of the reasons for open defecation and urination and the indiscriminate dumping of refuse in unauthorised places.[200] Poor urban planning[201] and the exclusion of the

---

[193]   Interviewees 2, 10, 11, 16, 17 and 19.
[194]   Interviewee 10.
[195]   Interviewee 11.
[196]   Interviewee 2.
[197]   Interviewees 2, 6 and 8.
[198]   Interviewees 6 and 8.
[199]   Interviewee 15.
[200]   Interviewees 1, 6, 8, 10, 11, 13 and 14.

population[202] are also some of main social drivers identified by stakeholders. One of the regulators interviewed acknowledged that informal settlements were excluded "because they are illegal. Capturing them would amount to sanctioning their existence" and even within formal settlements that are prioritised for service delivery, the "areas with poor access roads receive lower priority because of the difficulty of service delivery" and are sometimes excluded from sanitation service delivery due to poor accessibility.[203] The exclusion of informal settlements also has a political connotation, as noted by a media activist: "the government feels that they (informal settlements) should not be there, like the people are a threat… However, during elections, the government recognises them and installs polling booths in the community but after the elections, nothing happens."[204] Lack of access to land also inhibits the construction of sanitation facilities and disaster risk preparedness, to the extent that humanitarian actors are unable to construct camps for people who are affected by humanitarian crisis.[205] This is despite the fact that Section 1 of the Land Use Act 1978 vests all land in the government to hold in trust for the citizens, and the government can therefore obtain land for public uses subject to compensating persons holding any private ownership rights in the land. Other social drivers identified by stakeholders are non-prioritisation of sanitation by the government and the population;[206] poor maintenance of existing facilities and a culture of indifference and nonchalance which translates into poor sanitation habits;[207] tenure insecurity in informal settlements;[208] and negative cultural norms which support open defecation.[209] Experts also highlighted that sanitation receives scant attention in development planning and public discourse, because is not a "sexy" topic.[210]

Environmental Drivers

Stakeholders highlighted the following environmental drivers: an inaccessible topography[211] or special environmental features like coastal areas which may require specially designed floating toilets;[212] drought;[213] flooding events caused by natural or anthropogenic activities

---

[201] Interviewee 3.
[202] Interviewees 8, 10, 11 and 12.
[203] Interviewee 8
[204] Interviewee 10.
[205] Interviewee 6.
[206] Interviewees 10 and 13.
[207] Interviewees 8, 10 and 13.
[208] Interviewees 10, 11, 12, 16, 17 and 19.
[209] Interviewees 1, 2, 4, 7, 10, 13 and 16.
[210] Interviewees 1, 2, 10 and 13.
[211] Interviewees 8 and 11.
[212] Interviewee 10.

which lead to the destruction of sanitation infrastructure;[214] water pollution from oil exploration activities and faecal contamination;[215] and low rainfall which affects the generation of hydropower (the main source of electricity in Nigeria) for the operation of sanitation and water infrastructure like pumps.[216] However, the power supply problem in Nigeria is also due to the weak infrastructure for the transmission and distribution of the available power.[217]

Indirect

Economic Drivers

Stakeholders alluded to inadequate funding and the inefficient allocation of available funds for the provision of common goods like sewer networks;[218] the inefficiencies in public spending are sometimes due to corruption (see 8.2.2).[219] An estimated annual investment of USD 2.5 billion was required between 2007 and 2015 to attain the MDGs sanitation target while the actual spending though difficult to ascertain was only about a third of the necessary investment (AMCOW, 2011). The investments in sanitation are mainly investments in suasive instruments like trainings and advocacy (OSSAP-SDGs and NBS 2017), whereas investments in water are mostly in the form of hardware components like technologies for water service delivery or rural sanitation only.[220] For instance, the 2017 Appropriation Budget for the FMA, one of the line ministries in charge of sanitation (see 8.3.3), mentions sanitation only once in connection with rural roads and water sanitation (amounting to Naira (₦) 6,461,935,200), whereas there are four entries for the construction/provision of water facilities, three entries for the rehabilitation/repairs of water facilities, and one for supply/installation of water purification equipment and chemicals, and supply/installation of additional 2 No. 75KVA water pump head, respectively (all nine entries for water supply collectively amount to ₦126,641,333). This indicates more budgetary allocations to water supply than sanitation. The estimated cost of meeting basic WASH needs in the first year under the current SDGs framework is 0.63% of the national GDP, while the capital costs of meeting SDGs targets 6.1 and 6.2 is 1.7% of the GDP (Hutton & Varughese, 2016). This is above the eTkwini commitment made by the Nigerian government, under the auspices of the

---

[213] Interviewees 3 and 10.
[214] Interviewees 3, 10 and 13.
[215] Interviewees 3, 10 and 11.
[216] See Benue State Water Supply and Sanitation Policy, First Edition, January 2008, p. 12; interviewee 18.
[217] Interviewees 2 and 10.
[218] Interviewees 1, 2, 3, 4, 7, 8, 10, 13 and 16.
[219] Interviewee 10.
[220] Interviewee 2.

Africa San Conference on Sanitation and Hygiene, Durban, 2008, to allocate a minimum of 0.5% of GDP for sanitation and hygiene programs and monitor the implementation of this commitment. The eTkwini commitment and other similar international commitments for sanitation (see Chapter 5) have not even been met, due to lack of political will, corruption (see Box 8.1) and non-prioritisation of sanitation sector spending.[221] Despite some apparent evidence of political will (particularly, the various policies, strategies and programmes for rural sanitation and the MDGs and SDGs sanitation targets), the stakeholders decried the lead role of donors in the sanitation sector and the failure of the government to finalise the draft NWSP for over a decade since it was prepared (in 2004). The NWSP needs to be approved by the Federal Executive Council, in order for it to become a full-fledged policy, but it is still operational as the most comprehensive national policy on sanitation (including the management of both solid and liquid waste) (FMWR et al., 2016).

Social Drivers

Stakeholders highlighted the following indirect social drivers: (a) insurgency and insecurity, leading to retrogression in access to sanitation due to the destruction of sanitation infrastructure and disruption of law and order in the affected parts of Nigeria, particularly in the North East (due to Boko Haram) and the Niger Delta region (as a result of militancy) (see 8.2.2);[222] and (b) population growth which puts pressure on the existing sanitation infrastructure.[223] Further, urbanisation and poor planning of cities exacerbates harmful practices like indiscriminate sand filling and dumping of waste for land reclamation.[224] Stakeholders also stated that the lack of awareness and poor access to justice by vulnerable members of the population, such as IDPs and refugees (see for instance, The UN Refugee Agency 2016), also constrain the enforcement of the HRS for the unserved.[225]

Environmental Drivers

Some of the interviewees highlighted the effect of climate change and the related weather variability on access to sanitation, through exacerbating environmental hazards and affecting the operation of sanitation infrastructure.[226] This is especially significant because many sanitation and hygiene practices in Nigeria require water (see 3.5.1 – 3.5.4, 8.2.3).

---

[221] Interviewees 1, 10 and 13.
[222] Interviewee 13.
[223] Interviewees 2 and 8.
[224] Interviewees 8 and 10.
[225] Interviewees 10, 17 and 19.
[226] Interviewees 1, 2 and 11.

8.4   LEGAL BASIS FOR THE HUMAN RIGHT TO SANITATION IN NIGERIA

This section examines the sources of the human right to sanitation in Nigeria. It does so in order to establish that although the HRS is not expressly recognised as a fundamental right under the Federal Constitution of 1999, there are alternative sources which impose legal obligations for the realisation of the HRS on Nigeria. The section builds on the analysis of the legal bases for the HRS in international law in Chapter 5, and examines treaties to which Nigeria is a party (see 8.4.1), and national laws and judicial decisions (see 8.4.2) that provide the legal basis for the HRS in Nigeria.

8.4.1 Treaties

Nigeria is a State Party to various international and regional treaties which recognise the HRS (Table 8.2). However, treaty ratification does not automatically translate into domestication of the treaty provisions because the CFRN, section 12 requires all international treaties to be enacted into law by the National Assembly in other to take effect in Nigeria. This raises the question of the legal status of the treaties to which Nigeria is a State Party but which are yet to be domesticated by the National Assembly (see Table 8.2).

Arguably, by signing the treaty Nigeria has already expressed its intention to be bound by the provisions based on the principle of *pacta sunt servanda*, i.e. the principle of good faith. This argument is supported by the Vienna Convention on the Law of Treaties, article 26 which states that "every treaty in force is binding upon the parties to it and must be performed by them in good faith." The principle was reflected in the judgment of the Court of Appeal in the case of *Mojekwu and Ors. v. Ejikeme and Ors*,[227] where the court set aside the *Ili ekpe* custom for contravening the fundamental rights of women as guaranteed under the CFRN, and the Convention for the Elimination of All Forms of Discrimination against Women of 1979 (CEDAW) which had not been domesticated in Nigeria. Although on appeal, the Supreme Court refrained from ruling on the validity of the lower court's position on the protection of gender rights under the constitution and international law, it nonetheless overturned the earlier decision of the Court of Appeal stating that there was no justification for the lower court to have pronounced the relevant custom to be repugnant to natural justice, equity, and good conscience, as the repugnancy was not an issue before the court and the undermining of customary law in that way was not justifiable (Nwabueze, 2010).[228]

---

[227] (2002) 5 NWLR (Pt. 657) at 402.
[228] Mojekwu v. Iwuchukwu (2004) 11 NWLR Pt. 883, at 196.

Therefore, it would appear that HRS provisions in ratified treaties, including treaty provisions that form a part of customary international law, as evidence of a general practice accepted as law,[229] and provisions that reflect the general principles of law recognised by civilized nations[230] would be justiciable in Nigerian courts.

Table 8.2   Ratification and domestication of international and regional treaties (relevant to Nigeria) recognising the human right to sanitation

Year	Treaty	Ratification/ accession	Domestication
	HUMANITARIAN SITUATIONS		
1949	Geneva Conventions (I, III and IV) & Protocol I	1961 & 1988	N/A
	FORMAL & INFORMAL SETTLEMENTS & HUMANITARIAN SITUATIONS		
1966	International Convention on Economic, Social, and Cultural Rights	1993	N/A
1979	Convention on the Elimination of all Forms of Discrimination Against Women	1985	Some states have gender laws*
1989	Convention on the Rights of the Child	1991	21 out of the 36 states currently* have Childs' Right Laws
1981	African Charter on Human and Peoples' Rights	1983	African Charter on Human and Peoples' Rights (Ratification and Enforcement) Act, 1983
1990	African Charter on the Rights and Welfare of the Child	2001	21 out of the 36 states currently have Childs' Right Laws
2000	Protocol to the African Charter on Human and Peoples' Rights on the Rights of Women in Africa	2004	N/A
2006	Convention on the Rights of Persons with Disabilities	2010	N/A

*Issues concerning women and children fall under the residual legislative list within the purview of the House of Assembly of the respective states in Nigeria, by virtue of the provisions of the 1999 Constitution, section 4(7)

8.4.2 National Laws and Judicial Decisions

The NWRP states that: "[A]ccording to the Constitution of the Country, every Nigerian has the right to adequate water supply and sanitation, nutrition, clothing, shelter, basic education, and health care, as well as physical security and the means of making a living" (NWRP, p.2), and one of its strategies for improving the quantity and quality of water supply and sanitation

---

[229]   See the Statute of the International Court of Justice, 1945, article 38(b).
[230]   See the Statute of the International Court of Justice, 1945, article 38(c).

services is to "promote the right of access to clean water and basic sanitation for all citizens" (NWRP, p.24). However, the HRS is not expressly recognised within the fundamental rights and freedoms contained in the Constitution, but Sections 33 and 34 of the Constitution, respectively guarantee the fundamental human rights to life and human dignity which have been the basis for advocating for access to basic sanitation in other jurisdictions (see 5.2.2). The Constitution also protects persons from entering into or destroying property in order to safeguard sewage services.[231] The provisions are further significant because threats and actual violations of such rights contained in Chapter 4 are justiciable before the Federal High Court.

Chapter 2 (Sections 13-24) of the Constitution also contains fundamental objective and directive principles of State policy that may be linked to the economic, social and cultural rights forming the legal basis for the HRS at the international level (as discussed in Section 5.2). The directive principles essentially represent the high ideal which States aspire to attain;[232] but their enforcement is still largely political (Aguda, 2000). Table 8.3 expounds on the relevant provisions of Chapter 2, linked to the African Charter on Human and Peoples' Rights 1983 that is domesticated in Nigeria. Although Section 13 of the Constitution makes it: "… the duty and responsibility of all organs of government, and of all authorities and persons, exercising legislative, executive or judicial powers, to conform to, observe and apply the provisions of this Chapter of this Constitution," by virtue of Section 6(6)(c) the provisions of Chapter 2 are only justiciable in connection with other constitutional guarantees.[233] This differentiation between the (justiciable) civil and political rights and freedoms guaranteed in Chapter 4, and the non-justiciable provisions of Chapter 2 that are more closely related to the economic, social and cultural rights, is perhaps because civil and political rights are more relevant for foreign investors, and the country with its relative niche as an oil exporter does not have any additional incentive to engage in the global constitutional race to the top by expanding on socio-economic rights (Agbase, 1998). Nonetheless, due to the high level of poverty and inequities in access to resource in Nigeria the enforceability of Chapter 2 has been increasingly advocated by scholars and human rights activists.

---

[231] Section 44(2)(m) authorises any person or authority to cause damage to buildings, economic trees or crops, in order to enter, survey or dig any land, or to lay, install or erect poles, cables, wires, pipes, or other conductors or structures on any land, in order to provide or maintain the supply or distribution of public facilities or public utilities, including water supply and sewage services.

[232] Minerva Mills Ltd & Ors v Union of India & Ors (1980) SCC (3) 625 (India).

[233] Olasifoye v Federal Republic of Nigeria (2005) 51 WRN 52.

Table 8.3 Legal basis for the human right to sanitation under Chapter 2 of the Constitution of the Federal Republic of Nigeria 1999 and related provisions of the African Charter on Human and Peoples' Rights 1983

Chapter 2 Provisions relevant to the HRS (Section)	Details of the Directive Principles	Related economic, social and cultural rights under the African Charter on Human and Peoples' Rights 1983 (ACHPR)
14(2)(b)	The security and welfare of the people shall be the primary purpose of government	Dignity (ACHPR, article 5); liberty and security (ACHPR, article 6)
14(2)(c)	Participation by the people in their government shall be ensured in accordance with the Constitution	Participation (ACHPR, article 13(1))
16(2)(d)	Universal access to suitable and adequate shelter, and food, reasonable national minimum living wage, old age care and pensions, and unemployment, sick benefits and welfare for the disabled	Equal access to public services (ACHPR, article 13(2)); non-discrimination (ACHPR, articles 2 and 3); the best attainable state of physical and mental health (ACHPR, article 16)
17(3)(c)	Just and humane work conditions, adequate facilities for leisure and social, religious and cultural life; State policies to safeguard the health, safety and welfare of all workers	Work under equitable and favourable conditions (ACHPR, article 15)
17(3)(f)	Children, young persons and the aged are [entitled to be] protected against exploitation, and neglect	Protection of the rights of the child (ACHPR, article 18(3))
17(3)(g)	Public assistance in conditions of need	Equal access to public services (ACHPR, article 13(2))
20	The State shall protect and improve the environment and safeguard the water, air, land, forest and wild life	General satisfactory environment favourable to human development (ACHPR, article 24)

Proponents of non-justiciability generally rely on Section 6(6)(c) of the CFRN and the view that the Constitution is intended to state supreme rules of law, without opinions, aspirations, directives and policies (Popoola, 2010). This view by extension could mean expunging Chapter 2 from the constitution if it is not intended to be justiciable. Although Item 60(a) of the Exclusive Legislative List for the exercise of federal legislative powers under the CFRN provides *inter alia* for: "The establishment and regulation of authorities for the Federation or any part thereof - (a) To promote and enforce the observance of the Fundamental Objectives and Directive Principles contained in this Constitution;" there seems to be a mixed attitude by courts in Nigeria on the justiciability of the directive principles in Chapter 2 largely based on whether or not the directive principle are further contained in legislation or their enforcement violates Chapter 4 of the CFRN (Fagbohun, 2010; Popoola, 2010).

In cases where the laws hinged on Chapter 2 of the CFRN have been challenged, the attitude of the courts in Nigeria has been to uphold the power of the federal government to give effect to the relevant provisions through legislative enactments.[234] This is because the CFRN is not exhaustive on the measures for enforcing the human rights provisions.[235] In a case involving the right to education which is contained in the African Charter on Human and Peoples' Rights (ACHPR) already domesticated in Nigeria, the ECOWAS Court has held that the non-justiciability of the right under the CFRN does not oust its jurisdiction because the right is domesticated in municipal law.[236] Indeed, all the rights contained in the ACHPR are justiciable before domestic courts on the basis of the African Charter on Human and Peoples' Rights (Ratification and Enforcement) Act of 1983 (see Table 8.3 for socio-economic rights provisions under the ACHPR that are relevant to the HRS). Further, in cases where the implementation of the directive principles results in the violation of Chapter 4, the courts also exercise jurisdiction on the directive principles, like the Court of Appeal did in the case of *Archbishop Anthony Olubunmi Okogie & Ors v Attorney General of Lagos State*,[237] thereby presumably making Chapter 2 justiciable. International human rights mechanisms have also strengthened the justiciability of Chapter 2 provisions. For instance, the ECOWAS Court in the case of *Socio-Economic Rights and Accountability Project (SERAP) v. Nigeria*, dismissed Nigeria's contention that education is "a mere directive policy of the government and not a legal entitlement of the citizens" and held that the ECOWAS Court "has jurisdiction over human rights enshrined in the African Charter and the fact that these rights are domesticated in the municipal law of Nigeria cannot oust the jurisdiction of the Court."[238]

---

[234] Attorney General of Ondo State v. Attorney General of the Federation & Ors. (2002) NWLR concerning the enactment of the Independent Corrupt Practices and Other Related Offences Act; A.G. Lagos State v. A.G. Federation (2003) concerning the enactment of the Federal Environmental Protection Agency Act.

[235] I.N.E.C. v. Musa (2003).

[236] Socio Economic Rights and Accountability Projects (SERAP) v FRN and Universal Basic Education Commission (UBEC)

[237] (1981) 2 NCLR 350.

[238] No. ECW/CCJ/APP/0808.

8.5 SANITATION GOVERNANCE INSTRUMENTS

This section analyses whether the sanitation governance instruments in the case study address the lack of sanitation faced by actors, given the drivers already highlighted (see 8.3). It also proffers specific recommendations for redesign to improve the effectiveness of both the HRS instruments (see 8.5.1) and the non-HRS instruments (see 8.5.2) for sanitation governance, taking into account the drivers that affect the performance of the instruments within the specified context.

8.5.1 Human Rights Instruments

This sub-section discusses regulatory and economic instruments for implementing HRS.

Regulatory

Direct provision of access through public utilities

The 1999 Constitution, 4th Schedule, requires direct and universal provision and maintenance of public conveniences, sewage and refuse disposal by local governments and there are legal guidelines on the ratio of users to public sanitation facilities in the environmental sanitation laws and regulations.[239] There are also environmental health officers from the ministry of environment who monitor sanitation standards in public places.[240]. International organisations have also been directly involved in improving access to sanitation. For instance, the UN-Habitat Urban Renewal Scheme pursues sustainable upgrading of informal settlements and improving the liveability of the environment by ensuring access to sanitation services, *inter alia* in selected areas.

However, most local governments lack the financial and technical capacity to deliver this responsibility effectively (Asaju, 2010; Bello-Imam, 2007) and follow-up on sanitation projects in order to avoid retrogression For instance, one interviewee noted that although the UN-Habitat Urban Renewal Scheme has been effective to some extent, "there is more to be done in terms of ensuring sustainability. After the urban renewal project is completed, the responsibility returns to the government and they may be unable to continue providing services in such areas."[241] The environmental health officers who are supposed to conduct

---

[239] For instance, the Sanitation and Pollution Management Law of Edo State stipulates a ratio of 1:50 persons for mobile toilets at public functions and schools (section 20), and toilets at the ratio of 1:30 persons under other circumstances (section 26).

[240] Interviewees 2 and 3.

[241] Interviewee 11.

sanitary inspection are too few to cover the entire country.[242] Another constraint is that the involvement of the government in the delivery of sanitation services through public utilities affects the strict separation of politics and the management of services both unrealistic and less efficient (Schwartz and Schouten 2007). This makes the autonomy of service providers all the more important for ensuring that service providers operate without undue external influence, are in control of the resources for sanitation provision, and are accountable for their operations. Although the NWRP recognises the need for autonomy, it does not specify how this can be achieved. I also observed that the existing key performance indicators, which could form an objective quantifiable basis for the independent monitoring of service providers, have been developed by the regulators independently and sometimes in collaboration with international donors, but there is little or no input from the population in the policy process. This weakens the autonomy of the service providers and their accountability to the rightsholders. The direct provision of sanitation services by the government therefore requires political reforms to either strengthen the local governments with the resources that they need to implement their constitutional duties for sanitation services in their local communities or an amendment of the constitution to divest local governments of the role, in favour of the states. The local governments are however better positioned to deliver local sanitation services, based on the principle of subsidiarity and this makes the first option for political reforms more attractive.

Although the process of constitutional amendment is also very cumbersome and may cause unnecessary delays in the progressive implementation of the HRS, the local governments may be supported by the higher tiers of government within the existing legal framework to develop strong local monitoring mechanisms for sanitation to address the shortfall in environmental health officers. It is possible to design a local volunteer corps of environmental health inspectors to also improve the manpower resources for sanitation monitoring. This can be tailored after the National Environmental Standards and Regulations Enforcement Agency (NESREA) Green Corps initiative, through which private individuals are trained to support the enforcement of environmental laws in their communities, on a voluntary basis.

---

[242]   Interviewees 2, 3, 10 and 13.

Human Rights Monitoring Mechanisms

The National Human Rights Commission and NGOs like WaterAid monitor access to sanitation among vulnerable groups like prisoners and residents of informal settlements, respectively (WaterAid, 2015; WaterLex, 2015).[243] This provides additional information on access to sanitation among specific vulnerable groups, beyond the usual focus on access in urban and rural areas, as found in most official reports (see 8.2.3). The HR special procedures and monitoring mechanisms also attract international attention to otherwise local sanitation service problems and thereby pressure politicians, policymakers and technocrats into action, and provides an additional mechanism for HR enforcement after national remedies have been exhausted or where they are inadequate. For instance, in the case of *The Social and Economic Rights Action Center [SERAC] and the Center for Economic and Social Rights [CESCR] v. Nigeria*,[244] the Social and Economic Rights Action Centre (SERAC) instituted a case to the African Commission on Human and Peoples' Rights alleging that Nigeria had violated the rights of the Ogoni people through its involvement in Shell Petroleum Development Corporation's oil exploration activities in their community. The Commission found, on the merits that Nigeria was in violation of the African Charter on Human and Peoples' Rights of 1981, articles 2, 4, 14, 16, 18(1), 21 and 24. Hence, the international HR enforcement mechanisms may strengthen a State's accountability towards rightsholders, despite weak national HR enforcement mechanisms.

However, the current monitoring processes mainly focus on physical access to improved sanitation facilities, but do not directly monitor environmental impacts of the sanitation services/facilities (sustainability) or other HRS principles (see Chapter 5), and the most recent official reports on status to sanitation and compliance with the HRS obligations in Nigeria have either been authored by international NGOs like WaterAid, or by national agencies in response to calls by other international organisations or the epistemic community, rather than as an initiative from the government to determine the baseline of access to sanitation in Nigeria. The reports are also often limited to the same three components of sanitation: toilets, hand washing, and solid waste management, leaving out the other components of sanitation that are listed in paragraph 1.3.2 of the NESP (see 8.2.4). Nonetheless, additional components like medical waste management (for instance, Longe & Williams, 2006; Abah & Ohimain, 2011; Ogbonna, Chindah & Ubani, 2012; Nwachukwu,

---

[243] Interviewee 16.
[244] (2001) AHRLR 60 (ACHPR 2001).

Orji & Ugbogu, 2013), and school sanitation (Babalobi, 2013) are analysed in the scholarly literature with empirical evidence of the poor level of access to other sanitation components. Further, judicial decisions by themselves do not translate into positive actions from the State and additionally require political will and resources in order to ensure tangible benefit for the rightsholders (Ladan, 2009).

It is therefore important for investments in the sanitation sector to focus not only on behavioural changes and sanitation infrastructure, but also on strengthening local HRS monitoring mechanisms. The HRS monitoring may be linked to the existing sanitation monitoring mechanisms like the NDHS to ensure cross-sectoral learning and improvements in the quality of data on sanitation services. The monitoring process could also cover the full range of sanitation components in the NESP and inform national development strategies and plans.

Economic

Development Finance

Development finance (see 5.4.2) is an important source of funding for sanitation services; it augments public finance and tariffs from users. The WASH sector in Nigeria, as an emerging economy, attracts low development finance per capita at around USD 1.1, mostly in loans rather than grants (UN-Water, 2014). The available finance is often committed to rural sanitation services and scaling up strategy to meet universal access to sanitation (Federal Ministry of Water Resources [FMWR], 2016), while urban and peri-urban sanitation suffer from fragmented leadership and poor funding *inter alia* (see 8.2.4 and 8.3).[245] Nonetheless, heavy dependence on external funding of national development priorities like sanitation could result in the informal weakening of State structures (Burns et al. 2017). Corruption and political considerations may also affect the efficient utilisation of the available funds (Akpabio 2012).[246]

It is therefore important to explore additional options for funding sanitation services from internal sources, including non-monetary contributions from users. Strong accountability mechanisms and transparent processes can be institutionalised to ensure that available funds are efficiently committed to local sanitation priorities. Quasi-judicial mechanisms (like

---

[245] Interviewees 2, 3 and 13.
[246] Interviewees 10, 17 and 19.

administrative tribunals and petitions) may also be strengthened for the speedy sanctioning of petty corruption cases and to improve access to justice in the sanitation sector.

Subsidy

The NWRP provides for water charges to ensure economic efficiency, while guaranteeing affordable access for the poorest through a cross-subsidy. Following the shift away from subsidy approaches for toilets to non-subsidy approaches like community-led total sanitation (see 8.5.2), it is only waste collection services in some urban areas (like Lagos, Benin, and Port Harcourt) that are still subsidised in order to encourage speedy uptake of the services.[247] The subsidy is paid by the regulators to their respective private sector partners, under various public-private partnership (PPP) contract models (see 8.2.4).[248] The PPP can potentially improve the quality and quantity of public sanitation goods and services, enhance the performance orientation and responsiveness of the service providers to the needs of the public, improve creativity and innovation as a result of the combined resources of the different partners; engender greater public participation in the monitoring of the public services; and assist with building social capital and strengthening the capacity of civic groups to take collective action to solve their group problems, which prioritises social development (see generally Fiszbein, 2000; Ayee & Cook, 2003; Hawranek, 2000; Jones, 2000).

However, the beneficial effects of PPPs accrue where the institutional capacity of the State is already highly developed, the private sector well established, and public services were close to achieving full coverage in many cases (Plummer, 2002). Otherwise, there is little incentive for the private sector to make the necessary huge investments in service provision in informal settlements or other vulnerable areas and PPPs inadvertently create inequities in access to sanitation (Devas et al., 2001). In the absence of a strong regulatory framework, improvements in service performance standards are unlikely (Batley, 1996). PPP are also vulnerable to political changes (Fiszbein, 2000), and may collapse where the public sector does not meet its obligations to the private service providers. While subsidised PPPs have improved coverage (in terms of solid waste collection services) in the urban areas, they have failed to reduce the inequality in access to sanitation because of the exclusion of informal settlements.[249] The subsidised services are therefore a form of covert discrimination that

---

[247]   Interviewees 2, 7 and 8.
[248]   Interviewees 2, 7, 8, 9, 10, 11, 12, 13, 17 and 19.
[249]   Interviewees 7 and 8.

exacerbates the inequities in access to sanitation within the population.[250] In some cases, even where the regulator had provided additional funding to bridge the shortfall in tariffs and thereby motivate private service providers to extend coverage in informal settlements, the informal settlements were still being excluded.[251] Evidence from the case study, supported by the literature (Kariuki, 2014), also suggests that subsidies may not benefit the poorest due to insistence on other formal requirements like a registered address, or the lack of flexible payment mechanisms (see 8.2.4).[252]

In order to address the limitations of the current system of subsidies, it is preferable to institutionalise public ownership and control of sanitation goods and services with strong downward accountability mechanisms in favour of the population. The PPPs require strong regulatory mechanisms to protect the interests of vulnerable rightsholders. Where non-State actors are engaged as service providers, their operations ought to be regulated and controlled by the State to ensure full respect, protection, and fulfilment of the HRS, and the PPP contracts negotiations process ought to be transparent and involve the local population, including residents of informal settlements, who will depend on the services.

Suasive

Human rights advocacy and training materials

Human rights advocacy elevates improved sanitation from a local policy issue to become a universal priority, with increased opportunities for learning and knowledge transfer on different scales.[253] Advocacy at different levels of governance also places duty bearers under additional pressure to fulfil their HRS obligations and the manuals, reports and other publications of the Special Rapporteur (both former and current) are good examples of how suasive instruments improve the body of knowledge about the meaning, scope and how to implement the HRS.[254] The UN-Water Global Analysis and Assessment of Sanitation and Drinking-Water, for instance, promotes the national WASH sector review process and the results are useful for formulating commitments both nationally (such as changes in policy) and internationally (UN-Water, 2014). Suasive instruments focusing on vulnerable groups, like the Women Leaders for WASH which encourages the participation of women in sanitation governance and the WASH Pledge commitment by the affected corporations to

---

[250]   Interviewees 10, 11, 12, 134, 14, 16, 17 and 19.
[251]   Interviewees 10 and 15.
[252]   Interviewee 10.
[253]   Interviewees 24 and 43.
[254]   Interviewee 36.

implement safe access to WASH for their employees, highlights vulnerabilities and promotes inclusive sanitation governance.

However, some suasive instruments are written in a highly aspirational language which does not distinguish between the legally binding obligations and good practices. For instance, the manual on the rights to water and sanitation published by the former Special Rapporteur offers remedies for non-implementation of the rights, describes technology for implementing the rights, considers the challenges facing urban and rural areas separately and offers policy recommendations on how to address each context, but, in the analysis there is no distinction between obligations that are already legally binding and those that are evidence of good practice but non-binding. This creates some ambiguity for judicial enforcement. The involvement of a variety of actors and loose application of the HRS language may appear to improve the broad the implementation of the HRS norm or lead to a conflation of values and the hi-jack of the human rights agenda by technocrats in the long run, to the disadvantage of the poor, vulnerable and marginalised groups (Barlow 2009; Mehta, Allouche, Nicol & Walnycki, 2014).

In addition to showcasing examples of good practices for implementing the HRS, it is therefore essential for suasive instruments to clearly highlight the binding nature of HRS obligations at all levels of governance.

8.5.2 Non-human Rights Instruments

I discuss below the regulatory, economic, management, and suasive non-HRS instruments for sanitation governance.

Regulatory

Criminalisation of pollution

Criminal law is used to regulate the environment and has sanctioned offences against public health, as follows: "[A]ny person who corrupts or fouls the water of any spring, stream, well, tank, reservoir, or place, so as to render it less fit for the purpose for which it is ordinarily used, is guilty of a misdemeanour, and is liable to imprisonment for six months."[255] Offenders who dump sewage in unauthorized places may also be prosecuted by the relevant agencies.[256] Further, the enforcement agencies are equipped with the power of entry to

---

[255] Criminal Code Act 1916, Section 245.
[256] Interviewees 7, 8 and 9.

inspect premises, and the laws sanction open defecation and urination, water and land pollution with waste as well as non-payment of tariffs *inter alia* with punishments like closure of premises, fines and imprisonment terms.[257] With the recent reforms in sanitation laws and policies, criminal sanctions have been revised to better reflect` the modern realities and improve deterrence. Some states in Nigeria (like Abia, Adamawa, Anambra, Bayelsa, Delta, Edo, Kogi, Lagos, Ondo and Sokoto) have also established special courts to expedite the prosecution of sanitation offences (Ijaiya, 2013). In Lagos, in 2015 alone, 4449 suspects were arrested on suspicion of sanitation related offences (like contravening the restriction of movement during the designated hours for the observance of environmental sanitation day, discussed further under suasive instruments below); 3,178 were convicted and 376 discharged with 839 suspects still awaiting trial by the beginning of 2016 (Oyebade, 2016). Sanctions may also be imposed by a community, based on the collective ethos of the local population, and without recourse to the formal justice system.

However, criminal sanctions are sometimes imposed on small informal sanitation service providers who may not be able to afford the cost and formality of registration, but nonetheless augment the sparse formal sanitation services in many formal and informal settlements.[258] The regulatory authorities also often lack the resources to monitor and enforce compliance and prosecute the cases in a timely manner.

It is therefore necessary for criminal sanctions to be imposed against acts and omissions which threaten public health and the environment, but informal service providers may be supported with small loans or registration waivers from the state governments to enable them to obtain the necessary licenses for their operations. They could also be trained on best practises by the regulators, so that their operations do not inadvertently constitute an offence against public health. Minor sanitation offences may also be punished through quasi-judicial means, such as fines imposed by the regulators, and judicial mechanisms can be resorted to only in contentious cases or for serious offences, to reduce the pressure on the already stretched courts.

---

[257] Interviewees 4, 7 and 8. See for instance, National Environmental Standards and Regulations Enforcement Agency (Establishment) Act, section 30; Waste Management Agency Bill (Rivers State) 2013, section 53(5)(6)(7)(8)(9) and section 54; Sanitation and Pollution Management Law (Edo State) sections 6, 23, 31.

[258] Interviewees 7, 8, 13, 17 and 19.

Environmental protection laws, including licencing exemptions

Nigeria has environmental policies and regulations that are relevant for the HRS. The legal framework for sanitation under the British colonial administration was focused on land reclamation, and waste management for the control of vectors like mosquitoes and the reduction of public health risks, particularly for the Europeans, in order to safeguard British socio-economic interests in the local communities (His Majesty's Office 1906; 1909; 1910a; 1910b). Post-independence, the sanitation laws and policy reforms in Nigeria have been largely reactive, triggered by either unforeseen local circumstances, treaty obligations, or the international development agenda (Agbase, 1998; Akpabio, 2012). For instance, the discovery of 4,000 tons of toxic waste dumped by an Italian ship on the port of Koko, a city in the south of Nigeria, in 1987, led to the enactment of the Harmful (Toxic) Waste Criminal Provision Decree 42 of 1988. Subsequently, in 1989, the National Policy on the Environment was adopted with the policy goal of sustainable development, soon after sustainable development had emerged as a guiding principle for environmental protection and human development at the international level.

Earlier laws relevant for sanitation (like the Waterworks Act of 1915; Public Health Act of 1917; Oil in Navigable Waters Act of 1968 and Petroleum Act of 1969) focused on preventing the contamination of water resources and the environment by human activities, while subsequent laws (like the Land Use Act of 1978 and Water Resources Act of 2004) established water rights and licensing exemptions to ensure effective water resources management for domestic and socio-economic uses. Licensing exemptions where they cover personal and domestic sanitation and hygiene uses improve access for people who rely on wet sanitation systems without any additional financial costs. The NWRP recognises that no one can own water, but only a right for environmental and basic human needs, and that "…every citizen shall have a right to access water for sustenance while guaranteeing access for controlled socio-economic use."

The more recent laws (like the National Environmental Standards and Regulation Enforcement Agency 2007) adopt an integrated approach to prevent contamination of the environment by solid and liquid waste; prohibit open defecation or urinating in drains;[259] and require hygienically maintained toilets in commercial buildings,[260] as a way of improving access to sanitation services in public places. The National Environmental (Sanitation and

---

[259] Environmental Management (Miscellaneous Provisions) Law (Ogun State) 2005 section 34(1)(f).
[260] Environmental Laws Lagos State 2017, Section 69.

Wastes Control) Regulations[261] regulates effluent discharge, and it contains guidelines for recommended approved number of toilets and septic tank construction.

However, some parts of Nigeria are arid especially in the North and therefore face more challenges in accessing groundwater in sufficient quantity and good quality for their basic needs. Even in the water abundant parts of the country, indiscriminate abstraction may deplete finite water resources and damage environmental sustainability in the long run. There are also legal constraints to consider where the water rights and licensing exemptions only benefit landowners and exclude people living in informal settlements, or rightsholders are charged unaffordable rates for water abstraction without any support to ensure alternative safe public supply options. In Lagos, people are now required to obtain a permit before they can sink a borehole, hydraulic and other similar structures for water supply. Similarly, the Delta state government has imposed a borehole levy on private, commercial and industrial boreholes, respectively. The charges are exclusive of the high cost of manual and heavy duty drilling of boreholes (see 8.2.3). The neo-liberal policy reform is vehemently opposed by the population and even the current Special Rapporteur (with reference to Lagos) as a HR violation in the absence of public water and sanitation services for rightsholders. This is significant for many Nigerians who have to rely on water for their sanitation and hygiene uses. Further, the law enforcement is mostly confined to urban areas due to limited resources;[262] the survey respondents were mostly unaware of the guidelines for recommended approved number of toilets or the construction of septic tanks, for instance.

To protect access to water for basic sanitation in arid or semi-arid regions and during emergencies that affect the water supply infrastructure, potable water could be used only for satisfying basic human needs in such circumstances, except where there is express approval of regulatory authorities for non-related uses, based on the doctrine of necessity. Water may also be regulated as a common pool resource to check over-exploitation and degradation (Colin-Castilo & Woodward, 2015; Fisher et. al., 2010; Ostrom, 2008), in addition to operationalizing environmental principles like precaution and polluter pays (see 7.2.2). The economic good principle supports the valuation of water services and presumably improves financial sustainability and conservative use of water but it needs to be operationalized through pro-poor instruments (like cross-subsidies) to ensure inclusiveness (Jiménez & Pérez-Foguet, 2010; Obani & Gupta, 2016). The borehole levies can therefore be critically

---

[261]   2009 S.I. No. 29.
[262]   Interviewees 7, 8 and 11.

evaluated to prevent the violation of the HRS. This would benefit from a national threshold for affordability (taking into consideration the full the direct and indirect costs of sanitation and the economic realities), as proposed earlier (see 8.2.4). The regulatory capacity may also be expanded through engaging the services of volunteers, in addition to advocacy and awareness campaigns about the provisions of sanitation laws (see 8.5.1).

Economic

Budgeting, emergency programme fund, regular resources & thematic funding in emergencies

The budgeting process, of international humanitarian agencies especially, is subjected to financial standards which promote accountability, transparency, and better management of corporate financial risks in the humanitarian sector (UNICEF, 2017). This creates an opportunity for learning and improved monitoring of sanitation sector spending even outside humanitarian contexts. The emergency programme fund, regular resources, and thematic funding (see 6.4.1) also offer critical financial resources to augment national funding, and provide basic sanitation and other necessities for the recovery of the affected population.

However, international humanitarian actors may prioritise spending on hygiene promotion through hygiene studies and advertising over the direct provision of sanitation services, whereas the evidence suggests that hygiene awareness is already higher than anticipated among local people and the direct provision of essential hygiene items like soap and water and the use of interactive methods would better enhance learning and positive behaviour change (Fernando, Gunapala & Jayantha, 2009).[263] Beyond the provision of basic sanitation services, sanitation is often not addressed as a first order priority in humanitarian interventions (see generally Breau & Samuel, 2016; Klasing et al., 2011). Due to the perceived temporary nature of emergencies and humanitarian crisis, sanitation interventions are also built with minimal considerations for durability.[264] This is counterproductive in humanitarian crisis situations that become protracted, as often happens with severe natural hazards and conflicts.

To improve their impact, medium and long term economic instruments ought to involve representatives of the target community in the planning process, and reflect local priorities rather than a set of predefined sanitation and hygiene goods and services formulated by the humanitarian actors. The HRS principles (see 5.3) also offer a minimum standard that may be

---

[263] Interviewees 9, 21 and 22.
[264] Interviewee 21.

communicated to humanitarian actors and progressively implemented with the maximum available resources (see 5.2.3).

Private Investment

Private investment by individuals and households is the main source of funding for on-site and decentralised sanitation solutions and significantly reduces the funding gap for sanitation infrastructure. It is flexible and can include non-monetary investments like labour, and this empowers the poor. The opportunity for private investments also generates local business interest and enhances service expansion for users who can afford service tariffs.[265] The quality of the infrastructure and management differs widely based on the amount of financial and technical resources invested and the extent of regulation (see 3.5).

However, the recent neo-liberal reforms in the sanitation sector increases the focus on cost recovery,[266] and the private sector being profit-oriented cannot be mandated with the responsibility of providing public goods and fulfilling HR (like sanitation) without strong regulation and downward accountability mechanisms to empower the rightsholders from unaffordable tariffs and poor quality services.[267] Consequently, anti-privatisation campaigns have been hinged on HR (Bakker, 2007; Barlow, 2009; Corporate Accountability International [CAI], 2015; Uwaegbulam, 2016). Corruption and political instability may also hinder the regulation of private service providers (Ayee & Cook, 2003).

The foregoing reiterates the need for enhancing the regulatory and monitoring capacity of the relevant agencies for improved supervision of private investments to comply with the HRS. Similar recommendations under the discussion of subsidies (see Section 8.5.1) also apply to private investments.

Management

Community-Led Total Sanitation (CLTS)

Community-Led Total Sanitation (CLTS) was introduced in Nigeria in 2005 through triggering random communities, but triggering a complete LGA was considered to be a more effective approach and by 2008 the scope of CLTS in Nigeria was significantly broader in the run-up to the International Year of Sanitation (Federal Ministry of Water Resources et al.

---

[265] Interviewees 1, 6, 7, 8, 10, 13, 14 and 15.
[266] Nigeria Water Sector Road Map, 2011; interviewees 7, 8, 10 and 14.
[267] Interviewees 1 and 11.

2016).[268] There is now a wide spread subscription from government at all levels, and all donor assisted rural sanitation projects adopt CLTS as their main approach (Federal Ministry of Water Resources 2016).[269] CLTS has been initiated throughout the 36 states and the FCT Abuja, and by July, 2014, there were 19,467 communities triggered, including 9,728 certified open defecation free (ODF) and 3,276 that were close to being certified (Federal Ministry of Water Resources et al. 2016). Through focusing on triggering effective local participation to eradicate OD in communities, CLTS may promote inclusion.[270] As observed by one of the respondents, CLTS relies on the active participation of local communities to "…assess their sanitation situation, analyse it, and identify the problems they are facing, take action to address the problems and set strategies for reaching their goals … using locally sourced materials."[271] The relatively low cost of CLTS is considered to be more financially sustainable than other instruments that require subsidies and "end as soon as the subsidies dry up."[272]

However, while the selection of communities for CLTS was initially based on the level of vulnerability to water and sanitation-related diseases, there is no consistency in the selection criteria for new communities to participate in CLTS across Nigeria (United Nations Children's Fund [UNICEF], 2012). Some of the considerations are based on political factors, or the existence of a WASH department to locally facilitate community entry, a dedicated bank account for WASH, and evidence of local commitment to the CLTS approach (UNICEF, 2012). Some of these criteria (like a dedicated bank account for WASH) disenfranchise poor communities. CLTS may also violate the HRS without standards and resources to ensure quality infrastructure and mechanisms to protect vulnerable people against potential negative impacts of CLTS (like discrimination) (Galvin, 2015; Obani & Gupta, 2016).

It is therefore important for the NTGS to review the selection criteria for local communities participating in CLTS programmes, and publish the same in local languages for all interested communities to be aware. The criteria could be based on a needs assessment and devoid of discriminatory provisions. The NTGS may also commission a detailed meta-analysis of the

---

[268] Interviewees 2, 3 and 7.
[269] Interviewee 2.
[270] Ibid.
[271] Ibid.
[272] Ibid.

HR-impacts of CLTS across all communities that have been triggered, and develop strategies to address any unintended negative consequences.

Environmental health clubs in schools

Environmental health clubs in schools promote capacity building for children who may otherwise be unable to influence the sanitation governance process to address their special needs.[273] However, there are concerns that children could be easily exploited or overworked by poorly trained facilitators who do not fully understand the objectives of the programme. Further, the clubs may imbue the children with non-HR values that compete with the HRS except any rules incoherence is addressed to ensure compatibility with the HRS. It is therefore important for the facilitators to be properly trained and the School WASH projects and other child-led sanitation instruments can be carefully evaluated to ensure compliance with HRS norms and child rights standards.

Suasive

Monthly environmental sanitation exercise

In various parts of Nigeria, the federal and state government designate the first or last Saturday of every month as the monthly Environmental Sanitation Day (ESD). The practise was initially designed to compel citizens to remain at home during the designated hours and actively clean their homes, drains, and the surrounding environment. This could improve opportunities for local sanitation mapping initiatives, collective action for sanitation at the community level, and free access for sanitation service trucks; such local mapping initiatives have been an important source of WASH data and very useful for identifying vulnerable groups (de Albuquerque, 2014; dos Santos & Gupta, 2017).

Nonetheless, the household survey uncovered a general apathy amongst a vast majority of the population who see the practise as an infringement on their constitutionally guaranteed fundamental right of free movement. In 2016, the restriction was successfully challenged before the Court of Appeal, sitting in Lagos in the case of *Faith Okafor v. Attorney General of Lagos State*. The courts held that the restriction of movement was a violation of sections 34, 35 and 41 of the 1999 Constitution, which respectively preserve the rights to respect for the dignity of a person, personal liberty and freedom of movement. The failure of regulatory

---

[273] Interviewee 7.

authorities to conduct house inspections, due to limited resources, has also been a disincentive for the public to comply with the ESD.[274]

Hence, there needs to be a public reorientation on the importance of good sanitation and hygiene practices both during and beyond designated ESD. The enforcement capacity of regulatory agencies also needs to be strengthened through some of the measures already suggested (like establishing a well-trained volunteer sanitation enforcement corps), in addition to designing incentives for self-regulation within local communities.

Advocacy, Training, Assessment and Reporting Mechanisms

The NWSP recognises effective hygiene education and promotion as necessary for positive behavioural change and full sanitation coverage. Electronic and print media are useful for disseminating information on good sanitation and hygiene practices, publicising sanitation news and facts, and engaging in open discussion with the public on sanitation issues, particularly where the message is transmitted in the local language.[275] Advocacy may compel the government to prioritise sanitation investments due to internal pressure from the population.[276] Additionally, Nigeria has many domestic assessment and reporting mechanisms that cover access to sanitation services, *inter alia*, including the Water and Sanitation Monitoring Programme (WSMP); Core Welfare Indicator Survey (CWIS), and the National Demographic Health Survey (NDHS). However, there are inherent limitations in the various mechanisms which has resulted in the paucity of reliable data on the status of the HRS (see 8.2) The WSMP lacks a lead agency and has been plagued by tensions between the priorities of government, donors, and personnel, making it a weak and ineffective monitoring instrument (Odhiambo, 2010). The CWIS ought to be conducted by the government once every decade, as part of the national census, but the national population census figures have been historically controversial and the last census national population census was conducted in 2006. The most recent assessment is the NDHS already covered in Section 8.2. In addition, international mechanisms like the JMP also provide useful insight into sanitation access, subject to the limitations highlighted in Section 8.2 (see also 7.3.4) However, this does not fulfil the requirements of the HRS where the data is not disaggregated to show the situation of vulnerable and marginalised people, or where the data only measures one or a few HRS principles (like access) but other principles are being violated (like affordability),

---

[274] Interviewee 14.
[275] Interviewee 1, 2, 3, 4, 6, 7, 8, 12 and 15.
[276] Interviewees 1, 2, 6, 10, 11, 12, 13, 14, 16, 17, 18 and 19.

despite increased coverage. However, the new SDGs sanitation target shows the possibility of adapting non-HR mechanisms to integrate HRS principles; such integration may be pursued by all stakeholders in the sanitation sector towards improving knowledge transfer between the HRS and other disciplines and governance frameworks relevant for sanitation.

8.6   SANITATION GOVERNANCE PRINCIPLES, DRIVERS, INCLUSIVE DEVELOPMENT AND LEGAL PLURALISM

This section analyses whether the principles and instruments address the drivers of poor sanitation services (see 8.6.1), and their ability to achieve ID (see 8.6.2), and incoherence in the sanitation governance framework, using legal pluralism theory (see 8.6.3).

8.6.1 Sanitation Governance Principles, Instruments and Drivers

The sanitation governance framework contains a mix of twenty-seven HRS and non-HRS principles, including: (i) acceptability; (ii) access to information; (iii) accessibility; (iv) accountability; (v) affordability; (vi) autonomy of the service providers; (vii) availability; (viii) capacity building; (ix) cost sharing; (x) demand responsiveness; (xi) economic good; (xii) environmental offsetting; (xiii) equitable access and poverty reduction; (xiv) extra-territorial obligations; (xv) gender responsiveness; (xvi) integration; (xvii) participation; (xviii) polluter pays; (xix) precaution; (xx) prevention; (xxi) private sector participation; (xxii) right to a healthy environment; (xxiii) safety; (xxiv) social good; (xxv) subsidiarity; (xxvi) sustainability; and (xxvii) water rights for sustenance. The relevance of these principles are already covered in Chapter 5 (see 5.6.1) and Chapter 7 (see 7.4.1), and will not be repeated here. The case study further highlights an additional five principles for sanitation governance, namely: (a) autonomy of service providers; (b) policy making and regulatory role of the government; (c) respect for constitutional law; (d) demand responsiveness; and (e) no ownership, only a right to use water. These five principles have been covered in the discussion of key stakeholder roles (see 8.2.4) and governance instruments (see 8.5). This section therefore focuses on the extent to which the principles are effective in tackling the drivers of poor sanitation by briefly analysing the extent to which they are operationalized in practise (elaborating on Table 8.1).

The foregoing principles focus mainly on water, and address sanitation governance mainly in connection with water quality and quantity. Hence, the principles are poorly developed with reference to sanitation governance specifically. While water is recognised under the NWRP as a second order priority (following after minimum stream flow requirement as a first order priority), sanitation is not clearly assigned any priority of use. The prioritisation of water alone falls short of addressing the sanitation needs of users who rely on dry systems; and the case study did not show any significant evidence of prioritisation of sanitation in the allocation of maximum available resources as required under HR law (see 5.2.2 and 8.4.1), nor a strong HR-based approach in the delivery of services. Rather, the central goal of sanitation governance appears to be to improve financial flows to the sector with a goal to ensuring that households can in the long term, cover the full cost of their sanitation services (FMWR et al., 2016). This is essential for financial sustainability but requires safeguards to ensure that the HRS is not violated for the poor and vulnerable people who cannot make financial contributions for their basic needs.

Although the policy process mostly focuses on rural, semi-urban and urban areas, he government and stakeholders in the sector have instituted strategies like the Partnership for Expanded Water Supply, Sanitation and Hygiene (PEWASH), launched in 2016, designed to achieve universal sanitation coverage in rural areas, without an equivalent programme for other contexts with significant sanitation problems (like urban areas, humanitarian situations and informal settlements). Hence, the problem of lack of leadership for sanitation governance in urban areas (especially highlighted by the experts and stakeholders I interviewed, see 8.3) still persists. Further, it is the neo-liberal principles (like economic good, cost recovery and demand responsiveness) which have been the bases for recent sanitation sector reforms at various levels of governance, while the HRS principles and participatory governance instruments like the sanitation ladder and participatory budgeting are yet to be operationalized across Nigeria (FMWR et al., 2016).

The foregoing shows that though there is some potential for addressing the drivers of poor sanitation governance through the existing sanitation governance framework, there is a further need to redesign and strengthen the policy implementation process. This is in addition to the recommendations already made on redesigning the sanitation governance instruments in Section 8.5 and addressing specific drivers (in the next paragraph). The next sub-section further evaluates the impact of the existing principles and instruments on ID (see 8.5.2).

Instruments for addressing the drivers

In theory, the HRS can be satisfied through right-based instruments that impose an obligation on the State as primary duty bearer to respect, protect and fulfil the sanitation needs of the population, especially the poor, vulnerable and marginalised. In a society with high income levels, this can easily be achieved for households with the capacity to pay for their sanitation services. Otherwise, the HRS offers instruments like progressive pricing, cross-subsidy, and regulatory safeguards to protect the rights of the poor, vulnerable and marginalised. However, where there are scarce resources for expanding the sanitation network and a myriad of environmental, economic and social drivers, imposing a duty to respect, protect and fulfil the HRS together with accountability mechanisms are essential but not enough. Rather, tackling the drivers of poor sanitation services require a broad range of both HRS instruments and complementary non-HR instruments (including technology) to ensure progressive realisation of universal access to sanitation services that satisfy the HRS principles.

Challenging topography, space constraints and system failures like erratic power supply require the substitution of sanitation infrastructure with more suitable alternative technologies for the local context, and incentives for users to adapt the alternatives. Space constraint may also require the construction off shared facilities, where this does not offend local customs. It is therefore also important for sanitation regulations to accommodate shared facilities that meet sanitary standards, to improve the chances of external funding. It is further important to approach sanitation service delivery as a complete value chain which requires utilities as well as small service providers who may be better equipped with small equipment that can access small lanes than the big trucks used by utilities.

Climate change-related and natural hazards, **weather variability**, and **high temperatures and turbidity** require efficient disaster risk management to strengthen systems' resilience and prioritising sanitation services in emergency response planning. A reduction in greenhouse gas emissions and climate finance can also generally improve sanitation systems and climate-resilient communities (United Nations Development Program [UNDP], 2016). Additionally, climate change requires a strong regulatory regime for a legally binding obligation to reduce greenhouse gas emissions globally, in line with the precautionary principle and support for poor States to employ adaptation measures for climate-change induced hazards based on common but differentiated responsibility. The regulatory regime may integrate the latest scientific evidence on safe levels of CO_2 emissions, for instance and relevant indicators like the level of atmospheric CO_2 concentration, top-of-atmosphere radiative forcing (Rockström et al., 2009; Steffen et al., 2015) and wider developmental concerns in the regulation of economic activities contributing to climate change (Obani & Gupta 2016c; Stern 2011).

Discounting the future, especially among poor people and preference distortion affecting willingness to pay require educating the poor on the health, social and economic costs of OD and other unsanitary practices and protecting their property rights and investments in sanitation infrastructure. The legal protection of sanitation investments can be distinct from the question of informality because sanitation is a human right that imposes legal obligations to provide services to everyone in need (see 4.2.2 and 5.2.3). Sanitation is also a public good and the negative externalities from non-users cannot be confined easily; it is therefore essential to ensure universal access irrespective of the legal status of the users (see 3.3). The case study also highlighted an additional factor affecting WTP, proximity to waste management facilities. This requires regulators to ensure that waste management facilities are sited away from residential areas and well managed wherever they may be located, to avoid public nuisance. Waste managers may also be penalised for failure to meet the regulatory standards for the operation of their facilities.

Distance to the facility makes it necessary to provide sanitation facilities within a reasonable distance of the vicinity of users. What constitutes a reasonable distance is contextual and depends on both the physiology and circumstances of the users, as well as the environment where the facility is needed? For instance, children may be unable to use sanitation facilities that are far away from home unaccompanied at night, and an adult confined to a wheelchair may be unable to use a toilet right inside the room that lacks disability access. Hence, the suitability of the available facility is also important and this needs to be reflected in an indicator for measuring access.

Exclusion of minorities and other negative social practices that hamper access to sanitation services require all stakeholders to be educated on the public good nature of sanitation and the inefficiencies of excluding some members of the population from accessing services, because the negative externalities that result cannot be easily confined to the non-users alone. In addition to suasive instruments, addressing the sanitation laws can be expanded to protect vulnerable populations from discrimination, ensure equal access to sanitation resources and services and promote the effective participation of vulnerable people in the sanitation governance process.

Household poverty and unaffordable tariffs and connection fees require integrating sanitation service provision in poverty alleviation strategies, developing low-cost technology options (at the local level of governance) that advance the HRS principles and progressively deliver higher social, health and environmental functions (see 3.5.5), and building a strong sanitation marketing mechanism that is informed by a needs assessment for the vulnerable population and integrates non-monetary contributions, payments in instalment and cross-subsidies for the poor. Household poverty may also be addressed through micro-finance loans or other loans with minimal interest rates and flexible requirements for collateral. Although household poverty does not directly affect access for people living in humanitarian situations, the increasing number of humanitarian crisis situations and the limited resources for humanitarian assistance for sanitation raises questions of whether or not refugees and internally displaced persons can pay for their sanitation needs, and how (see Chapter 9).

Inefficient tariff collection system, including unaffordable tariffs and connection fees further require the depoliticising sanitation tariffs, understanding the economic capacity of users, developing a variety of HRS-compatible technology options to suit different wealth quintiles, and setting the right cost for services. Nonetheless, the proportion of the cost to be borne by the rightsholders needs to be affordable and flexible economic instruments like progressive pricing and cross-subsidy may be relevant. Otherwise, the current practise of imposing flat rates, irrespective of the quantity of waste generated, may impose additional cost on poor users even though they may not generate much waste.

Insecurity and conflict also make it expedient to strengthen systems' resilience and prioritising sanitation services in emergency response planning. **Poor social cohesion** further requires moving away from an emphasis on individual behavioural changes, prescription of a pre-defined sanitation technology, and the marketing of private sanitation goods, towards a social marketing model that addresses the structural causes of poor sanitation services, and markets technology options that are demanded by the community (based on a participatory mechanism like a functional sanitation ladder) and addresses communal concern. With improved social cohesion, there is likely to be a higher sense of commitment among the population to address the sanitation problem and maintain facilities. The success of management instruments like the CLTS (see 7.3. and 8.5.2) is evidence of the power of collective action for addressing poor sanitation services.

Insufficient funds require a baseline study to clarify how much is currently being spent, on what component of sanitation, and by whom? The different stakeholders in the sanitation sector, including government ministries, departments and agencies services can be encouraged to establish a clear heading for sanitation services within their budgets to improve the tracking of sanitation funds. The government can also be pressured into complying with its international funding commitments like the eTkwini (see 8.3) (for instance, through international and local advocacy efforts) and establish transparency and anti-corruption strategies to ensure the efficient utilisation of funds in the sector. **Poorly targeted funds** also require subsidiarity, participatory budgeting and a strong accountability mechanism through which the vulnerable population can express their sanitation needs and thereby influence the expenditure of sanitation funds for those who need it the most.

Low awareness makes it important to educate rightsholders on the design, functionality, use, operation and maintenance of different sanitation technology options; sources of information and technical advice about their sanitation services; and what to do if they have any problems with their sanitation services or providers, including judicial and non-judicial mechanisms for resolving disputes over services. Such education can be transmitted using different media and languages, including local languages and signs; displayed in public places like markets; and taught to children, to ensure that the message reaches a wider audience.

National poverty requires improved mechanisms for coordinated planning and financing of sanitation projects by different stakeholders at scale. The case study also supports the need for external partners to align their development finance and programmes with national priorities to ensure local ownership of the development agenda or programmes supported by the external partners. Consequently, it is important to integrate the aid effectiveness principles adopted in 2005 as part of the Paris Declaration on Aid Effectiveness and reaffirmed in the 2008 Accra Agenda for Action, which include: ownership, alignment, harmonization, result-based management and mutual accountability, while institutionalising safeguards against the eroding of national democratic processes by the donors (Bissio, 2013; McCourt, 2017). The HRS principles of affordability, equality and non-discrimination, transparency and access to information about public expenditure on sanitation, and extra-territorial obligations imposed on rich States to support the poorer States in realising the HRS can further mitigate national poverty and therefore need to be strengthened in the policy framework.

Non-acceptance based on culture require an understanding of cultural sanitation practices and how these interact with the HRS principles, investing in suasive instruments to improve good sanitation and hygiene practices as required, and imposing sanctions to deter negative practices. It further requires a negotiation of acceptable sanitation standards and technology, between the local community and the regulators, and enforcement of the negotiated standards without making the cost of sanitation unaffordable. Otherwise, poor sanitation standards may lead to informality, a vicious cycle of poor sanitation services despite the high technical standards specified in the legal framework, and discrimination against the poor.

Other social drivers require population planning measures and sexual and reproductive health awareness to manage **population growth and density**; urban planning and the upgrading of informal settlements and rural areas to address **mass migration and uncontrolled urbanisation**; mainstreaming of the HRS and the protection of sanitation infrastructure in the national security agenda; and complementary humanitarian principles which assure access to basic needs for vulnerable populations in humanitarian situations.

Pollution/water scarcity requires: (a) incorporating emission limits in sanitation governance rules, backed up with strong enforcement mechanisms to prevent pollution and ensure reparation where violations occur, based on the polluter pays principle; (b) replacing harmful components of the sanitation system with alternatives that are more environmentally sustainable; (c) promoting dry sanitation systems in the arid regions and wet systems in the more water abundant regions, and where wet sanitation systems are used, (d) adapting measures to protect against indiscriminate abstraction and the unsustainable of water or other natural resources for sanitation, in furtherance of the protection of ecosystems and water as a finite resource, heritage and common pool resource (CPR); and (e) water conservation through water efficiency standards for water-borne sanitation systems and retrofitting of existing water borne systems in arid integrated water resources management. To improve the uptake of conservation programmes, there needs to be incentives for the end-users and additional technical and financial support based on a needs assessment. It is also important for the sanitation ladder used in national strategies like PEWASH to be revised, by the regulators in conjunction with local communities and other stakeholders, and delinked from water. Rather, the stakeholders may jointly develop a variety of sanitation technology options that are suitable for the local environmental conditions and use locally available materials and skills. Further, the sustainability of sanitation systems may be enhanced through incorporating pollution prevention and Integrated Water Resource Management (IWRM) by integrating environmental functions in the design of sanitation policies and infrastructure, and enlisting the cooperation of communities in setting up sanitation service targets and the implementation of governance strategies through community-based (management/participatory) governance instruments.

Poor maintenance culture needs to be addressed by focusing on the root causes. For instance, where poor maintenance is as a result of lack of knowledge about how to operate and maintain the facility, the users need to be give the necessary training; if it is because of lack of resources like cleaning agents, low-cost alternative cleaning agents can be provided for users; and if it is due to general non-challans, then sanctions may be imposed for failure to maintain good sanitary standards and enforced against offenders.

Risk aversion was partly addressed in the case study through good practices like government-funded subsidies to private sector, in order to encourage private service providers to cover vulnerable populations. It also requires the State to retain its regulatory role and ensure compliance of service providers with HRS obligations through the mandatory extension of basic sanitation services to vulnerable populations, for instance.

Tenure insecurity requires guaranteed universal access to sanitation, as a public good, irrespective of legal title to property; and protecting property rights and investments in sanitation infrastructure from forced eviction or capture. In addition, communities in informal settlements can also explore on-site sanitation technology options to reduce the dependence on utilities.

8.6.2 Sanitation Governance Framework and Inclusive Development

In Section 8.6.1, I have made the case for effective operationalization of the sanitation governance principles contained in the legal and regulatory framework, towards the realisation of the HRS in Nigeria. Nonetheless, the implementation of these principles can lead to different outcomes for ID (i.e. see different results in different quadrants of Figure 8.2) and therefore requires further careful consideration. Figure 8.2 illustrates how the principle of cost recovery may either promote ID (Q4 in Figure 8.2) or exacerbate one or more forms of exclusion (Q1-Q3 in Figure 8.2); the outcome depends on whether or not there is a well-targeted cross-subsidy to ensure affordability for the poor and vulnerable people, and whether the sanitation services prevent environmental pollution. Figure 8.2 uses ODF as an indicator of the effectiveness of the instrument, showing that the separation of excreta from the immediate environment does not necessarily address social relational or ecological inclusion (for instance if the excreta is disposed of in a water source, without treatment or if poor people who cannot afford the cost of sanitation services are subjected to inhuman treatment in the process of executing CLTS).

Social and relational inclusion

Although the HRS is recognised by regulators, experts and other stakeholders in the sanitation sector in Nigeria, some of the existing sanitation governance instruments, including principles, may entrench overt discrimination against the poor and vulnerable populations like people living in informal settlements.[277] For instance, the exclusion of informal settlements from sanitation coverage (see 8.4.2) exacerbates social and relational exclusion, despite increasing coverage in the formal settlements that benefit from the services. Other regulatory instruments like the technical guidelines for constructing sanitation facilities would only contribute to social and relational inclusion where the maximum available resources are applied to support compliance by the vulnerable (that is, Q3 or Q4 in Figure 8.2 based on the environmental impact of the instrument). Generally, instruments promoting individual or communal responsibility for meeting the full cost of their sanitation needs, without support for the poor and vulnerable, fall within Q1 or Q2 in Figure 8.3, depending on the integration of environmental sustainability.

The increasing implementation of neo-liberal policies without cross-subsidies to cushion the effect on the poor and without effective public participation in the policy process has also entrenched structural inequities and weak downward accountability, in favour of the rightsholders. Nonetheless, exempting the private sector from participating in the sanitation sector or totally expunging economic principles would amount to throwing out the baby with the bath water because of the evidence from the case study that where properly regulated to ensure quality and affordability of services, the private sector can provide resources for the expansion of sanitation services.

The delivery of sanitation and hygiene awareness campaigns in local languages (see 8.5.2) strengthens participation and access to information. The disaggregation of data on access to sanitation could also highlight covert discrimination and the needs of vulnerable and marginalised groups, and is therefore preferable to monitoring the average increase in coverage, generally.[278]

Ecological inclusion

The right to abstract water without charge, for domestic purposes, may lead to environmental degradation and requires better regulation and enforcement of checks (see 8.5.2). In cases

---

[277]   Interviewees 8, 10, 14, 16, 17, 18 and 19.
[278]   Interviewee 3.

where economic instruments are used to regulate abstraction, it is important that these do not make services unaffordable. Further, given the ecological diversity of Nigeria, it may be necessary for the government to differentiate the regulatory standards (in terms of the quantity and quality) of water abstracted for socio-economic purposes. The NWSP requires that "[E]ach household in urban areas (population above 20,000) must own and have access to safe sanitary facility that uses suitable and affordable water conveyance systems (at least pour-flush toilet)" (NWSP, p. 7). However, such prescription does not take into account the geological differences and local environmental conditions in the arid parts of Nigeria with water scarcity and may hamper ecological inclusion (in Q1 or Q3 of Figure 8.2, depending on the further impact of the facility and governance process on social and relational inclusion). The non-HRS principles like common pool resource, cost sharing, economic good, integration and pollution prevention, in addition to economic instruments like pollution permits can potentially stimulate a green economy and environmental sustainability.[279] Nonetheless, this may not improve social and relational inclusion except where the revenue generated is used to cross subsidise the poor (see Q4 in Figure 8.2).

---

[279] Interviewee 1.

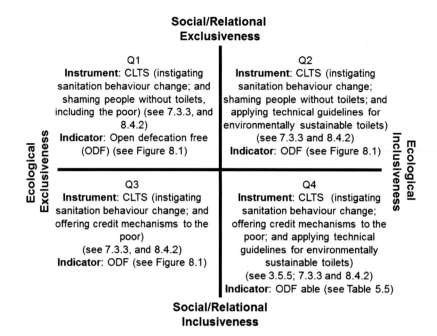

Figure 8.2 Assessing the principle of cost recovery for inclusive development

8.6.3 Legal Pluralism in the Nigerian Sanitation Governance Architecture

The objective of the NWSP is to ensure access to adequate, affordable and sustainable sanitation for all Nigerians, through the participation of various stakeholders, including the government, development partners and other international organisations, NGOs, the private sector, communities, households and individuals. There are also non-HRS principles contained in the NWSP and other relevant policies on sanitation governance (like the NEP and the NWRP). The plurality makes it important to analyse the interactions between the HRS principles and non-HRS principles, towards resolving any incoherence and fostering support for the progressive realisation of the HRS. This section shows that the introduction of the HRS into the policy framework for sanitation governance in Nigeria, *vis-a-vis* non-HR principles may result in accommodation, competition, indifference, or mutual support (see Table 8.4). Table 8.4 illustrates the four legal pluralism typologies, as they operate across the three tiers of government in Nigeria (federal level, and the states and local levels).

Competition

The case study shows that competition could result in any of the following three instances, where: (a) the policy process does not expressly prioritise the use of water to satisfy human sanitation needs, even though wet sanitation systems are at the top of the sanitation ladder, but drinking water is clearly prioritised as a second order water use; (b) neo-liberal policies like cost recovery and demand responsiveness are institutionalised, without cross-subsidy to protect the poor from being denied access because they cannot afford the cost of basic services; and (c) citizens are denied their rights to access water resources for their basic needs, due to high water abstraction charges or the criminalisation of service provision to vulnerable groups like residents of informal settlements. Competition also exists where the economic, social and cultural rights (forming the basis for the HRS in international law) are non-justiciable within the national system (see 8.4.2). Similarly, competition exists when various stakeholders, like the members of the NTGS, approach sanitation governance with different priorities that are primarily based on their respective core mandate areas.[280] This is reflected in the lack of a common definition of sanitation or uniform data on coverage, for instance (see 8.2.3 and 8.2.4). Excluding the informal settlements from accessing subsidised sanitation services and criminalising the informal service sector also competes with universal access.[281] At the sub-national levels (states and local), the exclusion of informal settlements from accessing sanitation services competes with the realisation of the HRS.

Indifference

The lack of formal recognition of the HRS in the Constitution (see 8.4.2) amounts to indifference. The HRS principles are also poorly developed and implemented in the policy process, even when compared to water. As a result, despite Nigeria's participation in various political processes and United Nations General Assembly and Human Rights Council resolutions that uphold the HRS, the HRS norm has not brought about a significant shift in the sanitation policy discourse. Rather, there is a continuity in practise wherein both HRS and non-HR instruments continue to operate in two parallel unconnected tracks. This is compounded by plurality because, as noted by one stakeholder: "sector policies and approaches are multiple and incoherent and are located with different government actors with no leadership and shared ownership of the problem."[282] Although many stakeholders have adopted a right-based language in discussing access to sanitation, their operations are largely

---

[280] Interviewees 1, 2, 4 and 9.
[281] Interviewees 17 and 19.
[282] Interviewee 16.

based on non-HRS principles like cost recovery. Indifference in sanitation governance is further marked by the limited application of the cost sharing principle to capital investment for sewerage, storm water, on-site sanitation systems in public places and rural sanitation, without covering vulnerable groups in urban areas; and situations where although the government policy is to ensure universal access to sanitation, the informal settlements are not covered by formal service providers in practice (a form of covert discrimination). Overall, there is lack of political will to progressively realise the HRS.[283] Most states do not expressly recognise the HRS in their sanitation laws, strategies and programmes, and are therefore indifferent to the HRS commitment in national sanitation policies.

Accommodation

The NWSP recognises participation as a guiding principle for the delivery of sanitation services; the principle is similarly recognised under the NWRP, and the NPES (see Table 8.1). The participation principle however applies equally to all stakeholders, including private businesses, based on the assumption that the sound business principles of the private sector will strengthen the sector. Accommodation exists where efforts are made to promote participation and access to information in the policy process; this offers opportunities for stakeholders to contribute their unique perspectives to the sanitation governance process, try to resolve any incoherence between the HRS and non-HR principles, and strengthening accountability (Obani & Gupta, 2014b; 2016a). Accommodation results where the regulators make an effort to promote cooperation in the management of transboundary waters and increase the mutual benefits of shared water resources, both within Nigeria and transboundary waters (NWRP), in order to promote access to water for domestic uses (including sanitation and hygiene).

At the sub-national levels, sanitation programmes which try to encourage the participation of the local population in the ownership and management of sanitation infrastructure thereby promote accommodation. It also encourages the participation of non-governmental organisations and other civil society groups because they contribute to improving the technical capacity and development skills of local communities and thereby strengthen the sanitation sector. Further, the NEP seeks to create public awareness on environmental matters and improve participation in the policy process. The Freedom of Information Act 2011 may support this objective by ensuring transparency in the sanitation governance, making public

---

[283] Interviewees 3, 10, 11, 13, 14 and 16.

records and information freely accessible, protecting public records and information to the extent consistent with the public interest, and protecting serving public officers from adverse consequences of disclosing certain kinds of official information without authorization. Nonetheless, the households surveyed were hardly aware of judicial and non-judicial mechanisms for obtaining information about their sanitation services, held by either public authorities or private entities, or seeking redress when they had complaints about the services. They mostly rely on the local media, particularly a radio programme called "Man around Town", hosted on ITV radio 92.3 FM, to air their grievances and contact the relevant authorities on their behalf. This is yielding positive results by empowering the local population to seek remedy against their waste managers, and the Commissioner of Environment in Edo state has established an open door policy for people with complaints about the services from their waste managers to contact him directly.

Mutual support

The national sanitation policies and regulatory framework includes the principles of pollution prevention, IWRM, integration of environmental management in the delivery of sanitation services, access to information, licensing and monitoring, and guidelines for safe sanitation facilities, *inter alia*, thereby supporting the HRS. The SDGs also supports the HRS (subject to the limitations discussed in Section 7.3.4) by further elevating the local sanitation problems on the international development agenda and providing a strong monitoring mechanism for tracking progressive realization of improvements in access to safely managed sanitation facilities (through the JMP). Similarly, CLTS supports the HRS through instigating communities to be ODF, provided that the vulnerable are afforded the additional resources that they need to realise ODF status (see 8.6.2).

Mutual support also results where international or regional organisations (like the African Commission) and courts (like the ECOWAS Court) promote the realisation of the socio-economic rights of Nigerians, which can further strengthen the enforcement of the HRS despite the limited justiciability of socio-economic rights within the Nigerian legal system (see 8.4.2). Further, the ICESCR prohibits denying people access to the Covenant rights due to their land tenure status and this protects informal settlements from discrimination.[284] At the sub-national levels, the regulation of open defecation and urination (ODR) supports the HR including the HRS and the right to a healthy environment. However, it is equally important

---

[284] CESCR, General Comment No. 20, 2009 (E/C.12/GC /20), paragraph 25

for the population to be provided with safer alternatives for their sanitation needs, such as hygienically maintained public sanitation facilities.

Table 8.4 Typology of relationships between the human right to sanitation and non-human rights principles for sanitation governance in Nigeria

Type of Relationship	Federal Level	States and Local Levels
Competition	The NWRP does not expressly recognise human sanitation and hygiene needs as a priority use of water, and it contains neo-liberal principles like cost recovery which can lead to the exclusion of poor and vulnerable rightsholders who cannot afford to pay for their basic needs without a subsidy	Where some states regulators exclude informal settlements from accessing public sanitation services, and criminalise the informal sanitation services (see 8.5.1 and 8.5.2)
Indifference	Environmental sanitation policies emphasize universal coverage although informal settlements continue to be excluded from services in practice	Where the sanitation laws, and strategies and programmes at the sub-national levels do not capture the HRS
Accommodation	The NWSP and the NWRP contain the participation principle to try to involve users, and all other stakeholders in the sanitation governance process, including the private sector	Where sanitation programmes, like the CLTS, try to encourage local ownership and management of sanitation facilities, with support for the effective participation of the local population (see 8.5.2)
Mutual Support	International quasi-judicial bodies and courts have upheld the socio-economic rights of citizens, whereas the rights may be non-justiciable based on the constitutional law, but are supported by national laws	Where sanitation laws, like the Sanitation and Pollution Management Law (Edo State), regulate open defecation and require operators of commercial buildings to provide public toilets (see 8.5.2)

8.7 IMPLICATIONS AND RECOMMENDATIONS FOR THE INCLUSIVE REALISATION OF THE HUMAN RIGHT TO SANITATION

Although Nigeria is a State Party to the ICESCR (which is the main legal basis for the HRS in international law) and international instruments (like the UNGA and HRC Resolutions recognising the HRS), this has not translated into an express recognition of the right in the 1999 Constitution of the Federal Republic of Nigeria. Remarkably, although the policy and regulatory framework generally support the HRS *de facto*, there are inherent contradictions which affect the practical realization of the right. The case study, including the household survey conducted in Benin City, highlights the practical complexities of realizing the HRS in an emerging economy which might not have been evident from a traditional desk review of the national policy and regulatory framework. It also showcases the performance of HRS and non-HR instruments for sanitation governance, including their impact on actors, given the prevailing drivers in the case study. It further proves that the HRS can be subject to the predominant discourses and governance approaches at any given level of governance, its formal recognition policies does not unequivocally assure ID, and the realisation of the HRS may require different instruments depending on the given context. Nonetheless, progressive implementation of the HRS towards the realisation of the SDGs sanitation target within the next 12 years means that the HRS needs to be fully institutionalised by 2030 and any inconsistencies between the HRS and other principles for sanitation governance would need to be resolved, in favour of universal access to safe and affordable sanitation services. Therefore, to address the question: how does the human right to sanitation (HRS) influence the normative framework for sanitation governance towards inclusive development (ID) outcomes across different levels of governance in Nigeria?, this section covers the harmonisation of the national sanitation policy process, formal recognition of the HRS, including the resolution of rules incoherence in the legal system, and the operationalization of the HRS norm.

Harmonisation of the national sanitation policy process

The case study reveals a high level of incoherence within the sanitation policy process. Sanitation governance remains strongly linked with water and the water-flush system is at the top of the sanitation ladder used for sanitation programmes, like PEWASH. The continued linking of sanitation and water is good for ensuring water quality, but it is not expedient for the development of the HRS norm specifically. The prescription of a wet sanitation system at the top of the sanitation ladder also does not reflect the wide geological differences and poor access to water either due to economic, environmental or socio-legal drivers. I therefore suggest that a national sanitation policy could be formulated to harmonise the existing sanitation policies and mainstream the HRS. At a minimum, the policy reform process may:

(a) Finalise the draft National Water-Sanitation policy pending the formulation of a sanitation policy by local stakeholders;

(b) Develop a new national sanitation policy that mutually supports but is not intricately linked with the water sector policy, define the scope of the HRS and the service levels and indications that are required to realise the right in different parts of the country (outlined in Chapter 5), and makes the HRS applicable to all persons, irrespective of their legal title to property or other social status;

(c) Require the effective participation of users in the budgeting and planning process for sanitation services, including the choice of technology. This would further improve accountability and transparency at various levels of governance;

(d) Establish an independent agency, made up of rightsholders and other stakeholders in the sanitation sector, to monitor public revenue and allocations to the sanitation sector, and harmonise the sanitation services monitoring systems to improve data validity for measuring the status of access to sanitation, based on the HRS principles;

(e) Align the implementation of international organisations' and donors' sanitation programmes for sanitation with local policies and priorities, in order not to erode the government's legitimacy for failure to deliver on local priorities;

(f) Develop a common formula for determining the full direct and indirect costs of basic sanitation services and setting a national threshold for affordability of services, taking into consideration any special local circumstances and individual vulnerabilities. In this regard, a threshold of ≤3% of a household's disposable income can be applied to people in the higher wealth quintiles, to cross-subsidise the basic sanitation needs of the poor, who can also make non-monetary contributions like labour. Progressive pricing and mechanisms for instalment payment could also be used to promote affordability for the poor.

Formal recognition of the human right to sanitation and resolution of incoherence in the national constitution and states' sanitation laws

In addition to policy reform, the strongest indication of the prioritisation of the HRS within a federal legal system like Nigeria is a constitutional guarantee (see, Obani & Gupta (2015) for a general discussion of the HRS in national constitutions, and Annex I for the details of States with HRS laws). The HRS may also be included in all new sanitation laws across the three tiers of governance.

The formal recognition of the HRS in the constitution, laws and policies, backed with effective instruments for implementation, would go a long way to lending a human face to sanitation governance, providing guiding norms for resolving the current inconsistencies in the policy framework, and improving social and relational inclusion, at the minimum. Although the economic good principle and pollution prevention through the use of economic instruments may generate additional funding for sanitation services, the public good nature of sanitation as a human right requires that: (a) the tariffs do not create a negative incentive for people to resort to open defecation and other unsafe alternatives to formal sanitation services, (b) users effectively participate in the design of the tariff structure and payment modalities, and (c) provisions are made to protect access and use of sanitation services by people who may be unable to pay the standard tariffs. Conversely, the motivations for uptake of market based instruments in the WASH sector include financial status, WASH needs, health goals and social relationships and these can be leveraged upon among users who have the capacity to pay an affordable price for sanitation rather than focusing solely on cost recovery and the marketing of predetermined sanitation goods and services (Barrington et al., 2016). The formal recognition of the HRS is necessary to address the incoherence in the policy framework for sanitation governance (see 8.6).

Further, at the national level, sanitation governance currently retains three parallel objectives, namely: (a) a focus on poverty eradication while eliminating subsidies; (b) promotion of public health and environmental protection while relying on predominantly economic instruments and principles like water as an economic good and cost sharing with beneficiaries without cross-subsidies for the poor; and (c) a recognition of water as a social good without modalities for ensuring public participation in the sanitation governance process in practise. At the sub-national levels, the evolution of sanitation governance has culminated in a shift away from subsidies towards commercialisation and the mainstreaming of human rights only at the policy level without any changes in the delivery of sanitation services to households and individuals. Therefore, the legal framework for sanitation needs to be reviewed to address inconsistencies between the current neo-liberal principles and HRS principles, and prioritise the satisfaction of the human sanitation needs of people in the lowest income quintiles (more than 50% of the Nigerian population). In addition to ensuring internal consistency, the guarantee of the HRS in legislation would provide a strong legal basis for rightsholders to enforce their right both within the national legal system and through international mechanisms (see 5.4.2 and 8.4.2). Nonetheless, a legislative provision and judicial declaration recognising the HRS would still require operationalization through complementary technology, and economic, management and suasive instruments.

Operationalizing the human right to sanitation principles

The case study highlights both the HRS and non-HR principles encountered in the literature (see Chapters 5, 6 and 7) and additional principles (like the autonomy of service providers, policy making and regulatory role of government and cost sharing with beneficiaries), included in the national sanitation policies and aimed at ensuring the sustainability of the sector. A majority of the HRS principles (all, excluding physical access) are not monitored in practice, and they are mainly operationalized through instruments which may not adequately reflect the HRS principles (see 8.2 and 8.5). For instance, the major sanitation assessments focus on the number of facilities available for users without indicating whether the facilities are conducive for users with special needs (like physical disability, children, or women who are menstruating). Similarly, the policies emphasise affordability for the poor, but there are no safeguards against disconnection from services due to inability to pay; rather, the government prosecutes the offenders irrespective of their reason for non-payment. The existing HRS instruments (like subsidies for waste management services, litigation and quasi-judicial mechanisms), may also be ineffectual against the drivers of poor sanitation services or may compound the existing inequities due to defects in their design or implementation (see 8.6.2). Given the foregoing, the following measures may be considered, in order to improve the implementation of the HRS principles in Nigeria:

(a) Re-educate all the stakeholders, particularly the regulators and service providers on the meaning of the HRS and the obligations its imposes on them to respect, protect and fulfil the right within their operations;

(b) Reallocate the existing subsidies to low-income households; and

(c) Resolve contradictions and incoherence in sanitation governance:

i.   competition may be addressed by guaranteeing access to basic sanitation services for poor and vulnerable users, cross-subsidising the poor, assisting the vulnerable populations to access judicial and quasi-judicial mechanisms like the complaint procedure under the ICESCR or regional courts like ECOWAS to enforce their HRS;

ii.   indifference would require reporting and monitoring systems to assess the progressive realisation of the HRS generally and its principles respectively, a clear budget line that amounts to at least 1.5% of the national GDP for personal sanitation and hygiene, and sanitation awareness and advocacy to address negative cultural practices. It is further important to institutionalise safeguards against arbitrary disconnection of users from sanitation services and unfair terms in service level agreements and contracts entered into with their service providers (whether in writing or by conduct), by aligning service delivery with the HRS principles;

iii. accommodation needs participatory mechanisms that promote mutual knowledge exchange and empower the vulnerable, consent rules to ensure that the outcomes of consultation processes are implemented, and co-production of sanitation goods and services by all stakeholders; and

iv.   mutual support is a desirable position which requires first order learning and greater cohesion to enhance support.

Chapter 9. Human Right to Sanitation and the Inclusive Development Imperative

9.1   REVISITING THE RESEARCH QUESTIONS

If the human right to sanitation (HRS) is incorporated within the human right to water, it will not get the special attention it needs for meaningful implementation (see 1.3.1). It is implied from the International Covenant on Economic, Social and Cultural Rights 1966 (ICESCR). It is also increasingly recognised in other legal instruments and political declarations at various levels of governance, both expressly and implicitly, as illustrated in Chapters 5 and 8. Nonetheless, over 4 billion people around the world lack access to safely managed sanitation facilities due to reasons that are not confined to sanitation laws and policy frameworks. Rather, the drivers of poor sanitation services also include economic, environmental and social factors that may affect access to sanitation services either directly or indirectly. Further, there are parallel sanitation governance principles operating alongside the HRS norm. In order to ensure that the HRS is not just rhetoric, especially for the people living without access to safely managed sanitation facilities, it is therefore important to go beyond a traditional legal analysis and explore the performance of the HRS and the drivers (see 3.4) which affect access to sanitation in practise; that is what I have attempted in this thesis. Given the potential contradictions between the HRS and other non-human rights instruments, including principles, for sanitation governance, as evident from the case study, it is also important for sanitation governance to be guided by an overarching norm that ensures universal access to sanitation without compromising on environmental sustainability. For this purpose, I selected inclusive development (ID) as the overarching norm for my analysis in this thesis, and my main research question is: *How can the human right to sanitation be interpreted and implemented to promote inclusive development?* I answer the main question based on my research findings on the five research questions already addressed in the previous Chapters:

(i)   What are the drivers of poor sanitation services and how are these currently being addressed in sanitation governance frameworks?

(ii)   How has the human right to sanitation evolved across different levels of governance, from international to local; how do the human right to sanitation principles address the drivers?

(iii)   Which humanitarian law and any other non-human rights instruments, including principles and indicators, for sanitation governance promote the progressive realisation of the human right to sanitation, through addressing the drivers of poor sanitation services?

(iv)   How does legal pluralism operate in sanitation governance, with the implementation of the human right to sanitation, alongside non-human rights instruments and principles?

(v)   How can the human right to sanitation institution be redesigned to advance ID outcomes across multiple levels of governance?

9.2   ADDRESSING THE DRIVERS OF POOR SANITATION SERVICES

This section partly cumulates the findings on the first three sub-research questions to show that the HRS does not currently address the main economic, social and environmental drivers of poor sanitation services. However, it is critical for the HRS to address these drivers because the problem of poor sanitation services is not a purely legal issue but arises from a combination of factors within the economy, the physical environment and society generally (see 3.4). The analysis in Chapter 3 highlighted twenty-six economic, environmental and social drivers, but the HRS potentially addresses sixteen drivers that are linked to poverty and discriminatory practices. It does this only in a formal legal sense, by obliging States, as the primary duty bearers, to respect, protect and fulfil the HRS for the unserved and underserved population within their jurisdictions while also supporting the fulfilment of the right for populations outside their jurisdictions. For instance, with increasing population and other anthropogenic factors creating pollution and exacerbate water scarcity, governance instruments enshrining HRS principles like sustainability and safety impose an obligation on duty bearers to avoid causing pollution and water scarcity (relevant to people who rely on wet sanitation systems). However, this addresses the drivers only to a limited extent because the HRS is mainly anthropocentric and does not sufficiently address sustainability and safety from an ecocentric perspective (Feris, 2015). This approach has practical limitations, and does not address important environmental drivers (challenging or inaccessible topography, natural disasters, high temperatures/turbidity in source water and climate variability and change), and partially addresses economic drivers (excluding discounting the future, preference distortion, risk aversion), and social drivers (excluding space constraints, insecurity, conflicts and poor social cohesion, mass migration/urbanisation) which affect the poor, marginalised and vulnerable populations lacking the resources to address these drivers privately.

The HRS, rooted as it is in the ICESCR, mainly focuses on economic and social (including political) drivers, as reflected in the HRS instruments, including the principles, examined in Chapter 5. For instance, while the HRS requires affordable access to sanitation services and this can be achieved through progressive pricing, the HRS neither specifies the threshold for affordability nor requires States to provide free sanitation services for the poor. The case study (Chapter 8) further illustrated the tensions between affordability of sanitation services for the poor and neo-liberal policies (like cost recovery) where there is no efficient system of cross-subsidies in place. Hence, the HRS as currently formulated does not sufficiently address the environmental drivers of poor sanitation services, and it only addresses the economic and social drivers to a limited extent (Interviewee 44, July 24, 2014). Table 9.1 illustrates the impact of the HRS principles and non-HR principles on the drivers and the outstanding drivers that are not addressed by the principles.

The HRS is very important for addressing drivers that are linked to poverty, vulnerability and marginalisation among individuals and households and offers mechanisms for legal redress for those whose basic sanitation needs are not being met. This makes it important for the HRS to be expressly recognised in national laws and policy documents, for the benefit of the local population (see 9.3.1). The drivers affect individuals and households in formal and informal settings, and humanitarian situations who generally lack the capacity to invest in sanitation infrastructure, and therefore require the support of the State and/or humanitarian actors (in the case of humanitarian situations) in order to realise their right. Hence, the HRS framework can be more effective against the drivers by adopting a comprehensive response to the various causes and forms of discriminatory practices which may include poverty and additional factors that may vary from one setting to another. In addition to the HRS, there are principles for humanitarian assistance and environment and water management that can address some of the drivers to some extent as illustrated in Chapters 6, 7 and 8. Technology is also very important for countering environmental drivers (like challenging topography and high temperatures and turbidity in source water), and social drivers (like weak power infrastructure and space constraints). Urban planning and demographic policies may also be crucial for addressing the social drivers. This means that in order to address the drivers of poor sanitation services, the HRS framework needs to be redesigned to also include or take into account a variety of instruments and principles for addressing the outstanding drivers and this may require complementary non-human rights instruments, including technology, as illustrated in the case study chapter (see 8.6). Hence, redesigning the HRS to be more effective for addressing a wider variety of drivers has implications for the definition of the HRS and pluralism in sanitation governance.

Table 9.1 Impact of the human right to sanitation and other principles on the drivers of poor sanitation services

Drivers		HRS (see 5.6)	Humani-tarian Assist-ance (see 6.6)	Non-human Rights (see 7.4)	Case Study (see 8.5)
	Drivers	**Impact of the Principles**			
	DIRECT				
Env.	Challenging or inaccessible topography	-	-	-	-
	High temperatures/high turbidity in source water	-	-	+	-
	Natural hazards	-	-	+	-
	Pollution/water scarcity	+	+	+	+
Eco.	Discounting the future, especially among poor people	-	-	+	-
	Household poverty	+	-	+	+
	Inefficient tariff collection system	+	-	+	+
	Preference distortion affecting WTP	-	-	+	-
	Risk aversion	-	-	-	+
	Unaffordable tariffs & connection fees	+	-	+	+
Soc.	Distance to the facility	+	+	+	+
	Epileptic power supply	-	-	-	-
	Space constraints	-	-	-	-
	Tenure insecurity	++	-	+	+
	Non-acceptance of sanitation facility based on culture	+	+	+	+
	Negative cultural practices	+	+	+	+
	Exclusion of minorities from accessing services	++	+	+	+
	Poor maintenance culture/improper use of facilities	+	+	+	+
	Nonchalance	+	-	+	+
	INDIRECT				
Env.	Climate variability/change	-	-	+	-
Eco.	Insufficient/poorly targeted funds	+	+	+	+
	Huge foreign debts that limit public spending	+	-	-	-
	Sanctions affecting the sanitation sector	+	-	-	-
	National poverty	+	+	+	+
Soc.	Population density/growth	-	-	-	+
	Low awareness about sanitation	+	+	-	+
	Mass migration	-	-	-	-
	Insecurity, conflicts and poor social cohesion	-	+	-	-
	Uncontrolled urbanisation	-	-	-	+

Env. = Environmental; Eco. = Economic; Soc. = Social

- = Driver is not addressed by the principles; + = Driver is partially addressed by the principles

++ = Driver is fully addressed by principles

9.3 GOING BEYOND THE CURRENT STATE OF THE LAW

This section expands on the current legal conception of the HRS in view of the limitations of the right in addressing the drivers (see 9.2). It expands on the emergence of the HRS as a distinct right (see 9.3.1), the meaning of the right (see 9.3.2), the economic nature of sanitation goods and services as it affects the implementation of the HRS (see 9.3.3), and the indicators for measuring the performance of the HRS (see 9.3.4).

9.3.1 Emergence of Sanitation as Distinct Human Right

My first argument is that sanitation has emerged as a distinct human right, at least in international law. This is because it can be implied from the ICESCR and it is expressly recognised and supported by various international law instruments. Chapter 1 of this thesis shows that although sanitation is historically linked to water quality, there are strong arguments for and against delinking the human right to sanitation from the human right to water (see 1.2.2). Chapter 5 showed three approaches to recognising the HRS in international law: (a) as an implied right linked to express International Covenant on Social and Cultural Right 1966 (ICESCR) provisions like the rights to health and adequate standard of living, (b) as an implied right necessary for ensuring water quality and the realisation of the human right to water, and (c) as an independent right with a legal basis in the ICESCR. It also showed alternative interpretations of the HRS at the international and national levels, with examples from Guinea Bissau, South Africa and the United Kingdom.

There are three advantages of formal recognition. First, the human rights framing creates an unarguable narrative with legal appeal which can be a useful tool for lobbyists, and gives more weight to legal arguments for enforcement or redress in case of violations especially when the right is formally recognised in hard law sources like national constitutions.[285] Although as stated by the High Court of South Africa in the case of *Mandla Bushula v Ukhahlamba District Municipality*,[286] socio-economic rights (issuing from the ICESCR, 1966 which is the main legal basis for the recognition of the HRS in international and national legal frameworks) do not entitle rightsholders to immediate access to core services, they oblige the State to take reasonable measures for progressive realisation of HRS norms and immediate fulfilment of HRS core obligations using the maximum available resources (see 4.4.1, 4.4.2, 5.3.1, 5.4.3, 5.4.5 and 5.5.1). The ICESCR prohibits denying people access to the

---

[285]   Interviewees 29, 32, 34, 37 and 42.
[286]   [2012] High Court (Eastern Cape Division) 2200/09, [2012] ZAECGHC 1.

Covenant rights due to their land tenure status, and this is important to protect people living in informal settlements, for instance (see 5.2.3). Second, the human rights framework sends a strong message for global attention to the sanitation crisis and introduces new actors that can both pressure States and provide support for progressive realisation.[287] For instance, as part of the human rights framework, international organisations like the WSSCC and various NGOs introduce more avenues for funding and technical support, information, private public partnerships, and technological innovation. Third, the HRS potentially improves the quality of life of the billions of people without access through its binding immediate and continuous obligations, especially vulnerable and marginalised people.[288] Hence, people living in otherwise 'less visible' conditions (like informal settlements and protracted crisis or humanitarian situations) may be empowered through the formal recognition of the HRS to seek judicial redress in case of actual or threatened violation of the right.

Chapter 8, the case study chapter, specifically demonstrated that in the absence of express recognition of the HRS in domestic laws, there are at least two alternative approaches for recognising the HRS within the domestic legal framework, namely: (a) as an implied right based on obligations imposed by international law and supported by the national legal framework in recognition of State sovereignty; and (b) as an implied right based on related economic, social and cultural rights that are expressly contained in national legislations. Taking a cue from the right to water, independent recognition causes weak enforcement and poor development of the normative aspects of the right (Obani & Gupta 2015). The case study showed three limitations of the implied recognition of the HRS: (a) fragmentation in sanitation governance, especially in urban areas, (b) lack of shared meaning about the HRS among actors, and (c) incoherence in the implementation of sanitation governance principles, sometimes leading to contradictory outcomes. The case study also illustrated how three theoretical arguments in support of the continued combination of the human rights to sanitation and water, namely, that the combination: elevates the HRS in the development agenda,[289] improves water quality,[290] and provides the water needed for sanitation and hygiene purposes,[291] fail. This is largely because wet sanitation systems are not suited to the local context due to drivers like pollution, drought, natural hazards and epileptic power supply affecting the operation of water pumps. There are also equity issues resulting from

---

[287] Interviewees 23, 25, 26, 27, 28, 30, 35, 37, 39 and 40.
[288] Interviewees 24, 31, 33 and 39.
[289] Interviewees 1 and 11.
[290] Interviewee 2.
[291] Interviewees 10 and 11.

trade practices like virtual water transfers which can create water stress for poorer exporters (Wang et al., 2014) and reduce availability for low price local personal and domestic sanitation needs among people who rely on wet sanitation systems (Feng et al., 2012); these issues are also not currently fully addressed by the sanitation governance framework and requires support through non-sanitation policies and governance frameworks.

Hence, I argue that if the HRS is to be meaningfully implemented, it needs to be expressly recognised as an independent right at different levels of governance, in order to facilitate further development of the normative aspects and clarity in implementation. Hence, if the HRS is to be successfully implemented, this could require:

(a) formulating a legal interpretation of the human right to sanitation, perhaps through a UN Human Rights Council Resolution, which clearly defines the scope and content of the right, as well as the role of various stakeholders in ensuring progressive realisation;

(b) recognising an independent human right to sanitation within national frameworks and the resolution of rules incoherence between the human rights principles and other existing sanitation governance principles, and the development agenda generally; and

(c) strengthening the synergies between the HRS and other aspects of law and development policy in order to counteract the negative taboo and other cultural drivers that are unique to sanitation.

9.3.2 Deconstructing the Meaning of the Human Right to Sanitation

My second argument stems from the fact that the HRS presupposes a clear understanding of the meaning of sanitation which is not the case in practise. Beyond excreta containment, there seems to be few similarities in the meaning of sanitation adopted by various stakeholders. Rather, the lack of synergy between human rights scholars, other sanitation experts, policymakers and technocrats has resulted in the development of parallel definitions of sanitation without a common understanding of what the terms used mean (see Chapters 3, 5, 6 and 7).

Chapters 3 and the case study (Chapter 8) already highlighted contestations in the meaning of sanitation which make the implementation of the HRS complex. The definitions range from simply 'sanitation' as mentioned under the Sustainable Development Goals Target 6.2, through 'basic sanitation', 'improved sanitation', or 'environmental sanitation'. At the micro-level, there is also little convergence in the usage of each of these terms by stakeholders as shown in the case study. There is also a prevalence of technocratic approaches to defining sanitation and sanitation service levels based on access to various technologies for excreta management, for instance (see 3.5), and the individualistic nature of the HRS appears more amenable to the narrow definition of basic sanitation or simply access to toilets for personal sanitation and hygiene needs. Conversely, international humanitarian law framework for WASH integrates water supply, excreta disposal, vector control, solid waste management, and drainage. States also interpret the HRS differently (see 5.3) and this affects the nature of their obligations for the fulfilment of the HRS both within their immediate jurisdiction and extraterritorially. Where national policies prioritise the safe collection, removal, disposal or purification of human excreta, domestic wastewater and sewage from households, including provision to informal settlements, as is the case with the Water Services Act 108 of 1997 (South Africa), this is more likely to support ID than a policy which focuses on a system for the treatment and disposal or reuse of human sewage and associated hygiene but does not require the collection and transport of human waste and also allows for cost recovery without equal emphasis on instruments that ensure access for the poor and other vulnerable group, for instance the UK's Sanitation Statement (see 5.2). Thus, national and local strategies may be at odds with the international HRS norms, thereby resulting in HRS violations (see 5.6 and 8.5). Similarly, in humanitarian situations, the lack of shared meanings between human rights actors and international humanitarian actors also increases the likelihood of HRS violations despite increase in sanitation coverage (see 6.6).

Further, within the Nigeria case study, national policies like the National Policy on Water Sanitation Policy 2004 and the National Environmental Sanitation Policy 2005 either define sanitation narrowly or broadly, influenced by external partners. Nonetheless, based on the predominantly technocratic definition of improved sanitation among key stakeholders involved in sanitation interventions, sanitation studies and statistics in Nigeria have been focused on either household access to excreta containment, hand washing and in some cases vector control, in rural and urban areas, or access to sanitation facilities in schools. My analysis shows that a technocratic definition of sanitation which narrowly focuses on a predefined set of technologies or limited sanitation components, rather than ensuring a sustainable sanitation system, puts the social, relational and ecological components of ID at risk (see 3.5, 4.4, 4.5, 5.5, 5.6, 6.6 and 8.5). The divergent interpretations of the HRS also reduce its normative value and require clearly defined indicators for measuring violations at multiple levels of governance, to avoid the paradox of social, relational and ecological exclusion following the recognition of the HRS by stakeholders.

In order to advance ID, the definition of the HRS could build on the strengths of broad definitions of sanitation, like the WSSD's (see 3.2) and the conceptions of environmental sanitation that were encountered in the case study (see 8.2.4), to expand the definition proffered by Catarina de Albuquerque (former Independent Expert, and former Special Rapporteur on the Human Right to Safe Drinking Water and Sanitation from 2008-2014), and reinforce the relevance of the HRS for addressing the drivers of poor sanitation services. The HRS principles may also be expanded by incorporating elements of sanitation governance from other fields, as illustrated in Box 9.1. Clearly the needs will be different in different contexts and to address these different needs would require a space for contextual elaboration through participatory instruments like a functional sanitation ladder (see Figure 3.2), other complementary rights regimes, and non-HR policies which advance social organisation, local justice struggles, and the renegotiation of state-citizen relations. It is therefore important for human rights scholars to clearly define the HRS to include not only legal norms but also to reflect the following elements at a minimum:

(a) define the scope of sanitation services to include both private and public spaces and humanitarian situations, in view of the importance of sanitary conditions outside the household in relation to human wellbeing and the integrity of ecosystems;

(b) emphasize equitable access to sanitation services, sustainable financing and participatory governance approaches; and

(c) integrate all the services required to ensure the safe management of wastewater, solid waste, stormwater, and all other waste streams, including containment, collection, transportation (including sewerage networks), treatment and disposal or reuse, as relevant, through environmentally sustainable instruments.

9.3.3 Establishing the Economic Characteristics of Sanitation Goods and Services

My third argument centres on the divergent understandings of the economic nature of sanitation which has inadvertently increased emphasis on neo-liberal policies for sanitation marketing and enabled a predominantly technocratic response to sanitation problems, thereby denying access to the poor and vulnerable and marginalised people. The literature describes sanitation as a public, private, and/or merit good (see 3.3), while the classification of water as an economic good in policy documents also affects sanitation users who rely on wet systems.

A public good presents a collective action problem because it is both non-excludable and non-rivalrous, while a private good is both excludable and rivalrous, and merit goods are goods that may be provided by the State directly or through a system of incentives to counter the inherent preference distortion which affects private investments. My analysis shows that while sanitation taken as a whole presents user-distortion problems because of the availability of unhygienic (but seemingly free) alternatives like open defecation, the various components of sanitation infrastructure may be further classified as public, common, toll or private goods (see 3.3). Although the rich users may be able to afford the necessary investments for private sanitation goods like toilets connected to sewage systems, vulnerable people like residents in informal settlements generally lack the (legal) capacity and financial or technical resources to make similar investments. The provision of public sanitation goods like sewer networks also demands the intervention of the State. Conversely, the privatisation of sanitation services in poor countries with weak regulatory systems also exacerbates inequities in access where full cost recovery is emphasized without legal protections for the users. This is because the governance of sanitation goods as purely private goods hinders poor households from assessing sanitation, just as the commodification of sanitation technology would hinder poor households and countries from assessing sanitation technology which they may need.

In the case study, national policies on sanitation are influenced by external partners to recognise sanitation as an economic good while the local people mainly accentuate the public good nature of sanitation and the need for State provision or regulation of non-state providers to ensure equitable access, at the very least (see 8.2). The commercialisation of low cost but limited sanitation services for the poor externalizes environmental pollution and limits the future access to clean water and environmental sustainability. The provision of more sophisticated technology for the rich improves access without reducing inequities in coverage and may also externalise negative ecological impacts (see 8.5). These are some examples of how the economic classification of sanitation goods can either create or exacerbate the drivers of poor sanitation services (see 3.4). I therefore argue that in the light of the negative externalities of poor access to sanitation services (see 1.2) and how these are capable of affecting users and non-users, as well as the environment, sanitation goods and services could be primarily regulated as public good to promote the human right. Although the HRS does not dictate any economic model for the provision of sanitation, the public goods nature of sanitation eschews full reliance on neo-liberal economic instruments to ensure universal access. The complexity of classifying sanitation as an economic good mirrors the pluralistic foundations of sanitation governance and the interactions between different principles (including the HR and neo-liberal principles). In order to advance a public goods discourse, sanitation policies and programming could as a minimum:

(a) delink the provision of sanitation services and universal service coverage from legal ownership and property rights in order to minimise negative externalities from non-users;

(b) target sanitation governance instruments to ensure service expansion and universal access, with a focus on the poor, vulnerable and marginalised populations within the society. This may be achieved through a combination of instruments like free access to basic sanitation services in informal settlements (see 5.4.1) and cross subsidies for poor users (see 5.4.2), rather than restricting service coverage and subsidies to formal settlements like I recorded in the case study (see 8.4). Nonetheless, an important instrument that emerged from the case study is the use of financial palliatives to bridge the shortfalls in tariff collection in informal settlements in the short term, to encourage service provision by the private sector; and

(c) prioritise access to sanitation as an immediate survival need in humanitarian situations and deemphasize cost recovery from victims living in humanitarian situations especially during the immediate aftermath of an emergency or other forms of disasters which affect human livelihoods.

9.3.4 Indicators for Measuring and Evaluating the Performance of the Human Right to Sanitation

This section builds on the literature review chapters and case study to propose indicators that can be adapted by local stakeholders, for monitoring compliance with the HRS at the national and sub-national levels of governance. It therefore builds on my findings in response to the question of: Which (human right to sanitation), humanitarian law and any other non-human rights principles, instruments and indicators for sanitation governance promote the progressive realisation of the HRS, through addressing the drivers of poor sanitation services? First the section provides an overview of the indicators I am proposing for monitoring compliance with the HRS (see 10.4.1), then it recommends how to adapt this proposal at the national and sub-national levels for developing countries with nascent structures for monitoring the HRS, using the outcome of my sector analysis in the case study as an example without intending to be prescriptive (see 10.4.2)

Methodology for developing the proposed indicators

I followed three steps in formulating the indicators which I propose below. First, I identified 13 principles of the HRS in Chapter 5. I regard these principles as the attributes of the HRS for the purpose of developing the indicators which I propose. Next, I elaborated on the specific content of each of the principles in practical terms, drawing from additional complementary principles which I encountered in the literature and my case study (see Table 9.2). Third, I evaluated the existing indicators used either directly or indirectly to measure assess to sanitation[292] and the development of indicators for the human right to water and sanitation by scholars, in order to determine their relevance for monitoring compliance with the elaborated HRS principles (see Table 10.2). This was a helpful starting point because sanitation already exists as an indicator for some other related economic, social and cultural rights, like health and adequate housing, with more advanced monitoring processes. In the process, I realised that some aspects of the HRS principles require contextual information which cannot be translated into percentages *per se* (like compliance with procedural and substantive safeguards that need to be met to justify service disconnections). Hence, my proposal includes a mix of qualitative and quantitative indicators. I cluster the indicators into three types below: structural, process, and outcome indicators, in order to reflect the tripartite obligations (respect, protect and fulfil) imposed by the HRS on States and relevant non-State actors.

Although indicators are critical at every stage of the lifecycle of policies, from formulation through legitimisation, implementation, evaluation and change, my preoccupation is with the implementation stage or compliance, given that the HRS is now widely recognised in formal and informal legal orders at multiple levels of governance (see Chapters 5, 6, 7 and 8). Nonetheless, the inclusion of the HRS and the prioritization of vulnerable groups in the legal framework may be a deceptive indicator of progress where the inclusion of the HRS in the law does not translate into improved services for rightsholders, including the poor, vulnerable and marginalized individuals and groups.

The potentials of the HRS may not be achieved despite formal recognition of the right in the legal framework, for a number of reasons. First, the fragmentation of responsibility for

---

[292] This refers to the indicators covered in Chapter 5 that are already in use by human rights bodies like the Office of the United Nations High Commissioner for Human Rights (OHCHR), and international development actors like the UN-Water Global Analysis and Assessment of Sanitation and Drinking-Water (GLASS) and the post-2015 World Health Organization and UNICEF Joint Monitoring Programme (JMP) for monitoring the Sustainable Development Goals indicators for sanitation.

sanitation between government ministries, departments, and agencies, and the resulting poor monitoring and enforcement hampers the implementation of the HRS at the national and sub-national levels (COHRE et al., 2008). Second, there is also the problem of legality and non-prioritisation of *informal settlements* in service expansion plans (COHRE et al., 2008; Katukiza et al., 2012). Third, the local and national sanitation sector may be too weak to provide the necessary leadership and support for sustainable services (COHRE et al., 2008). Fourth, non-sanitation policies such as prohibitive zoning policies may prevent the necessary investment in sanitation infrastructure, irrespective of the recognition of the HRS in the relevant legal frameworks (Solo et al., 1993). Fifth, even where the necessary investments are made, it is often on an ad hoc basis, subject to the availability of funds or even actual political will without promoting any clearly defined and detailed overarching long-term strategy (Parkinson et al., 1998). Sixth, tenure insecurity and the underlying power issues especially where service providers are not legally obliged to extend their coverage to people without legal title to their land indirectly limit household investment in and/or access to sanitation infrastructure and force the poor to rely on often unregulated and more expensive informal services for their basic needs in informal settlements (Chaplin, 1999, 2011; Scott et al., 2013). Seventh, in the context of emergencies and humanitarian situations, lack of a rapid assessment mechanism generally limits the ability of humanitarian organisations to provide high quality responses locally (Veeramany et al., 2016; Zakari et al, 2015).

Additionally, in the light of the prevailing drivers, monitoring financial flows and other resources dedicated to the vulnerable, inequities reduction, budgetary strategies, direct and indirect discrimination, and monitoring the proportion of the targeted population that was extended sustainable access are some aspects of the HRS principles which could require quantifiable indicators in addition to the formal recognition of the HRS in the legal framework. The number of vulnerable people who effectively participate in the sanitation governance process, and who are aware and capable of accessing mechanisms for complaining about sanitation services and the justiciability of the HRS also creates a clearer indication of progress towards the HRS (de Albuquerque, 2014).

Some authors have previously proposed the adaptation of existing monitoring mechanisms, like the JMP-post 2015 under the Sustainable Development Goals framework and/or the GLAAS reporting mechanism, in order to monitor the realisation of the human right to water and sanitation (Baquero et al., 2015; Giné-Garriga et al., 2017; Meier et al., 2017). While the existing platforms and supporting literature offer a pool of viable indicators that are technically sound and would enjoy a high level of national coverage with frequent updates, they do not sufficiently capture the unique aspects of HRS principles that are different from the human right to water. I also supplement the foregoing with additional information that is not sufficiently captured in the existing mechanisms for monitoring assess to sanitation or the proposed indicators in the scholarly literature (including an extended analysis of the direct and indirect drivers of poor sanitation services, the underlying attitudes of the stakeholders towards the HRS uncovered in the case study, a legal pluralism diagnostics of rules incoherence at different levels of governance).

The indicators for acceptability, accessibility, accountability, affordability, availability, dignity, extra-territorial obligations, safety and sustainability emerge from the literature review and content analysis and inductive analysis of the HRS framework. The remaining indicators are selected from the existing human development monitoring frameworks: (a) Gini coefficient which ordinarily measures the disparity in the distribution of income among individuals or households within a country from a perfectly equal distribution (represented by a value of 0) to absolute inequality (represented by a value of 100), with 40 as the threshold adopted by the United Nations Human Settlements Programme, could serve as a proxy indicator of fair distribution of resources to counter direct economic drivers like household poverty; (b) Human Development Index (HDI), a composite index that measures life expectancy, education and per capita income which are some of the real life issues affected by poor sanitation services (see Section 1.2); (c) The Economists Intelligence Unit Democracy Index, a composite index that measures participation in electoral process and pluralism, civil liberties, the functioning of government, political participation and political culture from around 167 countries across the world and thereby serves as a proxy indicator of participation in democratic processes which is important for realizing the HRS; and (d) the World Justice Project Rule of Law Index, which ordinarily scores and ranks the rule of law in different countries based on eight factors, including: constraints on government powers, absence of corruption, open government, fundamental rights, order and security, regulatory enforcement, civil justice, and criminal justice, can serve as a proxy indicator of HRS principles especially the rule of law.

Structural, process, threshold and outcome indicators

Following the methodology I have outlined, I propose a fuller set of indicators in this thesis that allow the measurement of structures (environmental factors or resources invested and the qualities of the affected population), processes (the approaches adopted), outputs (quantity or quality of goods and services produced and the efficiency of the production process) and outcomes (impact of the outputs) indicators. Without intending to be prescriptive, I also propose threshold indicators that can individually serve as a quick signal of poor compliance with the HRS principles and the exclusion of disadvantaged rightsholders within a given reporting period, for instance in the past year. Further, I deliberately cluster the indicators to enhance their use for evaluating: (a) general commitment of (State and non-State) duty bearers to realising human rights standards through structural indicators, (b) measures taken including instruments for translating the commitment into policies and interventions through process indicators, and (c) the impact of the measures and instruments on the un-served and underserved population through outcome indicators.

To ensure a viable number of indicators, I propose one cross-cutting structural indicator for the HRS, and one process and one outcome indicator for each of the thirteen HRS principles (see Table 9.2). The structural indicator is: recognition of the human right to sanitation and the prioritisation of vulnerable groups in the legal framework. This indicator can be measured through the GLAAS reporting mechanism. The remaining indicators are drawn from the literature and the Sustainable Development Goal (SDG) framework. I found the SDG indicators to be especially relevant for monitoring availability (through SDG Targets 6.2.1), extra-territorial obligations (Target 6.5.2, 6.a.1), participation (6.b.1), safety (Target 6.3.1) and sustainability (Target 6.6.1). The integration of indicators from the GLAAS and SDG framework improves the monitoring of access to sanitation in at least two ways. First, the GLAAS and SDG indicators are technically sound and (will) enjoy a high level of national coverage with frequent updates. Second, the wider set of indicators proposed for the HRS in this thesis would also enrich the monitoring of access to sanitation at the micro level, from a predominantly relational perspective rather than relying on averages.

Table 9.2 Proposed process and outcome indicators for the human right to sanitation

PRINCIPLES	PROCESS INDICATORS	THRESHOLD INDICATORS	OUTCOME INDICATORS
Acceptability	Involvement of disadvantaged rightsholders* in the choice, design & implementation of sanitation interventions	80% of disadvantaged rightsholders assured of complete privacy, comfort and dignity [a, b, d]	100% increase in the proportion of sanitation facilities that are adaptable to the special needs of disadvantaged rightsholders, compared to advantaged rightsholders
Accessibility	Design, operation & maintenance of sanitation facilities for full access by disadvantaged rightsholders	80% increase in favour of disadvantaged rightsholders [a,b,d]	100% increase in the proportion of sanitation facilities that are accessible for safe use at all times of the day and night
Accountability	Strong legal framework established for the justiciability of the HRS	80% increase in favour of disadvantaged rightsholders (adapting the World Justice Project Rule of Law Index score ≥ 8.0 corresponding to respect for the rule of law) [k]	100% increase in the number of disadvantaged rightsholders whose sanitation services-related complaints were resolved, compared to advantaged rightsholders
Affordability	Allocation of 0.5% - 1.5% of GDP to implement the HRS for disadvantaged rightsholders	$\leq 3\%$ of household income for disadvantaged rightsholders [g, h]	Maximum expenditure of 3% of household income on basic sanitation needs for the disadvantaged rightsholders
Availability	Safely managed sanitation services available for use by disadvantaged rightsholders	100% increase in access to basic sanitation (see 3.2) in favour of disadvantaged rightsholders	100% increase in access to safely managed sanitation facilities in favour of disadvantaged rightsholders
Equality & non-discrimination	Financial flows committed to realising the HRS for disadvantaged rightsholders	Maximum Gini coefficient of ≤ 40 in the distribution of resources for sanitation for the advantaged and the disadvantaged rightsholders [c, j]	Maximum Gini coefficient of ≤ 40 in the distribution of resources for sanitation for the advantaged & the disadvantaged rightsholders
Extra-territorial obligation	HRS mainstreamed in international policies & the development agenda	80% increase in the proportion of development finance dedicated to implementing the HRS in poor States [a, b, d]	100% increase in the proportion of development finance dedicated to implementing the HRS in poor States
Participation	Effective participatory mechanisms designed for sanitation governance	80% increase in favour of disadvantaged rightsholders (adapting Economist Intelligence Unit's Democracy Index score ≥ 8.0 for "full democracy") [f]	100% increase in the participation of disadvantaged rightsholders involved in local sanitation governance, compared to advantaged rightsholders
Safety	Assessment mechanisms designed to ensure the resilience & integrity of sanitation facilities used by disadvantaged rightsholders	0% morbidity and mortality [a, b, d]	Proportion of disadvantaged rightsholders using safely managed sanitation and hygiene services; 0% morbidity & mortality from the use of sanitation facilities

(continued on next page)

Table 9.2 Proposed Process and outcome indicators for the human right to sanitation (continued)

PRINCIPLES	PROCESS INDICATORS	THRESHOLD INDICATORS	OUTCOME INDICATORS
Sustainability	Safe and sustainable access to sanitation facilities extended to disadvantaged rightsholders	80% of facilities still in use; Maximum atmospheric CO2 concentration from the facilities = 350 ppm [b, d, e]	100% of facilities still in use; 100% of wastewater and waste safely treated
Transparency & empowerment	Disadvantaged rightsholders and civil society included in the HRS monitoring processes	70% increase in favour of disadvantaged rightsholders (adapting the Human Development Index score ≥ 0.7 for "high human development") [i]	100% increase in the proportion of disadvantaged rightsholders that influence sanitation governance processes, compared to advantaged rightsholders

Indicators shaded grey can be monitored through adapting the SDG framework to meet the criteria of the relevant HRS principles (see Table 10.2)
*Disadvantaged rightsholders refer to rightsholders who cannot fully enjoy their rights as a result of poverty, marginalisation or some other form of vulnerability

Key references:
a - Baquero et al., 2015; b - de Albuquerque, 2014; c - Meier et al., 2017; d – Sphere Project, 2011; e – Steffen et al., 2015; f - The Economist Intelligence Unit, 2017; g - UN, 2012; h – UNDP, 2006; i – UNDP, 2016; j - UN-Habitat, 2016; k - World Justice Project, 2016

Developing Local Indicators: Nigeria case study

Within Nigeria, the indicators commonly used by external partners and local policymakers and technocrats neither sufficiently monitors the components of sanitation and principles outlined in national sanitation priorities nor do they sufficiently monitor the attributes of the HRS (see 8.4.3). Hence, there are two main gaps in the use of indicators for monitoring compliance with the HRS in Nigeria, in the areas of: (a) poorly reflecting the wide range of HRS principles and national principles for sanitation governance within the national policies, and (b) poorly integrating the wide range of sanitation components recognised in domestic policies.

To address these gaps, I suggest three main strategies as a starting point. Below, I break these strategies into practical action points which largely reiterate each other, for ease of implementation.

(a) Revising the national sanitation policy, delinked from water, and based on broad consultations involving regulators, service providers, financial organisations, NGOs and community-based organisations, scholars, and rightsholders, with a view to mainstreaming the HRS principles and ID as a guiding norm for the national sanitation framework. Some of the aspects that may be considered for revision in the policy include:

i.   Moving from the current predominantly technocratic approach to defining sanitation and sanitation components to a broader functional approach that addresses the social, relational, and ecological dimensions of sanitation;

ii. Defining the economic character of sanitation to prioritise its public goods aspects and address the preference distortions associated with merit goods and reorienting all stakeholders accordingly;

iii. Mainstreaming environmental sustainability in the sanitation governance framework;

iv. Developing quantitative goals for universal provision of sanitation services, service standards, targets, and financing plans that are based on a sanitation needs assessment, with evaluation processes that involve the participation of the rightsholders;

v.   Formulating sanitation governance instruments that promote equitable access and the HRS standards, with the necessary adaptations to suit local geological, social, political and economic contexts. This requires the adaptation of locally available knowledge and technologies, enabled by decentralization and subsidiarity, and appropriate top-down and bottom-up accountability mechanisms;

vi. Capacity building for rightsholders and duty bearers, and establishing mechanisms for access to justice to ensure the full protection of the HRS for the poorest, most vulnerable and marginalised.

(b) Streamlining all service contracts, and agreements with both the State, rightsholders and internal and external partners within the sanitation sector with the HRS standard and indicators, while reflecting local geological differences and service needs and ensuring transparency of the contracts. At a minimum, this would require institutionalising operational overlaps/mutual support between the HRS and existing legal orders and resolving any existing conflicts through:

i.   Codifying the HRS principles in the national constitution, and laws and policy framework for sanitation governance;

ii. Establishing strong top-down and bottom-up accountability mechanisms, with consent rules and mutual exchange of knowledge in participatory processes to address accommodation and indifference;

iii.   Legal protection of universal access to basic services, with progressive pricing, cross-subsidies, etc. rather than instruments that create a perverse incentive for poor sanitation and hygiene habits such as subsidies for formal areas irrespective of the economic status and capability of users or the exclusion of informal settlements which makes the residents which hampers ID.

(c) Restructuring the sanitation governance architecture and educating all stakeholders on the implications of the sanitation policy revision for their operations. For a start, the process could involve:

i.   Promoting cooperation, knowledge transfer, co-budgeting and exchange of resources, including manpower, between the various government agencies involved in the sanitation sector;

ii.   Restructuring the national sanitation policies to extend beyond a focus on big infrastructure for the provision of wastewater and waste management services and include the provision of decentralised infrastructure for the use of vulnerable groups;

iii.   Establishing accessible mechanisms for the resolution of disputes over sanitation services and remedies for violations of the HRS.

The indicators suggested for the international level can also be adapted as a starting point for developing appropriate indicators for the HRS at the national and sub-national levels, through participatory approaches that involve policymakers, technocrats, financial institutions, NGOs and CBOs, rightsholders including households and individuals, and any key stakeholder whose operations affect the realisation of the HRS in any way. At the national and provincial levels, the National Task Group for Sanitation and the State Task Group for Sanitation respectively offer structures that can be leveraged on for effectively coordinating the participation of the rightsholders and duty bearers to determine sanitation service targets based on the HRS principles, and select appropriate indicators for monitoring compliance across multiple levels of governance.

9.4   CONTRADICTIONS AND INCOHERENCE FROM PLURALITY IN SANITATION GOVERNANCE

This section addresses the fourth sub-research question on legal pluralism in sanitation governance, relying on the four heuristic types of legal pluralism developed by Bavinck and

Gupta (2014), which capture the quality and intensity of the relationships between legal systems (see 2.4.2). My consideration of the drivers and the HRS institution in the previous sections shows that the HRS encounters multiple discourses, interpretations and legal principles operating simultaneously at each level of governance, and there is need to understand the resulting legal pluralism relation as a first step towards achieving integration and complementarity for mutual support between HRS and other related discourses/principles affecting sanitation. The section builds on the competition, indifference and accommodation outlined in Table 9.3 to discuss two indications of contradictions (see 9.4.1) and fours indications of incoherence (see 9.4.2) that were significant from my analysis of legal pluralism in the thesis (see 5.6.3, 6.6.3, 7.4.3 and 8.5.3).

Table 9.3 Types of legal pluralism relationship between different rules in sanitation governance

	Type of Relationship			
	Competition	Indifference	Accommodation	Mutual Support
International human rights law	The ICESCR requires States to apply maximum available resources for HR implementation but it does not clarify how much of the resources may be allocated to the HRS as a basic necessity	Where the HRS is recognised in international law instruments, like the CEDAW and CRC, without being captured or implemented in the national legal system	Where the right to participation, included in the ICESCR, is used to try to encourage the rightsholders to participate in the policy process and contribute their unique perspectives on the prospects and challenges for realising their HRS	Where the decisions of international courts expand on the meaning and principles of the HRS, in consonance with the ICESCR and other international law instruments recognising the right
International humanitarian law	Where sanitation facilities are destroyed as a military necessity	Where the HRS principles like accountability are captured in the humanitarian framework but humanitarian situations continue to be governed as an exception to the application of human rights principles	Where efforts are made to incorporate the HRS principles in the domestic humanitarian framework	When the humanitarian framework adopts a broad definition of sanitation which integrates environmental sustainability and thereby enhances the HRS

(continued on next page)

Table 9.3 Types of legal pluralism relationship between different rules in sanitation governance (continued)

	Type of Relationship			
	Competition	Indifference	Accommodation	Mutual Support
International environmental law & development	Where the polluter-pays principle results in the exclusion of the poor from accessing sanitation services due to their inability to pay; the 'no priority' of use in article 10 of the UN Watercourses Convention competes with the priority of human sanitation and drinking needs among water uses	Where the law and policy framework captures HRS principles but excludes low cost and shared sanitation facilities needed, for instance under the MDGs where shared facilities were strictly considered unimproved	Where environmental laws and development policies adopt participatory approaches to try to encourage the local stakeholders to engage with the policy process, for instance through the SDGs sanitation ladder	Where non-human rights instruments like the UNECE Water Protocol contain obligations for sanitation services; protecting and preserving the marine environment and ecosystems (Part IV of the UN Watercourses Convention) promotes pollution prevention and the elimination of open defecation
National laws and policies (Nigeria)	The NWRP does not expressly recognise human sanitation and hygiene needs as a priority use of water, and it contains neo-liberal principles like cost recovery which can lead to the exclusion of poor and vulnerable rightsholders who cannot afford to pay for their basic needs without a subsidy	The 1999 Constitution does not expressly guarantee the HRS within the national legal system, although it grants the right to enter property in order to ensure public sewage services	The NWSP and the NWRP contain the participation principle to try to involve users, and all other stakeholders in the sanitation governance process, including the private sector	International quasi-judicial bodies and courts have upheld the socio-economic rights of citizens, whereas the rights may be non-justiciable based on the constitutional law but are supported by national laws

(continued on next page)

Table 9.3 Types of legal pluralism relationship between different rules in sanitation governance (continued)

	Type of Relationship			
	Competition	Indifference	Accommodation	Mutual Support
Local laws and policies	Where some states regulators exclude informal settlements from accessing public sanitation services, and criminalise the informal sanitation services (see 8.5.1 and 8.5.2)	Where the sanitation laws, and strategies and programmes at the sub-national levels do not capture the HRS	Where sanitation programmes, like the CLTS, try to encourage local ownership and management of sanitation facilities, with support for the effective participation of the local population (see 8.5.2)	Where sanitation laws, like the Sanitation and Pollution Management Law (Edo State), regulate open defecation and require operators of commercial buildings to provide public toilets (see 8.5.2)

Source: The table builds on Table 2 in Obani & Gupta, 2014b

9.4.1 Contradictions

Two contradictions emerge from implementing the HRS within frameworks that simultaneously recognize water as an economic good, and excluding vulnerable and marginalised groups like informal settlements from assessing sanitation services. Considering the first contradiction, the recognition of water as an economic good is significant for realizing the HRS for two main reasons, namely, the literature review, content analysis and the case study showed that: (a) sanitation is still closely linked to water quality across multiple levels of governance, and (b) water is critical for many personal sanitation and hygiene processes although on-site dry sanitation systems may sometimes be more viable than wet sanitation systems. At the international level, the four Guiding Principles on Water and Development contained in the Dublin Statement complement the HRS participation principle and the need for equality and non-discrimination against women in the governance of water and sanitation to some extent, but contradictions arise during implementation as a result of plurality. The recognition of water as an economic good has been widely adopted in support of the commodification of water and related sanitation services, while the human rights construct has been the basis for movements to counter the predominant neo-liberal basis underlying commodification and various forms of water and sanitation sector reform involving the private sector in many developing countries (Bakker, 2007; Barlow, 2009). Nonetheless, the economic good principle is not recognized in any of the UN resolutions recognizing the HRS either linked to the right to water or as an independent right. Rather the resolutions mostly urge States and international organisations to commit financial resources to the realisation of the right without clarifying the contradictions between the human right and the dominant neo-liberal underpinnings of sanitation sector reforms in the developing countries where the majority of people without access to sanitation currently live. The case study also practically illustrated the tensions that arise from recognising the human rights to sanitation and the economic good principle simultaneously; many actors at the national and sub-national levels were inclined to adopt the economic good principle and pursue cost recovery without instruments to ensure that the HRS was not violated for the poor and other vulnerable group, as a result of inability to pay.

The second contradiction is a practical one that occurs where inequities in accessing sanitation services are perpetuated by sanitation governance policies, resulting in retrogression in realising the HRS. The literature review and content analysis showed that the HRS principles are predominantly social and relational but do not guarantee the protection of vulnerable and marginalised groups due to *de facto* and *de jure* practices, such as the exclusion of informal settlements from accessing sanitation networks. At the international level, the UN resolutions recognising the HRS mainly expound on the need to ensure the realisation of the HRS for vulnerable and marginalised groups like women and girls, especially, but do not directly address the issue of tenure insecurity or the status of informal settlements. In the absence of an international treaty protecting the rights of residents of informal settlements, similar to the treaties on the rights of the child, elimination of discrimination against women and the protection of persons living with disabilities, the tensions between the HRS and the exclusion of informal settlements through overt and covert discriminatory practices requires legal clarification. The recognition of the rights to housing and adequate standard of living may not suffice, especially in developing countries where the provision of housing is still largely considered a private responsibility borne by the population and subject to the availability of private finance. The case study further illustrated that even where there is no *de facto* exclusion of informal settlements from accessing formal sanitation services, structural factors and the prevailing neo-liberal discourse serve to exclude the informal settlements without deliberate instruments like financial palliatives and mandating service expansion to informal settlements.

9.4.2 Incoherence

Four areas of incoherence emerge in relation to: (a) defining the human right to sanitation; (b) implementing the principles; (c) designing the approaches for implementing the HRS; and (d) the economic aspects of the HRS. First, there is no coherence in the definition of the HRS across the different levels of governance. While at the international level, the UN resolutions recognising the HRS often adopt the definition proffered by the former Special Rapporteur (see 3.2), the various sources of the HRS including supporting treaties, like the Convention on the Elimination of All Forms of Discrimination Against Women 1979 (CEDAW) and the Convention on the Rights of the Child 1989 (CRC), use different terms including 'sanitation', 'adequate sanitation', 'improved sanitation' and 'environmental sanitation' without sufficient conceptual clarification of the meaning of the terms used. Nonetheless, the definition by the former Special Rapporteur has been criticised for its limited consideration of environmental sustainability (Feris, 2015). The case study also illustrated the incoherence in the meaning of the HRS as different domestic actors adopted different terms in connection with operationalizing the HRS, largely influenced by external actors, especially donors. The predominant use of 'improved sanitation' by domestic actors in the cause study was also inconsistent with the broad stipulation of the components of sanitation in domestic policies, as well as the perception of users as revealed by the households survey. This resulted in a high level of indifference and in some cases strong opposition to national sanitation instruments and non-compliance by members of the public, compounded by the low monitoring and enforcement capacity of the relevant regulatory agencies.

Second, the HRS framework does not stipulate any economic model or a finite list of instruments through which its principles are to be operationalized. As a result of this, the implementation of the HRS depends on how the principles are interpreted by actors at different levels of governance. At the international level, the HRS has been significantly influenced by dominant development programmes for improving access to sanitation such as the Millennium Development Goals (MDG) sanitation target that was focused on halving the number of people without access to sanitation by 2015. Nonetheless, the 2030-bound SDG water and sanitation goal complements the HRS with indicators as well and a monitoring process that could either directly monitor progress on the realisation of HRS principles like availability (Targets 6.1.1, 6.3.1, 6.4.1, 6.4.2), participation (6.b.1), extra-territoriality and cooperation (Target 6.5.2, 6.a.1); complementary principles like integrated water resources management (Target 6.5.1); or the state of drivers of poor sanitation services like pollution or water scarcity (Target 6.6.1) (see generally 10.4.1). The other SDGs can also be broadly construed to either address the related drivers of poor sanitation services and/or offer instruments and indicators that can enhance the realisation of the HRS, thereby promoting mutual support between the SDGs and the HRS (see Table 10.5). The case study showed that actors at the national and sub-national levels often conflated the language and principles of international development programmes like the MDGs with adopting the HRS in practise. This, combined with the dominant influence of external actors, resulted in the monitoring of access to sanitation in terms of access to toilets and the rates of progress in urban and rural areas only. Whereas, the HRS would require disaggregating the relevant data to show the progress in access for all vulnerable groups within any given context, like the poor, women, girls, children and people living in informal settlements. Further, private service providers would only extend sanitation services within formal areas, thereby exacerbating the inequities between formal settlements and informal settlements, for instance, except the government offered financial palliatives targeting the informal settlements.

Third, there is incoherence in the approaches to recognising and implementing the HRS across multiple levels of governance. At the international level, the literature review and content analysis showed three main approaches to recognising the human right to sanitation, namely: (a) as an implied right relevant for the realisation of other economic, social and cultural rights, (b) as a combined right linked with the right to water, and (c) as an independent right (Obani & Gupta, 2015). The implied recognition of the HRS in human rights treaties that are of limited scope either *ratione loci* or *ratione personae* like the CEDAW, the CRC and the Convention on the Rights of Persons with Disabilities 2006 (CRPD) has nonetheless resulted in fragmentation and incoherence in the normative framework of the HRS (Obani & Gupta, 2015; 2016a). For instance, the CEDAW imposes an obligation on States to ensure the right of women in rural areas to an adequate standard of living, including sanitation and water.[293] Though this falls short of recognising the right to sanitation for all women, especially those living in vulnerable conditions in (non-rural) formal settlements, informal settlements, and emergency situations, the CEDAW significantly recognises the right to sanitation without linking it to water. The CRC obliges

---

[293] CEDAW, article 14(2)(h)

State parties to conduct hygiene and environmental sanitation education as a measure to promote the right of a child to the "highest attainable standard of health".[294] The CRC provision relating to environmental sanitation presumably requires more than excreta removal as conceived by the concept of basic sanitation. In this regard, it is broader than most other treaty provisions on the HRS. Although the CRC only obliges States to provide sanitation education rather than sanitation services, the treaty is also remarkable because it again does not recognise the HRS in direct connection with water, which it also guarantees as a right.[295]

This raises the question of whether there is need for an international treaty recognising the HRS universally, without any restrictions either *ratione loci* or *ratione personae* (see 9.3.1). While such a treaty would unequivocally clarify the legal basis for the HRS in international law, the process of making treaties is highly political and there is limited evidence that treaties automatically improve human rights conditions within State Parties (see for instance, Magesan, 2013). The case study also showed how the HRS can be recognised in the constitution, laws, policies or regulations within the domestic legal framework. Each of these legal instruments can have complementary legal status where for instance the national policy constitution guarantees the HRS; enabling laws are passed to specify the standards for fulfilling the right based on the HRS principles; the regulator passes a regulation outlining how the HRS will be implemented through their operations; and national policy generally conveys the plan of action of the government for the realisation of the HRS. It is however important to expressly provide for the HRS in the constitution, which is the *grund norm* in many jurisdictions, or in a higher level legal norm like a sanitation law to ensure that the right is justiciable.

The HRS is also internally focused and poorly integrated with other areas of law and development; this creates incoherence in practice, as illustrated in Chapters 6, 7 and 8. For instance, despite the participation norm, contracts and arbitration terms for sanitation services are often still secretly negotiated between the government and its private partners or international organisations without the participation of the intended beneficiaries (Barlow, 2009). As a result, Bilateral Investment Treaties (BITs) and the arbitral awards of the International Centre for Settlement of Investment Disputes (ICSID) over service contracts in developing countries have sometimes worked against the fulfilment of human rights obligations either directly through stabilisation clauses that prevent new laws or necessary

---

[294] Convention on the Rights of the Child 1989 (CRC).
[295] CRC, articles 24(1) & 24(2)(c)

amendments to the existing legal framework, or indirectly by preventing the States' direct investment in service provision due to indemnities (Thielbörger, 2009).

Fourth, the literature review, content analysis and case study showed that there is incoherence in the funding of sanitation services and this affects the availability of resources for realising the HRS across multiple levels of governance. At the international level, although the eThekwini commitment was made since 2008 by African States, to invest 0.5% of their respective GDPs in the sanitation and hygiene sector, only Equatorial Guinea was reported to be on track while the other African countries were mostly off track in the spending commitments for sanitation. The poor funding of the sector in developing countries may be attributed to four factors, including: (a) low government investment which may be motivated by the perception that the sector is of interest to donors and therefore requires less domestic investment; (b) shortfalls in implementation caused by delays in donor funding or absorptive capacity by government agencies; (c) low donor spending on the maintenance of existing infrastructure and a perceived preference for capital investment in new projects, and (d) fiscal constraints by government and donor agencies (Martin & Watts, 2013). The case study also showed indications of poor funding of urban sanitation compared to rural sanitation, partly as a result of donor interest in rural sanitation; lack of accessibility of data on budgeted and actual spending due to fragmentation and budgeting practices; and low spending on sanitation by donors and the government compared to spending on water. This implies that despite the obligation on States and international organisations to commit financial resources to the sanitation, the sector is still poorly funded and there are no strong indications of improvements by actors across multiple levels of governance.

Nonetheless, the complex nature of the sanitation problem (see 10.2) and the limits of human rights principles for sanitation governance in addressing the drivers (see 10.3) require a multi-disciplinary and multi-dimensional governance framework that spans beyond the present confines of the human rights construct. For instance, there are non-human rights principles which can potentially address some of the main environmental drivers of poor sanitation services and in many other ways augment the HRS framework. Additionally, given the contextual nature of the drivers and the need for localised solutions, it may be necessary to adopt divergent principles in sanitation governance at different levels of governance and in different locations. The pluralistic nature of sanitation governance therefore offers potential for strengthening the HRS framework. To achieve this, it is necessary to: (a) mainstream an overarching norm like ID that can provide a standard for addressing social, economic and environmental priorities through sanitation governance processes and instruments, (b) resolve definitional issues and contradictions in the design of instruments and implementation of the HRS in line with the prevailing drivers and overarching norm, building on the existing knowledge on sanitation produced by various disciplines, and (c) expand the obligations of non-State actors involved in the sanitation sector based on the HRS standards. If the HRS is to be meaningfully implemented, sanitation interventions could be evidence-based and informed by assessments that cover the following at a minimum:

(a) identify the existing sanitation governance instruments and their outcomes for different un-served and underserved segments of the population;

(b) identify the different causes of vulnerability and/or exclusion as these may require different policy measures even where they intersect;

(c) analyse the impact of governance instruments from other sectors on individuals and households without access to sanitation services;

(d) identify additional instruments to reduce the inequities in access to sanitation for the un-served and underserved segments of the population, and

(e) analyse the impact of the instruments on the different causes of vulnerability/exclusion, in order to avoid contradictory outcomes.

9.5 RECOMMENDATIONS FOR INTERPRETING AND IMPLEMENTING THE HUMAN RIGHT TO SANITATION

Building on the foregoing conclusions and answers to the research questions, this section makes the following recommendations for HRS interpretation and implementation in order to address the overarching research question for this thesis: *How can the human right to sanitation be interpreted and implemented to promote inclusive development?*

Legal scholars tend to interpret the HRS through a purely legal sense and canvas for recognition and enforcement of the right, in order to progressively realise universal access to basic sanitation services. As stated by Meier et al. (2017:3), "[H]uman rights offer a universal framework to advance justice in water and sanitation policy. Rather than viewing safe drinking water and adequate sanitation as only basic needs, human rights implicate specific responsibilities to realize water and sanitation as legal entitlements." This can improve the accountability of States (to the rightsholders) and ensure sanitation service expansion in accordance with the social, economic and environmental criteria that are necessary for human wellbeing. However, a purely legalistic approach does not address the wide range of economic, environmental and social drivers of poor sanitation (see 9.2), and is complex to frame in the context of populations that rely on decentralised sanitation solutions (perhaps due to the failure of public utilities in low-income countries) (see 8.2) and non-State actors who support the local sanitation governance process (for instance during humanitarian crisis) (see 6.6).

Sanitation is a public good and the HRS can be strengthened through a broad interpretation that is situated within the broader scope of environmental justice, nature rights and the public goods discourse (as discussed in 9.3.2), and an expanded implementation of the eleven HRS principles discussed in this thesis (building on Box 9.1). The expansion of the HRS framework through complementary non-HRS instruments, including principles, as proposed in this thesis can address the drivers of poor sanitation services and strengthen the social, relational and ecological dimensions of ID in the process where universal access to sanitation is pursued within ecological limits and disadvantaged rightsholders are provided the necessary support to enable them enjoy the right. For instance, internally displaced persons and refugees need to be afforded the necessary support to live in dignity, especially where the host country can afford to provide basic sanitation with international support. Similarly, to protect people living in informal settlements, their access to sanitation needs to be distinguished from land rights issues, because human rights are inherent and would otherwise be violated if it is made dependent on the legal status or the relationship between the individual and the State.

Box 9.1   Elaborating the human right to sanitation through human rights and non-human rights principles

1. Acceptability
   - Design facilities to match the users' needs and preferences within ecological limits
   - Eliminate taboo and discriminatory practices against vulnerable groups like discrimination against women and girls during menstruation
   - Educate users on the harmful effects of negative attitudes and practices concerning sanitation through capacity building and awareness

2. Accessibility
   - Situate sanitation facilities in safe places that are easy to reach, within the immediate vicinity of users or as close to the users as possible, to avoid causing harm
   - Design sanitation facilities for easy use by everyone, including people with special needs
   - Ensure that pathways to the facilities that are located outside the household are well lit and safe
   - Ensure that menstrual hygiene and other needs of vulnerable groups are reflected in the design of sanitation services

3. Accountability
   - Clarify the roles of stakeholders in the sanitation sector and how they intersect
   - Establish mechanisms for monitoring the operations of duty bearers and the impacts on rightsholders for instance by establishing safeguards against service disconnections and monitoring compliance with service standards
   - Ensure accessible and independent review mechanisms are available to aggrieved rightsholders
   - Compensate aggrieved rightsholders and prevent retrogression

4. Affordability
   - Explore cost-effective alternatives to ensure that the direct and indirect costs of sanitation services neither exceed 3% of household income nor interfere with other basic needs like food and shelter
   - Establish mechanisms to cover the cost of sanitation for the poor, people living in humanitarian situations and other vulnerable groups that cannot otherwise afford the cost of their basic sanitation needs, like cross-subsidies funded through the polluter-pays principle and prevention principle involving corporations
   - Provide flexible payment options, like payments in kind or instalment cash transfers, based on need in order to ameliorate the effect of the economic good principle on the poor
   - Establish legal safeguards against service disconnections for people who genuinely cannot afford to pay for their basic sanitation needs

5. Availability
   - Ensure a sufficient number of facilities in households (including those in informal settlements) and public places, in order to avoid long waiting times
   - Expand sanitation services to include collection, transport, treatment, disposal and/or reuse of human excreta and other waste streams that affect human wellbeing
   - Ensure functional sanitation services in the households and everywhere people spend a considerable amount of time
   - Tailor services to meet special needs of all categories of users, to ensure universal access, irrespective of tenure security and other legal restrictions

(continued on next page)

Box 9.1   Elaborating the human right to sanitation through human rights and non-human rights principles (continued)

6. Equality and non-discrimination
- Investigate all the forms of inequality *de facto* and *de jure* and ensure that the procedures or instruments for addressing inequality do not further stigmatize users
- Decriminalise service provision in informal settlements and expand the sanitation policy process to include all social contexts, including humanitarian situations, and urban and rural areas
- Develop mechanisms for addressing the various causes of inequality and reducing the disparities in access to sanitation among disadvantaged groups
- May require differentiated support, such as affirmative action, to reduce existing inequality
7. Extra-territorial obligation
- Respect, protect and fulfil the HRS both within their territories and extra-territorially
- Prevent third parties within their control from violating the HRS extra-territorially, based on the attribution of State responsibility for the conduct of non-State actors
- Prioritise the HRS in humanitarian assistance, trade and other international affairs
- Provide effective mechanisms for accountability in the discharge of extraterritorial obligations
8. Participation
- Establish mechanisms for full, free and meaningful participation for everyone affected by decisions about sanitation services
- Include consent rules in participatory mechanism (to guard against) to foster mutual exchange of knowledge and tackle accommodation
- Inform everyone about the participatory mechanism and how they operate and eliminate the barriers to accessing the mechanisms
- Involve local stakeholders in the decision making process including the designing sanitation programs, indicators and other processes for operationalizing the HRS
9. Safety
- Ensure that the design of sanitation facilities promotes the safety of users
- Promote hygienic maintenance of the facilities
- Ensure that sanitation facilities protect the environment from pollution or contamination by waste using principles like prevention and precaution
- Educate users on the correct means of using and hygienically maintaining the facilities
10. Sustainability
- Prioritise sanitation in disaster risk management and humanitarian assistance
- Integrate environmental sustainability in sanitation planning and programmes and minimise harmful environmental impact from sanitation services
- Design services to be resilient enough to meet both the present needs and the needs of the future generation
- Ensure local knowledge about how to operate and maintain sanitation facilities through capacity building in order to ensure the sustainability of the existing services
11. Transparency and access to information
- Inform rightsholders about their HRS and related enforcement mechanisms through the local languages and common communication media
- Provide open access to information about sanitation services
- Ensure that mechanism for accessing information and enforcing the HRS are both physically and economically accessible for all
- Strengthen the capacity of the population to access and use the available information through capacity building

The broad interpretation of the HRS and implementation alongside other complementary discourses imposes wider obligations on States and other stakeholders whose operations affect the realisation of the HRS (beyond the traditional ambit of respect, protect, and fulfil). This provides an opportunity to cooperation between the State and stakeholders, including human rights practitioners and other members of the epistemic community, to integrate solutions from different fields for tackling the diverse drivers of poor sanitation services. However, the resulting plurality may link the HRS with contradictory and incoherent rules that compound HRS implementation, if left unaddressed (see 5.6.3, 6.6.3, 7.4.3, 8.6.3 and 9.4). There are tools for strengthening the positive outcomes as discussed in this thesis. Further, pluralism makes it essential to develop indicators (building on Table 9.1) for an objective assessment of progressive realisation and enhanced accountability in the HRS implementation process. Without the foregoing, the formal recognition of the HRS would remain tokenistic at best without translating into progressive realisation of universal access to safely managed, accessible, acceptable and affordable sanitation services, and a participatory governance process that empowers the poorest, most vulnerable and marginalised rightsholders.

9.6   REFLECTIONS ON METHODS

I adopted a multi-disciplinary perspective (using both legal and social science research methods) because the rules, decision-making processes, and programs that define acceptable sanitation standards, allot roles to key actors for achieving the standards, and steer interactions among the actors stem from law and many other disciplines (including the social sciences, physical sciences and engineering). I also integrated both legal and non-legal publications to present the current state of knowledge on HRS governance and built upon current research on the evolution of the legal framework for the HRS by identifying relationships of indifference, competition, accommodation, and mutual support in the legal framework across multiple levels of governance. As a result, I was able to reach beyond current legal research by combining quantitative and qualitative methods (from law and social sciences) to evaluate the HRS framework against the drivers of poor sanitation services and the need for ID, proffering recommendations for redesigning HRS instruments, where necessary.

Adopting ID as an overarching norm (rather than sustainable development, which is often operationalized by prioritising economic growth over social and ecological sustainability) enriched my analysis of the ecological dimension of the HRS that is otherwise poorly addressed within a narrowly defined HRS framework (Feris, 2015). Combining these methods allowed me to gain useful knowledge from both legal authorities and sources from other disciplines (like international relations, governance, natural sciences, and economics) that were relevant for addressing my research questions. Nonetheless, in the process of combining the various methods I drew on my knowledge of the types of legal authorities to determine the weight I attached to primary, secondary, mandatory and persuasive legal authorities in my analysis. My reflections on HR law also made me realise that the relational dimension of ID needs to be strengthened through prioritising disadvantaged individuals and groups rather than de-emphasising the differences between the haves and the have-nots in the development process.

There are some key areas in which my methodology can be strengthened for future research. To start with, I had to rely on some unofficial translations of laws, policy documents and case reports that were not written in English Language originally because the official English translations were not accessible. Second, although I set out to conduct household surveys in three contexts (formal and informal settlements, and humanitarian situations), I was only able to cover the formal context due to legal and political issues, and personal safety concerns that I encountered in the field when trying to access informal settlements. The potential respondents in the internally displaced persons' camps which I visited were also not accessible because of a number of factors including physical and psychological vulnerability, and other personal considerations which they may not have disclosed. Consequently, I relied on experts and other stakeholders as well as literature review and content analysis but could not get the direct input of households in informal settlements and humanitarian situations. Third, I encountered difficulties in tracking sanitation spending due to the variety of actors, definitions and interventions encountered in the sanitation governance process. As a result, I could not provide a big-picture analysis of the actual spending on sanitation by different actors across multiple levels of governance even though this would have strengthened my analysis further. Fourth, I compiled a rich set of HRS (structural, process, threshold and outcome) indicators, by drawing from otherwise distinct disciplines to support the assessment of the HRS. The development of indicators for the HRS can be further enriched through strengthening local democratic processes to adapt the indicators that I have proposed in this thesis to local circumstances. Finally, the reliability of my research outcomes and recommendations can be reinforced by further studies involving multiple researchers from various disciplines, with additional systematically designed multi-level case studies and ethnographic research to validate official reports, where possible.

Nonetheless, I had the privilege of discussing my methodology with other researchers from multiple disciplines during the course of my PhD, including law, engineering, political science, environmental science and anthropology. I also received feedback from reviewers who assessed the journal articles and book chapters that I published on the basis of my PhD. I was able to adapt my methodology using all the feedback I received and they were generally positive that combining the methods the way I did enhanced the understanding of the HRS and the instruments for addressing the drivers of poor sanitation for both lawyers and non-lawyers involved in sanitation governance. I found that addressing the non-legal aspects of the HRS implementation process, which are usually not covered in a pure law research, also made the non-lawyers more interested in exploring HRS instruments and synergies with human rights practitioners. Hence, I hope that other researchers can build on these methods to improve multi-disciplinary learning and mutual support between the HRS and other disciplines working on sanitation governance. I also hope that future reports of the Special Rapporteur on the human rights to safe drinking water and sanitation can explore the impact of the HRS on both direct and indirect economic, environmental and social drivers that have not previously been addressed by the human rights assessment, as well as the development of measurable process, threshold and outcome indicators, as discussed in this thesis.

9.7 RECOMMENDATIONS FOR FURTHER RESEARCH

Based on my research experience, findings and reflection, I have identified the following five gaps for further research:

(a) investigating the political economy of sanitation at multiple levels of governance, from the international to the local, in order to demine an affordable rate for users in different social contexts and wealth quintiles;

(b) linking the HRS to the food, water, and energy nexus discourse;

(c) investigating effective instruments for affordability and accountability of sanitation in humanitarian situations;

(d) analysing the economic aspects of ID components in sanitation policy and programming; and

(e) evaluating the import of power politics for HRS interpretation and implementation across multiple levels of governance.

References

Abah, S. O., & Ohimain, E. I. (2011). Healthcare waste management in Nigeria: A case study. *Journal of Public health and Epidemiology, 3*(3), 99-110.

Abeysuriya, K., Mitchell, C., & White, S. (2007). Can corporate social responsibility resolve the sanitation question in developing Asian countries? *Ecological Economics, 62*(1), 174-183. doi:10.1016/j.ecolecon.2006.06.003.

Abosede, A., & Onakoya, A. (2013). Entrepreneurship, economic development and inclusive growth. *International Journal of Arts and Entrepreneurship, 1*(3), 375-387.

Abraham, W. (2011). Megacities as Sources for Pathogenic Bacteria in Rivers and Their Fate Downstream. *International Journal of Microbiology, 2011*, Article ID 798292, 13 pages. doi:10.1155/2011/798292.

Aburdene, P. (2007). *Megatrends 2010 : The rise of conscious capitalism.* Charlottesville, VA: Hampton Roads.

ACF-France. (2009). *The human right to water and sanitation in emergency situations the legal framework and a guide to advocacy.* New York: Global WASH Cluster.

African Ministerial Conference on Water [AMCOW]. (2011). *Water supply and sanitation in Nigeria: Turning finance into services for 2015 and beyond.* Nairobi: Water and Sanitation Program.

Agbase, P. O. (1998). Environmental challenges facing Africa: Governmental response. In J. G. Jabbra, O. P. Dwivedi & International Association of Schools and Institutes of Administration (Eds.), *Governmental response to environmental challenges in global perspective.* Amsterdam: IOS Press.

Aguilar-Barajas, I., Mahlknecht, J., Kaledin, J., Kjellén, M., & Mejía-Betancourt, A. (Eds.). (2015). *Water and cities in Latin America: Challenges for sustainable development.* Abingdon, UK: Routledge.

Akhter, S. R., Sarkar, R. K., Dutta, M., Khanom, R., Akter, N., Chowdhury, M. R., & Sultan, M. (2015). Issues with families and children in a disaster context: A qualitative perspective from rural Bangladesh. *International Journal of Disaster Risk Reduction, 13*, 313-323.

Akpabio, E. M. (2012). Water meanings, sanitation practices and hygiene behaviours in the cultural mirror: A perspective from Nigeria. *Journal of Water Sanitation and Hygiene for Development, 2*(3), 168-181.

Akpabio, E. M., & Subramanian, S. V. (2012). Water supply and sanitation practices in Nigeria: Applying local ecological knowledge to understand complexity. *ZEF Working Paper Series, 94*, 1-27.

Akpan, N. (2015). Floating toilets that clean themselves grow on a lake. *Social Entrepreneurs: Taking on World Problems.* [Article] Retrieved from http://www.npr.org/blogs/goatsandsoda/2014/12/23/371862099/floating-toilets-that-clean-themselves-grow-on-a-lake.S

Albu, M. (2010). *Emergency market mapping and analysis toolkit.* Warwickshire: Practical Action Publishing.

Alebeek, R. v., & Nollkaemper, A. (2012). The legal status of decisions by human rights treaty bodies in national law. In H. Keller & G. Ulfstein (Eds.), *UN human rights treaty bodies: Law and legitimacy* (pp. 356-413). Cambridge: Cambridge University Press.

Alexy, R. (2000). On the structure of legal principles. *Ratio Juris, 13*(3), 294-304.

Alston, P. (2009). A third generation of solidarity rights: Progressive development or obfuscation of international human rights law? *Netherlands International Law Review, 29*(3), 307-322. doi:10.1017/S0165070X00012882.

Alston, P., & Goodman, R. (2012). *International human rights: The successor to international human rights in context*. (2nd. Ed). Oxford: Oxford University Press.

Altheide, D. L., & Schneider, C. J. (2013). *Qualitative Media Analysis: Qualitative Research Methods Volume 38*. Thousand Oaks, California: SAGE Publications, Inc.

Amnesty International. (2010). *Risking rape to reach a toilet: Women's experiences in the slums of Nairobi, Kenya*. London: Amnesty International.

Arai, Y. (2011). The principle of humanity under international humanitarian law in the" is/ought" dichotomy: Surreptitious, capricious but conscientious meandering. *Japanese Yearbook of International Law, 54*, 333-364.

Arden, M. (2015). *Human rights and European law: Building new legal orders*. Oxford: Oxford University Press.

Arimah, B. C. (1996). Willingness to pay for improved environmental sanitation in a Nigerian city. *Journal of Environmental Management, 48*(2), 127-138.

Asaju, K. (2010). Local government autonomy in Nigeria: Politics and challenges of the 1999 Constitution. *International Journal of Advanced Legal Studies and Governance, 1*(1), 98-113.

Arthurson, K. (2002). Creating inclusive communities through balancing social mix: A critical relationship or tenuous link?. *Urban Policy and Research 20*(3), 245–261.

Arts, K. (2017). Inclusive sustainable development: A human rights perspective. *Current Opinion in Environmental Sustainability, 24,* 58-62.

Arts, K. (2014). Twenty-five years of the United Nations Convention on the Rights of the Child: Achievements and challenges. *Netherlands International Law Review, 61*(3), 267-303.

Aven, T. (2011). On different types of uncertainties in the context of the precautionary principle. *Risk Analysis, 31*(10), 1515-1525. doi:10.1111/j.1539-6924.2011.01612.x.

Ayee, J., & Cook, R. (2003). *"Toilet wars": Urban sanitation services and the politics of public-private partnerships in Ghana*. Brighton: Institute of Development Studies.

Babalobi, B. (2013). Water, sanitation and hygiene practices among primary-school children in Lagos: A case study of the Makoko slum community. *Water International, 38*(7), 921-929. doi:10.1080/02508060.2013.851368.

Babbie, E. (2012). *The practice of social research* (13th ed.). Belmont: Wadsworth

Baer, M. (2015). From water wars to water rights: Implementing the human right to water in Bolivia. *Journal of Human Rights, 14*(3), 353-376.

Baillat A, 2013; Corruption and the human right to water and sanitation: Human right-based approach to tackling corruption in the water sector. Geneva: WaterLex; Berlin: Water Integrity Network.

Bain, R., Cronk, R., Wright, J., Yang, H., Slaymaker, T., Bartram, J., & Hunter, P. R. (2014). Fecal contamination of drinking-water in low- and middle-income countries: A systematic review and meta-analysis. *PLoS Medicine, 11*(5), e1001644.

Baker, B. (2009). Hague peace conferences (1899 and 1907) *Max Planck Encyclopedia of Public International Law*. Oxford: Oxford University Press.

Bakker, K. (2003). Archipelagos and networks: Urbanization and water privatization in the south. *The Geographical Journal, 169*(4), 328-341.

Bakker, K. (2007). The "commons" versus the "commodity": Alter-globalization, anti-privatization and the human right to water in the global south. *Antipode, 39*(3), 430-455.

Bank, T. W. (1992). *WDR 1992: Development and the environment*. Oxford: World Bank and Oxford University Press.

Baquero, Ó. F., Jiménez, A., & Foguet, A. P. (2015). Reporting progress on the human right to water and sanitation through JMP and GLAAS. *Journal of Water, Sanitation and Hygiene for Development, 5*, 310-321.

Barber, N. W., & Ekins, R. (2016). Situating subsidiarity. *The American Journal of Jurisprudence,   61*(1),   5-12.   doi:https://doi-org.proxy.uba.uva.nl:2443/10.1093/ajj/auw002

Bardosh, K. (2015). Achieving "total sanitation" in rural african geographies: Poverty, participation and pit latrines in Eastern Zambia. *Geoforum, 66*, 53-63. doi:http://dx.doi.org/10.1016/j.geoforum.2015.09.004

Barlow, M. (2009). *Blue covenant : The global water crisis and the coming battle for the right to water*. New York: New Press

Barnard, S., Routray, P., Majorin, F., Peletz, R., Boisson, S., Sinha, A., & Clasen, T. (2013). Impact of Indian Total Sanitation Campaign on latrine coverage and use: A cross-sectional study in Orissa three years following programme implementation. *PLoS ONE, 8*(8), e71438.

Barrington, D. J., Sridharan, S., Shields, K. F., Saunders, S. G., Souter, R. T., & Bartram, J. (2017). Sanitation marketing: A systematic review and theoretical critique using the capability approach. *Social Science & Medicine, 194*, 128-134. doi:DOI: 10.1016/j.socscimed.2017.10.021.

Barrington, D. J., Sridharan, S., Saunders, S. G., Souter, R. T., Bartram, J., Shields, K. F., . . . Hughes, R. K. (2016). Improving community health through marketing exchanges: A participatory action research study on water, sanitation, and hygiene in three Melanesian countries. *Social Science & Medicine, 171*, 84-93. doi:http://dx.doi.org/10.1016/j.socscimed.2016.11.003

Basu, A., & Shankar, U. (2015). Balancing of competing rights through sustainable development: Role of Indian judiciary. *Jindal Global Law Review, 6*(1), 61-72. doi:10.1007/s41020-015-0003-6.

Bates, B., Kundzewicz, Z. W., Wu, S., & Palutikof, J. (Eds.). (2008). *Climate change and water: Technical paper of the Intergovernmental Panel on Climate Change.* Geneva, Switzerland: Intergovernmental Panel on Climate Change Secretariat.

Bates, R. (2010). The Road to the well: An evaluation of the customary right to water. *Review of European Community and International Environmental Law, 19*(3), 282-293.

Batley, R. (1996). Public-private relationships and performance in service provision. *Urban Studies, 33*(4-5), 723-751.

Baud, I. (2004). Markets, partnerships and sustainable development in solid waste management; raising the questions. In I. Baud, J. Post, & C. Furedy (Eds.), *Solid waste management and recycling: Actors, partnerships and policies in Hyderabad, India and Nairobi, Kenya.* New York, Boston, Dordrecht, London, Moscow: Kluwer Academic Publishers.

Baud, I., Pfeffer, K., & Scott, D. (2016). Configuring knowledge in urban water-related risks and vulnerability. *Habitat International, 54*, 95-99.

Baum, R., Luh, J., & Bartram, J. (2013). Sanitation: A global estimate of sewerage connections without treatment and the resulting impact on MDG progress. *Environmental Science & Technology, 47*(4), 1994-2000.

Bavinck, M., & Gupta, J. (2014). Legal pluralism in aquatic regimes: a challenge for governance. *Current Opinion in Environmental Sustainability, 11*, 78–85. http://dx.doi.org/10.1016/j.cosust.2014.10.003.

Bayefsky, A. F. (2001). *The UN human rights treaty system: Universality at the crossroads.* The Hague: Kluwer Law International.

Bello-Imam, I. B. (2007). *The local government system in Nigeria.* Ibadan, Nigeria.

Benda-Beckmann, F. v. (2001). Legal pluralism and social justice in economic and political development. *IdS Bulletin, 32*(1), 46-56.

Bennett, K., Bilak, A., Bullock, N., Cakaj, L., Clarey, M., Desai, B., . . . Yonetani, M. (2017). *Global report on internal displacement* (J. Lennard Ed.). Geneva, Switzerland: International Displacement Monitoring Centre.

Bernal, C. (2015). The right to water: Constitutional perspectives from the Global South. In S. Alam, S. Atapattu, C. G. Gonzalez, & J. Razzaque (Eds.), *International environmental law and the Global South* (pp. 277-293). New York: Cambridge University Press.

Bhullar, L. (2013). Ensuring safe municipal wastewater disposal in urban India: Is there a legal basis? *Journal of Environmental Law, 25*(2), 235-260.

Biran, A., Jenkins, M., Dabrase, P., & Bhagwat, I. (2011). Patterns and determinants of communal latrine usage in urban poverty pockets in Bhopal, India. *Tropical Medicine and International Health, 16*(7), 854-862.

Bisset, A. (Ed.) (2016). *International human rights docuemnts.* 10th edition. Oxford: Oxford University Press.

Bissio, R. (2013). The Paris Declaration on Aid Effectiveness. In United Nations Human Rights Office of the High Commissioner (Eds.), *The Paris Declaration on Aid Effectiveness: Realizing the right to development: Essays in commemoration of 25 years of the United Nations Declaration on the Right to Development* (pp. 233-247). New York and Geneva: United Nations.

Black, M., & Fawcett, B. (2008). *The last taboo: Opening the door on the global sanitation crisis.* London; Sterling, VA: Earthscan.

Borda, A. Z. (2013). A formal approach to article 38 (1)(d) of the ICJ Statute from the perspective of the international criminal courts and tribunals. *European Journal of International Law, 24*(2), 649-661.

Bos, K., & Gupta, J. (2016). Inclusive development, oil extraction and climate change: A multilevel analysis of Kenya. *International Journal of Sustainable Development & World Ecology, 23*(6), 482-492. doi:http://dx.doi.org/10.1080/13504509.2016.1162217.

Botting, M. J., Porbeni, E. O., Joffres, M. R., Johnston, B. C., Black, R. E., & Mills, E. J. (2010). Water and sanitation infrastructure for health: The impact of foreign aid. *Globalization and Health, 6*(1), 12-20.

Boulding, C., & Wampler, B. (2010). Voice, votes, and resources: Evaluating the effect of participatory democracy on well-being. *World Development, 38*, 125-135.

Breau, S. C., & Samuel, K. L. (Eds.). (2016). *Research Handbook on Disasters and International Law.* Cheltenham, UK; Northampton, MA, USA: Edward Elgar Publishing.

Brölmann, C. (2012). The Oxford guide to treaties symposium: Contractual and institutional elements in the treaty process. Retrieved from http://opiniojuris.org/2012/11/12/the-oxford-guide-to-treaties-symposium-contractual-and-institutional-elements-in-the-treaty-process/.

Brölmann, C. (2011). Transboundary aquifers as a concern of the international community. *International Community Law Review, 13*, 189-191.

Brookings-Bern Project on Internal Displacement. (2011). *IASC operational guidelines on the protection of persons in situations of natural disaster.* Washington, DC: The Brookings-Bern Project on Internal Displacement.

Brownlie, I. (2003). *Principles of Public International Law* (6th ed.). Oxford: Oxford University Press.

Burns, S. L., Krott, M., Sayadyan, H., & Giessen, L. (2017). The World Bank Improving Environmental and Natural Resource Policies: Power, Deregulation, and Privatization in (Post-Soviet) Armenia. *World Development, 92*, 215-224. doi:https://doi.org/10.1016/j.worlddev.2016.12.030.

Cairncross, S., & Valdmanis, V. (2006). Water supply, sanitation, and hygiene promotion. In D. T. Jamison, J. G. Breman, V. R. Measham, G. Alleyne, M. Cleason, & D. B. Evans (Eds.), *Disease control priorities in dveloping countries* (pp. 771-792). Washington, DC: The World Bank.

Cairns-Smith, S., Hill, H., & Nazarenko, E. (2014). Working paper - Urban sanitation: Why a portfolio of solutions is needed. The Boston Consulting Group. Retrieved from https://www.bcgperspectives.com/content/articles/development_health_sanitation_sol utions_urban_growth/

Carley, K. (1993). Coding choices for textual analysis: A comparison of content analysis and map analysis. *Sociological Methodology, 23*, 75-126.

Carreño, M. L., Cardona, O. D., & Barbat, A. H. (2007). A disaster risk management performance index. *Natural Hazards, 41*(1), 1-20. doi:10.1007/s11069-006-9008-y.

Castro, J. E. (2004). Urban water and the politics of citizenship: The case of the Mexico City Metropolitan Area during the 1980s and 1990s. *Environment and Planning A, 36*(2), 327-346.

Cavallo, G. A. (2013). The human right to water and sanitation: Going beyond corporate social responsibility. *Merkourios, 29*, 39-64.

Chambers, R. (2009). *Going to scale with community-led total sanitation: Reflections on experience, issues and ways forward. IDS Practice Paper 1*. Brighton: IDS.

Chaplin, S. E. (1999). Cities, sewers and poverty: India's politics of sanitation. *Environment and Urbanization*.

Chaplin, S. E. (2011). Indian cities, sanitation and the state: The politics of the failure to provide. *Environment and Urbanization, 23*(1), 57-70.

Chayes, A., & Chayes, A. H. (1993). On compliance. *International Organization, 47*(2), 175-205.

Chimni, B. S. (2006). Third world approaches to international law: A manifesto. *International Community Law Review, 8*(1), 3-27.

Chowdhury, A. (2016). *Financing poverty eradication*. Canberra: Australian National University.

Centre on Housing Rights and Evictions [COHRE], Swiss Agency for Development and Cooperation [SDC] & UN-HABITAT. (2007). *Manual on the right to water and sanitation*. Geneva: COHRE.

Centre on Housing Rights and Evictions [COHRE], UN-HABITAT, WaterAid & Swiss Agency for Development and Cooperation [SDC]. (2008). *Sanitation: A human rights imperative* Geneva: COHRE.

Colin-Castillo, S. & Woodward, R.T., (2015). Measuring the potential for self-governance: An approach for the management of the common-pool resources. *International Journal of the Commons, 9*(1), 281–305. DOI:http://doi.org/10.18352/ijc.490.

Collingsworth, T. (2002). The key human rights challenge: Developing enforcement mechanisms. *Harvard Human Rights Journal, 15*, 183-203.

Commission on Growth and Development. (2008). *The growth report: Strategies for sustained growth and inclusive development*. Washington, DC: The International Bank for Reconstruction and Development/The World Bank.

Committee on Oversight and Government Reform. (2011). *Hearing before the Committee on Oversight and Government Reform: House of representatives one hundred eleventh congress second session on H.R. 4869 to provide for restroom gender parity in federal buildings*. Washington: U.S. Government Printing Office Retrieved from http://www.gpoaccess.gov/congress/index.html

Conti, K. I., & Gupta, J. (2014). Protected by pluralism? Grappling with multiple legal frameworks in groundwater governance. *Current Opinion in Environmental Sustainability, 11*, 39-47.

Conti, K. I., & Gupta, J. (2016). Global governance principles for the sustainable development of groundwater resources. *International Environmental Agreements: Politics, Law and Economics, 16*(6), 849-871.

Cook, C. (2001). *Laying down the law*. Sydney: Butterworths.

Cordell, D., Drangert, J.-O., & White, S. (2009). The story of phosphorus: Global food security and food for thought. *Global Environmental Change, 19*(2), 292-305. doi:10.1016/j.gloenvcha.2008.10.009.

Cornwall, A., & Scoones, I. (2011). *Revolutionizing development: Reflections on the work of Robert Chambers*. London; Washington, DC Earthscan.

Corporate Accountability International. (2015). Lagos Summit rejects World Bank water privatization, demands access for all. Retrieved from https://www.stopcorporateabuse.org/press-release/lagos-summit-rejects-world-bank-water-privatization-demands-access-all

Cotton, A. (2013). Reporting aid flows for water supply and sanitation: Official development assistance. *Journal of Water, Sanitation and Hygiene for Development, 3*(3), 441-450. doi:10.2166/washdev.2013.058.

Coyle, S., & Morrow, K. (2004). *The philosophical foundations of environmental law: Property, rights and nature*. Oregon: Bloomsbury Publishing.

Craig, P. (2012). Subsidiarity: A political and legal analysis. *Journal of Common Market Studies, 50*(1), 72-87.

Creswell, J. W. (2012). *Qualitative inquiry and research design: Choosing among five approaches*. Thousand Oaks, London, New Dehli, Singapore: Sage Publications.

Crutzen, P. J., & Brauch, H. G. (Eds.). (2016). *Paul J. Crutzen:A pioneer on atmospheric chemistry and climate change in the anthropocene*. Switzerland: Springer International Publishing.

Cullet, P. (2013). Right to water in India: Plugging conceptual and practical gaps. *The International Journal of Human Rights, 17*(1), 56-78.

Daemane, M. M. M. (2014). Neo-liberalism and globalism, failing ideas for sustainable development and poverty alleviation for the developing world: The experiences of Asia, Sub-Saharan Africa (SSA) and Latin America. *Journal of Emerging Trends in Educational Research and Policy Studies, 5*(8), 196-202.

Dagdeviren, H., & Robertson, S. A. (2009). *Access to water in the slums of the developing world*. Brasilia: International Policy Centre for Inclusive Growth.

D'Amato, A. (2009). Softness in international law: A self-serving quest for new legal materials - A reply to Jean D'Aspremont. *European Journal of International Law, 20*(3), 897-910.

Dangour, A. D., Watson, L., Cumming, O., Boisson, S., Che, Y., Velleman, Y., . . . Uauy, R. (2013). Interventions to improve water quality and supply, sanitation and hygiene practices, and their effects on the nutritional status of children. *The Cochrane Database of Systematic Reviews, 2013*(8). doi:http://dx.doi.org/10.1002/14651858.CD009382.pub2.

Daniel. (2015). Environmental sanitation: Cleanest district in Sokoto to get N1m – Tambuwal. *Information NG*. Retrieved from http://www.informationng.com/2015/08/environmental-sanitation-cleanest-district-in-sokoto-to-get-n1m-tambuwal.html

Danilenko, A., van den Berg, C., Macheve, B., & Moffitt, J. L. (2014). *The IBNET Water Supply and Sanitation Blue Book 2014: The International Benchmarking Network for Water and Sanitation Utilities databook.* Washington DC: International Bank for Reconstruction and Development/The World Bank.

Darrow, M. (2012). The Millennium Development Goals: Milestones or millstones? - Human rights priorities for the post-2015 development agenda. *Yale Human Rights and Development Law Journal, 15*(1), 55-127.

d'Aspremont, J. (2008). Softness in international law: A self-serving quest for new legal materials. *European Journal of International Law, 19*(5), 1075-1093.

Dayal, R., van Wijk, C., & Mukherkee, N. (2000). *Methodology for participatory assessments: with communities, institutions and policy makers: Linking sustainability with demand, gender and poverty.* Washington DC, Delft: Water and Sanitation Program, IRC International Water and Sanitation Centre.

de Albuquerque, C. (2009). *Promotion and protection of all human rights, civil, political, economic, social and cultural rights, including the right to development: Report of the Independent Expert on the issue of human rights obligations related to access to safe drinking water and sanitation.* United Nations Human Right Council, A/HRC/12/24.

de Albuquerque, C. (2010). *Report of the Independent Expert on the Issue of Human Rights Obligations Related to Access to Safe Drinking Water and Sanitation.* United Nations General Assembly, A/65/254.

de Albuquerque, C. (2012). *On the right track back: Good practices in realising the rights to water and sanitation.* Lisbon: Textype.

de Albuquerque, C. (2014). *Realizing the human rights to water and sanitation: A handbook by the UN Special Rapporteur Catarina de Albuquerque.* Lisbon: UN.

de Beco, G. (2010). The interplay between human rights and development the other way round: The emerging use of quantitative tools for measuring the progressive realisation of economic, social and cultural rights. *Hum. Rts. & Int'l Legal Discourse, 4*, 265-287.

de Beco, G. (2011). Monitoring corruption from a human rights perspective. *International Journal of Human Rights, 15*(7), 1107-1124.

de Sadeleer, N. (2002). *Environmental principles: From political slogans to legal rules.* Oxford: Oxford University Press.

De Schutter, O., Eide, A., Khalfan, A., Orellana, M., Salomon, M., & Seidermanf, I. (2012). Commentary to the Maastricht Principles on Extraterritorial Obligations of States in the Area of Economic, Social and Cultural Rights. *Human Rights Quarterly, 34*, 1084-1169.

Dearden, L. (2016). Germany follows Switzerland and Denmark to seize cash and valuables from arriving refugees. Retrieved from http://www.independent.co.uk/news/world/europe/germany-follows-switzerland-and-denmark-to-seize-cash-and-valuables-from-arriving-refugees-a6828821.html

Dembour, M. (2006). *Who believes in human rights?: Reflections on the European Convention.* New York: Cambridge University Press.

Department for International Development. (2007). Sanitation policy background paper: Water is life, sanitation is dignity. London: DFID.

Dessy, S. E., & Vencatachellum, D. (2007). Debt relief and social services expenditure: The African experience, 1989–2003. *African Development Review, 19*(1), 200-216.

Deutsche Gesellschaft für Technische Zusammenarbeit (GTZ). (2009). *The human right to water and sanitation: Translating theory into practice.* Eschborn: Deutsche Gesellschaft für Technische Zusammenarbeit (GTZ) GmbH.

Devas, N., Amis, P., Beall, J., Grant, U., Mitlin, D., Rakodi, C., & Satterthwaite, D. (2001). *Urban governance and poverty: Lessons from a study of ten cities in the south.* Birmingham: The School of Public Policy, University of Birmingham.

Devine, J. (2010). *Global scaling up sanitation project: Introducing SaniFOAM - A framework to analyze sanitation behaviors to design effective sanitation programs.* Water and Sanitation Program. Retrieved from http://www.wsp.org/sites/wsp.org/files/publications/GSP_sanifoam.pdf

DFID. (2012). *Water, sanitation and hygiene portfolio review March 2012.* Retrieved from http://www.dfid.gov.uk/Documents/DFID%20WASH%20Portfolio%20Review.pdf

Dodane, P. H. (2012). Capital and operating costs of full-scale fecal sludge management and wastewater treatment systems in Dakar, Senegal. *Environmental Science & Technology, 46*(7), 3705-3711.

Donnelly, J. (2003). *Universal human rights in theory and practice* (2nd ed.). Ithaca; London: Cornwell University Press.

Donnelly, J. (2009). *Human dignity and human rights: Swiss Initiative to Commemorate the 60th Anniversary of the UDHR Protecting Dignity - An Agenda for Human Rights.* USA: University of Denver.

Donnelly, J. (2013). *Universal human rights in theory and practice* (3 ed.). Ithaca, NY: Cornell University Press.

dos Santos, R., & Gupta, J. (2017). Pro-poor water and sanitation: operationalising inclusive discourses to benefit the poor. *Current Opinion in Environmental Sustainability, 24*, 30-35. doi:https://doi.org/10.1016/j.cosust.2017.01.004.

Doumani, F. (2014). The cost of environmental degradation: A decade later. Retrieved from http://www.sweep-net.org/cost-environmental-degradation

Douzinas, C. (2000). *The end of human rights: Critical legal thought at the turn of the century.* Oxford: Hart Publishing.

Douzinas, C. (2013). The paradoxes of human rights. *Constellations, 20*(1), 51-67.

Dryzek, J. (1997). *The politics of the earth.* Oxford: Oxford University Press.

Eade, D. (1997). *Capacity-building: An approach to people-centered development.* Oxford, UK and Ireland: Oxfam.

Easton, G. (2010). Critical realism in case study research. *Industrial Marketing Management, 39*(1), 118-128. doi:https://doi.org/10.1016/j.indmarman.2008.06.004.

Edwards, J., & Deakin, N. (1992). Privatism and partnership in urban regeneration. *Public Administration, 70*(3), 359-368.

Eisenhardt, K. M. (1989). Building theories from case study research. *The Academy of Management Review, 14*(4), 532-550.

Ellis, K., & Feris, L. (2014). The right to sanitation: Time to delink from the right to water. *Human Rights Quarterly: A Comparative and International Journal of the Social Sciences, Philosophy, and Law, 36*(3), 607-629.

Enderlein, H., Wälti, S., & Zürn, M. (Eds.). (2010). *Handbook on multi-level governance.* Cheltenham, Massachusetts: Edward Elgar.

Ensink, J. H., Bastable, A., & Cairncross, S. (2015). Assessment of a membrane drinking water filter in an emergency setting. *Journal of Water and Health, 13*(2), 362-370.

Ensink, J. H., Blumenthal, U. J., & Brooker, S. (2008). Wastewater quality and the risk of intestinal nematode infection in sewage farming families in Hyderabad, India. *The American Journal of Tropical Medicine and Hygiene, 79*(4), 561-567.

Ersel, M. (2015). Water and sanitation standards in humanitarian action. *Turkish Journal of Emergency Medicine, 15*, 27-33.

Escobar, A. (2012). *Encountering development: The making and unmaking of the third world.* Princeton, New Jersey: Princeton University Press.

Evans, B. (2005). *Securing sanitation: The compelling case to address the crisis - Report*. Stockholm: Stockholm International Water Institute.

Evans, B. E., & Bartram, J. (2004). *The sanitation challenge: Turning commitment into reality*. Geneva, Switzerland: World Health Organization.

Evans, B., Colin, C., Jones, H., & Robinson, A. (2009). Sustainability and equity aspects of total sanitation programmes: A study of recent WaterAid-supported programmes in three countries - Global synthesis report. London: WaterAid.

Evans, W. D., Young, S., Bihm, J. W., Pattanayak, S. K., Rai, S., & Buszin, J. (2014). Social marketing of water and sanitation products: A systematic review of peer-reviewed literature. *Social Science and Medicine, 110*, 18-25.

Exley, J. L. R., Liseka, B., Cumming, O., & Ensink, J. H. J. (2015). The Sanitation Ladder, What Constitutes an Improved Form of Sanitation? *Environmental Science & Technology, 49*(2), 1086-1094.

Eyler, J. M. (2001). The changing assessments of John Snow's and William Farr's cholera studies. *Sozial- und Präventivmedizin, 46*(4), 225-232. doi:10.1007/bf01593177.

Ezeah, C., & Roberts, C. L. (2014). Waste governance agenda in Nigerian cities: A comparative analysis. *Habitat International, 41*, 121-128. doi:https://doi.org/10.1016/j.habitatint.2013.07.007.

Fagbohun, O. (2010). Legal arguments for non-justiciability of Chapter 2 of the 1999 Constitution. In E. Azinge & B. Owasanoye (Eds.), *Justiciability and constitutionalism: An economic analysis of law* (pp. 150-190). Lagos: Nigerian Institute of Advanced Legal Studies.

Fain, J. A. (2017). *Reading, understanding, and applying nursing research.* (5th Edition). Philadelphia: F.A. Davis Company.

Fan, H., Zhou, H., & Wang, J. (2014). Pyrolysis of municipal sewage sludge in a slowly heating and gas sweeping fixed-bed reactor. *Energy Conversion and Management, 88*(15), 1151-1158.

Farrall, J. (2009). Impossible expectations? The UN Security Council's promotion of the rule of law after conflict. In B. Bowden, H. Charlesworth, & J. Farrall (Eds.), *The role of international law in rebuilding societies after conflict: Great expectations* (pp. 134-157). Cambridge: Cambridge University Press.

Farran, S. (2006). Is legal pluralism an obstacle to human rights?: Considerations from the South Pacific. *The Journal of Legal Pluralism and Unofficial Law, 38*(52), 77-105.

Farrar, J. H., & Dugdale, A. M. (1990). *Introduction to legal method*. London: Sweet & Maxwell.

Fatoni, Z., & Stewart, D. E. (2012). Sanitation in an emergency situation: A case study of the eruption of Mt Merapi, Indonesia, 2010. *International Journal of Environmental Protection, 2*, 1-5.

Federal Minstry of Water Resources. (2016). *Partnership for expanded water aupply, sanitation & hygiene (PEWASH): Programmed strategy (2016 – 2030)*. Abuja: Federal Ministry of Water Resources>

Federal Ministry of Water Resources, European Union, UKAid, & UNICEF. (2016). *Making Nigeria open-defecation-free by 2025: A national road map*. Abuja: Federal Ministry of Water Resources.

Felsinger, K. (2010). *The public-private partnership handbook*. Manila: Asian Development Bank.

Feng, K., Siu, Y. L., Guan, D., & Hubacek, K. (2012). Assessing regional virtual water flows and water footprints in the Yellow River Basin, China: A consumption based approach. *Appl. Geogr., 32*, 691–701.

Fenner, R. A., Guthrie, P. M., & Piano, E. (2007). Process selection for sanitation systems and wastewater treatment in refugee camps during disaster-relief situations. *Water and Environment Journal, 21*(4), 252-264.

Feris, L. (2015). The human right to sanitation: A critique on the absence of environmental considerations. *Review of European, Comparative & International Environmental Law, 24*(1), 16-26.

Fernando, W., Gunapala, A., & Jayantha, W. (2009). Water supply and sanitation needs in a disaster: Lessons learned through the tsunami disaster in Sri Lanka. *Desalination, 248*(1-3), 14-21.

Finnemore, M., & Sikkink, K. (1998). International norm dynamics and political change. *International Organization, 52*(4), 887-917.

Fisher, B., Kulindwa, K., Mwanyoka, I., Turner, R. T., & Burgess, N. D. (2010). Common pool resource management and PES: Lessons and constraints for water PES in Tanzania. *Ecological Economics, 69*, 1253–1261.

Fisher, E. (2002). Precaution, precaution everywhere: Developing a common understanding of the precautionary principle in the European Community. *Maastricht Journal of European & Comparative Law, 9*, 7.

Fiszbein, A. (2000). Public-private partnerships as a strategy for local capacity building: Some suggestive evidence from Latin America. In P. Collins (Ed.), *Applying public administration in development: Guideposts to the future.* Chichester: Wiley.

Fiszbein, A., & Lowden, P. (1999). *Working together for a change: Government, civic, and business partnerships for poverty reduction in Latin America and the Caribbean.* Washington DC: The World Bank.

Flick, U., Kardorff, E. v., & Steinke, I. (2010). *A companion to qualitative research.* London, Thousand Oaks, New Delhi: Sage Publications.

Flores Baquero, O. s., Giné Garriga, R., Pérez Foguet, A., & Jiménez Fdez de Palencia, A. (2013). Post-2015 WASH targets and indicators: A review from a human rights perspective. Bareclona, Spain: University Research Institute for Sustainability Science and Technology (IS.UPC), Universitat Politècnica de Catalunya & ONAGAWA, Engineering for Human Development.

Flynn, J. (2012). Human rights in history and contemporary practice: Source materials for philosophy. In C. Corradetti (Ed.), *Philosophical dimensions of human rights: Some contemporary views* (pp. 3-22). Dordrecht: Springer.

Fonseca, C. (2014). Affordability of WASH services: Rules of thumb and why it's difficult to measure.   Retrieved from http://www.ircwash.org/blog/affordability-wash-services-rules-thumb-and-why-it%E2%80%99s-difficult-measure

Freitas, R. A. (1999). *Nanomedicine volume I: Basic capabilities.* Austin, TX: Landes Bioscience.

Fritz, D., Miller, U., Gude, A., Pruisken, A., & Rischewski, D. (2009). Making poverty reduction inclusive: Experiences from Cambodia, Tanzania and Vietnam. *Journal of International Development, 21*(5), 673-684.

Galvin, M. (2015). Talking shit: Is community-led total sanitation a radical and revolutionary approach to sanitation? *Wiley Interdisciplinary Reviews: Water, 2*(1), 9-20.

Garn, J. V., Caruso, B. A., Drews-Botsch, C. D., Kramer, M. R., Brumback, B. A., Rheingans, R. D., & Freeman, M. C. (2014). Factors associated with pupil toilet use in Kenyan primary schools. *International Journal of Environmental Research and Public Health, 11*(9), 9694-9711. doi:10.3390/ijerph110909694

Garn, J. V., Sclar, G. D., Freeman, M. C., Penakalapati, G., Alexander, K. T., Brooks, P., . . . Clasen, T. F. (2017). The impact of sanitation interventions on latrine coverage and latrine use: A systematic review and meta-analysis. *International Journal of Hygiene*

and   Environmental   Health,   220(2,   Part   B),   329-340.
doi:https://doi.org/10.1016/j.ijheh.2016.10.001

Gathii, J. T. (2010). Defining the relationship between human rights and corruption. *University of Pennsylvania Journal of International Law, 31*(1). 125-202.

Gavouneli, M. (2011). A human right to groundwater? *International Community Law Review, 13*(3), 305-319. doi:doi:https://doi.org/10.1163/187197311X582403

Gearty, C. A. (2006). Can human rights survive? *The Hamilyn Lectures 2005* (Vol. 57). Cambridge: Cambridge University Press.

Geels, F. W. (2002). Technological transitions as evolutionary reconfiguration processes: A multi-level perspective and a case-study. *Research Policy, 31*(8–9), 1257-1274. doi:http://dx.doi.org/10.1016/S0048-7333(02)00062-8

Geels, F. W. (2013). The impact of the financial–economic crisis on sustainability transitions: Financial investment, governance and public discourse. *Environmental Innovation and Societal Transitions, 6*(0), 67-95. doi:http://dx.doi.org/10.1016/j.eist.2012.11.004

Giles, M. (2012). The disappointments of civil society: The politics of NGO intervention in Northern Ghana. *Political Geography, 21*, 125-154.

Giné-Garriga, R., Flores-Baquero, Ó., Jiménez-Fdez de Palencia, A., & Pérez-Foguet, A. (2017). Monitoring sanitation and hygiene in the 2030 agenda for sustainable development: A review through the lens of human rights. *Science of the Total Environment, 580*, 1108-1119. doi:https://doi.org/10.1016/j.scitotenv.2016.12.066

Goldblatt, M. (1999). Assessing the effective demand for improved water supplies in informal settlements: A willingness to pay survey in Vlakfontein and Finetown, Johannesburg. *Geoforum, 30*(1), 27.

Gonçalves, S. (2014). The effects of participatory budgeting on municipal expenditures and infant   mortality   in   Brazil.   *World   Development,   53*,   94-110. doi:https://doi.org/10.1016/j.worlddev.2013.01.009

Goodbich, L. M. (1947). From League of Nations to United Nations. *International Organization 1*(1), 3-21.

Goodman, R., & Jinks, D. (2003). Measuring the effects of human rights treaties. *European Journal of International Law, 14*(1), 171-183.

Goulding, C. (2002). *Grounded theory: A practical guide for management, business and market researchers*. London: Sage.

Government of Belgium. (2007). Study on human rights and the access to water contribution of   the   Government   of   Belgium.   Retrieved   from http://www2.ohchr.org/english/issues/water/contributions/Belgium.pdf

Green, H. (2014). Use of theoretical and conceptual frameworks in qualitative research. *Nurse Researcher, 21*(6), 34-38.

Grey, D., & Sadoff, C. W. (2007). Sink or swim?: Water security for growth and development. *Water Policy, 9*, 545-571.

Griffin, J. (2008). *On human rights*. Oxford; New York: Oxford University Press.

Gross, E., & Günther, I. (2014). Why do households invest in sanitation in rural Benin: Health, wealth, or prestige? *Water Resources Research, 50*(10), 8314-8329.

Group insists 15 persons maximum to share household toilet (2017, August 1). Retrieved from   https://www.vanguardngr.com/2017/08/%EF%BB%BF%EF%BB%BFgroup-insists-15-persons-maximum-share-household-toilet/

Guimarães, E. F., Malheiros, T. F., & Marques, R. C. (2016). Inclusive governance: New concept of water supply and sanitation services in social vulnerability areas. *Utilities Policy, 43, Part A*, 124-129. doi:https://doi.org/10.1016/j.jup.2016.06.003

Guiteras, R., Levinsohn, J., & Mobarak, A. M. (2015). Encouraging sanitation investment in the developing world: A cluster-randomized trial. *Science, 348*(6237), 903-906. doi:10.1126/science.aaa0491

Gupta, J. (2014). *The history of global climate governance*. Cambridge: Cambridge University Press.

Gupta, J. & Arts, K. (2017). Achieving the 1.5°C objective: Just implementation through a right to (sustainable) development. Int. Environ. Agreements. doi:10.1007/s10784-017-9376-7.

Gupta, J., & Bavinck, M. (2014). Towards an elaborated theory of legal pluralism and aquatic resources. *Current Opinion in Environmental Sustainability, 11*, 86-93. doi:https://doi.org/10.1016/j.cosust.2014.10.006

Gupta, J., Pouw, N. R., & Ros-Tonen, M. A. (2015). Towards an elaborated theory of inclusive development. *The European Journal of Development Research, 27*(4), 541-559.

Gupta, J., & van der Grijp, N. M. (2010). *Mainstreaming climate change in development cooperation: Theory, practice and implications for the European Union*. Cambridge, UK: Cambridge University Press.

Gupta, J., & Vegelin, C. (2016). Sustainable development goals and inclusive development. *International Environmental Agreements: Politics, Law and Economics, 16*(3), 433-448.

Hafner-Burton, E. M., Helfer, L. R., & Fariss, C. J. (2011). Emergency and escape: Explaining derogations from human rights treaties. *International Organization, 65*(4), 673-707.

Hall, R. P., van Koppen, B., & van Houweling, E. (2014). The human right to water: The importance of domestic and productive water rights. *Science and Engineering Ethics, 20*(4), 849-868.

Hamlin, C. (1998). *Public health and social justice in the age of Chadwick: Britain, 1800-1854*. Cambridge: Cambridge University Press.

Hamner, S., Tripathi, A., Mishra, R., Bouskill, N., Broadaway, S., Pyle, B., & Ford, T. (2006). The role of water use patterns and sewage pollution in incidence of water-borne/enteric diseases along the Ganges river in Varanasi, India. *International Journal of Environmental Health Research, 16*(2), 113-132.

Hanley, N., & Spash, C. (1993). *Cost-benefit analysis and the environment*. Hampshire: Edward Elgar.

Hanson, S. (2003). *Legal method and reasoning*. Australia: Cavendish Publishing.

Hardin, G. (1968). The tragedy of the commons. *Science, 162*(3859), 1243-1248.

Harlow, C. (2006). Global administrative law: The quest for principles and values. *European Journal of International Law, 17*, 187-214. doi:10.1093/ejil/chi158

Harroff-Tavel, M. (1989). Neutrality and impartiality: The importance of these principles for the International Red Cross and Red Crescent Movement and the difficulties involved in applying them. *International Review of the Red Cross, 29*(273), 536-552.

Harvey, P. (2007). *Excreta disposal in emergencies: A field manual*. Loughborough: Water, Engineering and Development Centre (WEDC) Loughborough University of Technology.

Harvey, P., Baghri, S., & Reed, B. (2002). *Emergency sanitation: Assessment and programme design*. Loughborough: Water, Engineering and Development Centre, Loughborough Univ.

Hathaway, O. A. (2002). Do human rights treaties make a difference? *The Yale Law Journal, 111*(8), 1935-2042.

Haugen, H. M. (2011). Human rights principles: Can they be applied to improve the realization of social human rights? *Max Planck Yearbook of United Nations Law, 15*, 419-444.

Hawranek, P. M. (2000). Governance of economic development: The institutional and policy framework for public-private partnerships - a normative approach. *International Journal of Public Private Partnerships, 3*, 15-30.

Hayat, S., & Gupta, J. (2016). Kinds of freshwater and their relations to ecosystem services and human well-being. *Water Policy, wp2016182.* doi:10.2166/wp.2016.182

Head, J. G. (1974). *Public goods and public welfare.* Durham: Duke University Press.

Heeger, J. (2011). *Explorative study into the provision or emergency rehabilitation assistance by the Dutch Water sector.* The Netherlands: Netherlands Water Partnership.

Heijnen, M., Routray, P., Torondel, B., & Clasen, T. (2015a). Neighbour-shared versus communal latrines in urban slums: A cross-sectional study in Orissa, India exploring household demographics, accessibility, privacy, use and cleanliness. *Transactions of the Royal Society of Tropical Medicine and Hygiene, 109*(11), 690-699.

Heijnen, M., Routray, P., Torondel, B., & Clasen, T. (2015b). shared sanitation versus individual household latrines in urban slums: A cross-sectional study in Orissa, India. *The American journal of tropical medicine and hygiene, 93*(2), 263-268.

Helfer, L. R., & Slaughter, A. (1997). Toward a theory of effective supranational adjudication. *The Yale Law Journal, 107*(2), 273-391.

Henckaerts, J., & Doswald-Beck, L. (2005). *Customary international humanitarian law: Volume I - Rules.* New York: Cambridge University Press.

Hilhorst, D. (2005). Dead or alive?: Ten years of the Code of Conduct for Disaster Relief. *Humanitarian Practice Network Humanitarian Exchange, 29.*

His Majesty's Office. (1906). Colonial Reports – Annual: No. 507 – Southern Nigeria (Lagos). Report for 1905. London: Darling & Son.

His Majesty's Office. (1909). Colonial Reports – Annual: No. 594 – Nothern Nigeria. Report for 1907-8. London: Darling & Son.

His Majesty's Office. (1910a). Colonial Reports – Annual: No. 633 – Nothern Nigeria. Report for 1908-9. London: Darling & Son.

His Majesty's Office. (1910b). Colonial Reports – Annual: No. 665 – Southern Nigeria. Report for 1909. London: Darling & Son.

Hobbes, T., Gaskin, J. C. A. (1998). Leviathan. Retrieved from http://search.ebscohost.com/login.aspx?direct=true&scope=site&db=nlebk&db=nlabk &AN=12309

Holden, E., Linnerud, K., & Banister, D. (2016). The imperatives of sustainable development. *Sustainable Development.* doi:10.1002/sd.1647

Holdren, J. P., Daily, G. C., & Ehrlich, P. R. (1995). The meaning of sustainability: Biogeophysical aspetcs. In M. Munasingha & W. Shearer (Eds.), *Defining and measuring sustainability.* Washington, D C: The World Bank.

Holzer, V. (2012). *The 1951 Refugee Convention and the protection of people fleeing armed conflict and other situations of violence.* Geneva, Switzerland: United Nations High Commissioner for Refugees.

Hood, C. (2007). Intellectual obsolescence and intellectual makeovers: Reflections on the tools of government after two decades. *Governance: An International Journal of Policy, Administration, and Institutions, 20*(1), 127-144.

Horsley, T. (2012). Subsidiarity and the European Court of Justice: Missing pieces in the subsidiarity jigsaw? *Journal of Common Market Studies, 50*(2), 267-282.

Howard, G., & Bartram, J. (2003). *Domestic water quantity, service level, and health.* Geneva, Switzerland: World Health Organization.

Howlett, M. (2000). Managing the "hollow state": Procedural policy instruments and modern governance. *Canadian Public Administration, 43*(4), 412-431.

Howlett, M., & Rayner, J. (2007). Design principles for policy mixes: Cohesion and coherence in 'new governance arrangements'. *Policy and Society, 26*(4), 1-18. doi:http://dx.doi.org/10.1016/S1449-4035(07)70118-2

Hulland, K., Martin, N., Dreibelbis, R., Valliant, J. D., & Winch, P. (2015). *What factors affect sustained adoption of safe water, hygiene and sanitation technologies?: A systematic review of literature.* London: EPPI-Centre, Social Science Research Unit, UCL Institute of Education, University College London.

Human Rights Council. (2010). *Resolution adopted by the Human Rights Council 15/9: Human rights and access to safe drinking water and sanitation.*

Humanitarian Accountability Partnership. (2008). The guide to the HAP standard: Humanitarian accountability and quality management. Cowley, UK: Oxfam.

Hutton, G. (2012). Global costs and benefits of drinking-water supply and sanitation interventions to reach the MDG target and universal coverage. Retrieved from http://whqlibdoc.who.int/hq/2012/WHO_HSE_WSH_12.01_eng.pdf

Hutton, G. (2013). Global costs and benefits of reaching universal coverage of sanitation and drinking-water supply. *Journal of Water and Health, 11*(1), 1-12.

Hutton, G., & Varughese, M. (2016). *The costs of meeting the 2030 Sustainable Development Goal targets on drinking water, sanitation, and hygiene. Water and Sanitation Program: Technical paper 103171.* International Bank for Reconstruction and Development / World Bank.

IDGEC. (2005). *Science Plan: Institutional Dimensions of Global Environmental Change – IHDP Report No. 16.* Bonn: International Human Dimensions Programme (IHDP).

Ijaiya, H. (2013). The legal framework for solid waste disposal and management in Kwara State, Nigeria. *Journal of Environmental Protection, 4*(11), 1240-1244. doi:http://dx.doi.org/10.4236/jep.2013.411143.

Internal Displacement Monitoring Centre [IDMC] & Norwegian Refugee Council [NRC]. (2017). *Global report on internal displacement (GRID) 2017.* Geneva: IDMC.

International Committee of the Red Cross. (2002). *The law of armed conflict: Neutrality.* Geneva: International Committee of the Red Cross.

International Committee of the Red Cross. (2014). *International humanitarian law: Answers to your questions.* Geneva: International Committee of the Red Cross.

International Council on Human Rights Policy. (2009a). *When legal worlds overlap: Human rights, state and non-state law.* Versoix, Switzerland: ICHRP.

International Council on Human Rights Policy. (2009b). *Corruption and Human Rights: Making the Connection* - Working Paper. Geneva, Switzerland: ICHRP.

International Council on Human Rights Policy & Transparency International. (2010). *Integrating Human Rights in the Anti-Corruption Agenda: Challenges, Possibilities and Opportunities*, Geneva, Switzerland: ICHRP.

IRIN. (2012). Madagascar: Addressing toilet taboos to improve sanitation. *News.* Retrieved from http://www.irinnews.org/report/95136/madagascar-addressing-toilet-taboos-to-improve-sanitation

Irish, S., Aiemjoy, K., Torondel, B., Abdelahi, F., Ensink, J. H. J., & Shiff, C. (2013). Characteristics of latrines in central tanzania and their relation to fly catches. *PLoS ONE, 8*(7), e67951.

Isunju, J. B., Schwartz, K., Schouten, M. A., Johnson, W. P., & van Dijk, M. P. (2011). Socio-economic aspects of improved sanitation in slums: A review. *Public Health, 125*(6), 368-376.

Jacobs, M. (1999). Sustainable development as a contested concept. In A. Dobson (Ed.), *Fairness and futurity: Essays on environmental sustainability and social justice.* Oxford: Oxford University Press.

Jahre, M., Ergun, O., & Goentzel, J. (2015). One size fits all? Using standard global tools in humanitarian logistics. *Procedia Engineering, 107*, 18-26.

Jamali, D. (2004). Success and failure mechanisms of public private partnerships (PPPs) in developing countries. *International Journal of Public Sector Management, 17*(5), 414–430. doi:http://doi.org/10.1108/09513550410546598

Janse, R. (2013). A turn to legal pluralism in the rule of law promotion. *Erasmus Law Review, 6*(3/4), 181-190.

Jaramillo, M., & Alcázar, L. (2013). *Does participatory budgeting have an effect on the quality of public services?: The case of Peru's water and sanitation sector.* Retrieved from http://EconPapers.repec.org/RePEc:idb:brikps:79661

Jasper, C., Le, T.-T., & Bartram, J. (2012). Water and sanitation in schools: A systematic review of the health and educational outcomes. *International Journal of Environmental Research and Public Health, 9*(8), 2772-2787. doi:10.3390/ijerph9082772.

Jenkins, M. W., & Scott, B. (2007). Behavioral indicators of household decision-making and demand for sanitation and potential gains from social marketing in Ghana. *Social Science & Medicine, 64*(12), 2427-2442.

Jennings, R. Y. (1996). The judiciary, international and national, and the development of international law. *International and Comparative Law Quarterly, 45*(01), 1-12.

Jensen, M. H., Villumsen, M., & Petersen, T. D. (2014). *The AAAQ framework and the right to water: International indicators for availability, accessibility, acceptability and quality. An issue paper of the AAAQ Toolbox.* Copenhagen: Danish Institute for Human Rights.

Jeppsson, U., & Hellström, D. (2002). Systems analysis for environmental assessment of urban water and wastewater systems. *Water Science and Technology, 46*(6-7), 121-129.

Jewitt, S. (2011). Geographies of shit: Spatial and temporal variations in attitudes towards human waste. *Progress in Human Geography, 35*(5), 608-626.

Jewitt, S., & Ryley, H. (2014). It's a girl thing: Menstruation, school attendance, spatial mobility and wider gender inequalities in Kenya. *Geoforum, 56*, 137-147. doi:http://dx.doi.org/10.1016/j.geoforum.2014.07.006

Jiménez, A., & Pérez-Foguet, A. (2010). Building the role of local government authorities towards the achievement of the human right to water in rural Tanzania. *Natural Resources Forum, 34*(2), 93-105.

Joaquín, R., & Muñiz, T. (1997). Legal principles and legal theory. *Ratio Juris, 10*(3), 267-287.

Johannessen, A., Patinet, J., Carter, W., Lamb, J. (2012). *Sustainable sanitation for emergencies and reconstruction situations - Factsheet of Working Group 8. Sustainable Sanitation Alliance (SuSanA).* Retrieved from http://www.susana.org/en/knowledge-hub/resources-and-publications/library/details/797

Jones, R. (2000). A role for public-private partnerships in an enlarged European Union. *International Journal of Public Private Partnerships, 3*(1), 31-44.

Joshi, A., & Amadi, C. (2013). Impact of water, sanitation, and hygiene interventions on improving health outcomes among school children. *Journal of Environmental and Public Health, 2013*, 10. doi:10.1155/2013/984626.

Joshi, D., Fawcett, B., & Mannan, F. (2011). Health, hygiene and appropriate sanitation: Experiences and perceptions of the urban poor. *Environ. Urban. Environment and Urbanization, 23*(1), 91-111.

Joss, A., Zabczynski, S., Gobel, A., Hoffmann, B., Loffler, D., McArdell, C. S., . . . Siegrist, H. (2006). Biological degradation of pharmaceuticals in municipal wastewater treatment: Proposing a classification scheme. *Water Research, 40*(8), 1686-1696.

Juuti, P. S. (2007). First innovations of water supply and sanitation. In P. S. Juuti, T. S. Katko, & H. S. Vuorinen (Eds.), *Environmental History of Water* (pp. 17 - 43). London: IWA Publishing.

Kaddar, M., & Furrer, E. (2008). Are current debt relief alternatives an option for scaling up health financing in beneficiary countries? *Bulletin of the World Health Organization, 86*, 877-883.

Kaïka, M. (2003). Constructing scarcity and sensationalizing water politics: 170 days that shook Athens. *Antipode, 35*, 919-954.

Kälin, W. (2008). *Guiding principles on internal displacement: Annotations*. Washington, DC: American Society of International Law.

Kamga, S. D. (2013). The right to basic sanitation: A human right in need of constitutional guarantee in Africa. *South African Journal on Human Rights, 29*, 615-650.

Kar, K. (2003). *Subsidy or self-respect?: Participatory total community sanitation in Bangladesh* Brighton: Institute of Development Studies.

Kar, K., & Milward, K. (2011). *Digging in, spreading out and growing up: Introducing CLTS in Africa* (Vol. 2011). Brighton: Institute of Development Studies.

Kar, K., & Pasteur, K. (2005). *Subsidy or self-respect?: Community led total sanitation; an update on recent developments*. Brighton: Institute of Development Studies.

Kariuki, M. (2014). *Do pro-poor policies increase water coverage?: An analysis of service delivery in Kampala's informal settlements*. Washington: The World Bank.

Karpouzoglou, T., & Zimmer, A. (2016). Ways of knowing the wastewaterscape: Urban political ecology and the politics of wastewater in Delhi, India. *Habitat International, 54, Part 2*, 150-160. doi:http://dx.doi.org/10.1016/j.habitatint.2015.12.024.

Kassim, H., & Le Galès, P. (2010). Exploring governance in a multi-level polity: A policy instruments approach. *West European Politics, 33*(1), 1-21.

Katukiza, A. Y., Ronteltap, M., Niwagaba, C. B., Foppen, J. W. A., Kansiime, F., & Lens, P. N. L. (2012). Sustainable sanitation technology options for urban slums. *Biotechnology Advances, 30*(5), 964-978. doi:http://dx.doi.org/10.1016/j.biotechadv.2012.02.007

Katukiza, A. Y., Ronteltap, M., Oleja, A., Niwagaba, C. B., Kansiime, F., & Lens, P. N. L. (2010). Selection of sustainable sanitation technologies for urban slums: A case of Bwaise III in Kampala, Uganda. *Science of The Total Environment, 409*(1), 52-62. doi:http://dx.doi.org/10.1016/j.scitotenv.2010.09.032

Kennedy, D. (1987). The sources of international law. *American University International Law Review, 2*(1), 1-96.

Kenny, S., & Clarke, M. (Eds.). (2010). *Challenging capacity building: Comparative perspectives*. Hampshire: Palgrave Macmillan.

Keraita, B., Drechsel, P., & Konradsen, F. (2010). Up and down the sanitation ladder: Harmonizing the treatment and multiple-barrier perspectives on risk reduction in wastewater irrigated agriculture. *Irrigation and Drainage Systems, 24*(1-2), 23-35.

Kiefer, T., & Brölmann, C. (2005). Beyond state sovereignty: The human right to water. *Non-State Actors and International Law, 5*(3), 183-208.

Klasing, A. M., Moses, P. S., & Satterthwaite, M. L. (2011). Measuring the way forward in Haiti: Grounding disaster relief in the legal framework of human rights. *Health and Human Rights, 13*(1), 15-35.

Koh, H. H. (1999). How is international human rights law enforced? *Indiana Law Journal, 74*(4), 1397-1417.

Koskenniemi, M., & Leino, P. (2002). Fragmentation of international law?: Postmodern anxieties. *Leiden Journal of International Law, 15*(3), 553-579.

Kramek, N., & Loh, L. (2007). The history of Philadelphia's water supply and sanitation system. lessons in sustainability of developing urban water systems. *Master of Environmental Studies. Philadelphia, University of Pennsylvania, Philadelphia Global Water Initiative, junio.*

Krippendorff, K. (2013). *Content analysis: An introduction to its methodology* (3rd ed.). Los Angeles, London, New Delhi, Singapore: Sage Publications.

Kühl, S. (2009). Capacity development as the model for development aid organizations. *Development and Change, 40*(3), 551-577.

Kvarnström, E., McConville, J., Bracken, P., Johansson, M., & Fogde, M. (2011). The sanitation ladder - a need for a revamp? *Journal of Water, Sanitation and Hygiene for Development, 1*(1), 3-12. doi:doi: 10.2166/washdev.2011.014

Kwiringira, J., Atekyereza, P., Niwagaba, C., & Günther, I. (2014). Descending the sanitation ladder in urban Uganda: Evidence from Kampala slums. *BMC Public Health, 14*(1), 624.

Labib, A., & Read, M. (2015). A hybrid model for learning from failures: The Hurricane Katrina disaster. *Expert Systems with Applications, 42*(21), 7869-7881.

Lacey, A. (2010). The research process. In K. Gerrish & A. Lacey (Eds.), *The research process in nursing* (6th ed., pp. 13-26). West Sussex: Wiley-Blackwell.

Ladan, M. (2009). Nigeria. In L. J. Kotze & A. R. Paterson (Eds.), *The role of the judiciary in environmental governance: Comparative perspectives* (pp. 527-556). The Netherlands: Kluwer Law International.

Lahiri, S., & Chanthaphone, S. (2000). *Consumers choice.......The sanitation ladder: Rural sanitation options in Lao PDR.* Vientiane, Lao PDR: Nam Saat, WSP-EAP, UNICEF – Water and Environmental Sanitation Section, Lao PDR.

Lancet Editorial. (2010). Water and sanitation become human rights, albeit turbidly. *The Lancet, 376*(9739), 390. doi:doi:10.1016/S0140-6736(10)61203-2.

Lascoumes, P., & Le Galès, P. (2007). Introduction: Understanding public policy through its instruments-From the nature of instruments to the sociology of public policy instrumentation. *Governance: An International Journal of Policy, Administration, and Institutions, 20*(1), 1-21.

Lawrence-Agbai, I. (2016). "Nigeria loses N500bn on poor sanitation yearly – Reckitt Benckiser". *Today*, May 12, 2016.

Lawson, V. (2010). Reshaping economic geography?: Producing spaces of inclusive development. *Economic Geography, 86*(4), 351–360.

Laven, A. C. (2010). *The risks of inclusion: Shifts in governance processes and upgrading opportunities for cocoa farmers in Ghana.* Amsterdam: KIT Publishers.

Lee, P., & George, R. P. (2008). The nature and basis of human dignity. *Ratio Juris, 21*(2), 173-193.

Lensu, M. (2003). *Respect for culture and customs in international humanitarian assistance: Implications for principles and policy.* [Manuscript submitted in fulfilment of a Ph.D. in Government]. (UMI U615845). Ann Arbor, MI.

Lenton, R., Wright, A. M., & Lewis, K. (2005). *Health, dignity, and development: What will it take?* London and Sterling, Va.: Earthscan.

Limon, M. (2010). Human rights obligations and accountability in the face of climate change. *Georgia Journal of International and Comparative Law, 38*, 543-592.

Lixil, WaterAid Japan & Oxford Economics. (2016). *The true cost of poor sanitation*. Tokyo: Lixil.

Locke, J., Laslett, P. (1988). *Two treatises of government*. Cambridge [England]; New York: Cambridge University Press.

London, L., & Schneider, H. (2012). Globalisation and health inequalities: Can a human rights paradigm create space for civil society action?. *Social Science & Medicine, 74*, 6-13.

Longe, E., & Williams, A. (2006). A preliminary study of medical waste management in Lagos metropolis, Nigeria. *Journal of Environmental Health Science & Engineering, 3*(2), 133-139.

Lopes, A. M., Fam, D., & Williams, J. (2012). Designing sustainable sanitation: Involving design in innovative, transdisciplinary research. *Design Studies, 33*(3), 298-317. doi:10.1016/j.destud.2011.08.005

Luber, G., & Lemery, J. (2015). *Global climate change and human health: From science to practice*. San Francisco: Jossey-Bass.

Mader, P. (2012). Attempting the production of public goods through microfinance: The case of water and sanitation. *Economic Research-Ekonomska Istraživanja, 25*(sup1), 190-214.

Magesan, A. (2013). Human rights treaty ratification of aid receiving countries. *World Development, 45*(0), 175-188. doi:http://dx.doi.org/10.1016/j.worlddev.2012.11.003

Majoor, S., & Schwartz, K. (2015). Instruments of urban governance. In J. Gupta, K. Pfeffer, H. Verrest, & M. Ros-Tonen (Eds.), *Geographies of urban governance: Advanced theories, methods and practices*. Heidelberg: Springer International Publishing.

Malebo, H. M., Makundi, E. A., Mussa, R., Mushi, A. K., Munga, M. A., Mrisho, M., . . . Tenu, P. (2012). *Outcome and impact monitoring for scaling up Mtumba sanitation and hygiene participatory approach in Tanzania*. Dar es Salaam: Tanzanian National Institute for Medical Research.

Malkin, J., & Wildavsky, A. (1991). Why the traditional distinction between public and private goods should be abandoned. *Journal of Theoretical Politics, 3*(4), 355-378.

Mara, D. D. (1996). *Low-cost sewerage*. Chichester: Wiley.

Mara, D. D., & Guimarães, A. S. P. (1999). Simplified sewerage: Potential applicability in industrialized countries. *Urban Water, 1*(3), 257-259. doi:https://doi.org/10.1016/S1462-0758(99)00015-1

Marin, P. (2009). *Public-private partnerships fpor urban water utilities: A review of experiences in developing countries*. Washington DC: The World Bank

Marks, G. (1992). Structural policy in the European Community. In A. Sbragia (Ed.), *Europolitics : Institutions and policymaking in the "new" European Community* (pp. 191-225). Washington: Brookings Institute.

Martin, L. (2011). International legal discourse on the human right to water and sanitation from the Latin American point of view. *Inter-American and European Human Rights Journal, 4*(1-2), 136-154.

Martin, M., & Walker, J. (2015). *Financing the Sustainable Development Goals: Lessons from government spending on the MDGs - Government Spending Watch Report*. Oxfam and Development Finance International. Retrieved from http://policy-practice.oxfam.org.uk/publications/financing-thesustainable-development-goals-lessons-from-government-spending-on-556597

Martin, M., & Watts, R. (2013). Putting progress at risk?: MDG spending in developing countries. GB: Oxfam.

Martini, A., & Spataro, L. (2016). The principle of subsidiarity and the ethical factor in Giuseppe Toniolo's thought. *Journal of Business Ethics*, 1-15. doi:10.1007/s10551-016-3407-0

Martin-Simpson, S., Parkinson, J., & Katsou, E. (2017). Measuring the benefits of using market based approaches to provide water and sanitation in humanitarian contexts. *Journal of Environmental Management*. doi:https://doi.org/10.1016/j.jenvman.2017.03.009

Maru, M. T. (2014). *The Kampala Convention and its contributions to international law: Legal analyses and interpretations of the African Union Convention for the protection and assistance of internally displaced persons*. The Hague: Eleven International Publishing.

Mathew, K., Zachariah, S., Shordt, K., Snel, M., Cairncross, S., Biran, A., & Schmidt, W.-P. (2009). The sustainability and impact of school sanitation, water and hygiene education in southern India. *Waterlines, 28*(4), 275-292.

Maunganidze, D. (2016). International humanitarian law violations: The causes and ways of improving compliance. *International Journal of Humanities and Social Science Research, 2*, 30-33.

Maxwell, J. A. (2012). *Qualitative research design: An interactive approach*. Thousand Oaks: Sage.

Mayer, B. (2015). The applicability of the principle of prevention to climate change: A response to Zahar. *Climate Law, 5*(1), 1-24. doi:doi:https://doi.org/10.1163/18786561-00501001

Mbonu, C. (2009). Human rights bodies and mechanisms: Report of Ms. Christy Mbonu, Special Rapporteur on corruption and its impact on the full enjoyment of human rights. A/HRC/11/CRP.1. Human Rights Council Eleventh session 2 – 19 June 2009 Agenda item 5. Retrieved from http://www.ohchr.org/Documents/Issues/Development/GoodGovernance/A_HRC_11_CRP_1.pdf

McCourt, W. (). Lost in translation: The World Bank and the Paris Declaration. *Dev Policy Rev*. Accepted Author Manuscript. Doi:10.1111/dpr.12297.

McCrudden, C. (2008). Human dignity and judicial interpretation of human rights. *European Journal of International Law, 19*(4), 655-724.

McDade, P. V. (1986). The effect of Article 4 of the Vienna Convention on the Law of Treaties 1969. *International and Comparative Law Quarterly, 35*(3), 499-511. doi:10.1093/iclqaj/35.3.499

McDougal, L., & Beard, J. (2011). Revisiting Sphere: New standards of service delivery for new trends in protracted displacement. *Disasters, 35*(1), 87-101.

McFarlane, C. (2008). Sanitation in Mumbai's informal settlements: State,'slum', and infrastructure. *Environment and planning A, 40*(1), 88-107.

McGoldrick, D. (2009). Accommodating Muslims in Europe from adopting Sharia Law to religiously based opt outs from generally applicable laws. *Human Rights Law Review, 9*(4), 603-645.

McGranahan, G. (2013). *Community-driven sanitation improvement in deprived urban neighbourhoods: Meeting the challenges of local collective action, co-production, affordability and a trans-sectoral approach*. London: SHARE London School of Hygiene and Tropical Medicine.

McIntyre, O. (2012). The human right to water as a creature of global administrative law. *Water International, 37*, 654-669. doi:10.1080/02508060.2012.710948

McKay, J. (2014). General jurisprudence, empirical legal theory, epistemic fruit, and the ontology of 'law': Scope, scepticism, demarcation, artefacts, hermeneutic concepts, normativity and natural kinds. *Osgoode Legal Studies Research Paper Series, 36.*

McLellan, M., & Porter, T. (2007). *News, improved: How America's newsrooms are learning to change.* Washington, DC: CQ Press.

Mehmetoglu, M., & Altinay, L. (2006). Examination of grounded theory analysis with an application to hospitality research. *International Journal of Hospitality Management, 25*(1), 12-33. doi:https://doi.org/10.1016/j.ijhm.2004.12.002

Mehta, L., & Movik, S. (2011). *Shit matters: The potential of community-led total sanitation.* Rugby, UK: Practical Action Publishing.

Mehta, L., Allouche, J., Nicol, A., & Walnycki, A. (2014). Global environmental justice and the right to water: The case of peri-urban Cochabamba and Delhi. *Geoforum, 54*, 158-166.

Mehta, M. (2008). *Assessing Microfinance for Water and Sanitation: Exploring opportunities for scaling up.* Seattle: Bill and Melinda Gates Foundation.

Meier, B. M., Cronk, R., Luh, J., Bartram, J., & de Albuquerque, C. (2017). Monitoring the progressive realization of the human rights to water and sanitation: Frontier analysis as a basis to enhance human rights accountability. In K. Conca & E. Weinthal (Eds.), *The Oxford Handbook of Water Politics nd Policy* (25 pages). London: IWA Publishing. DOI:10.1093/oxfordhb/9780199335084.013.21. (forthcoming). Retrieved from http://bmeier.web.unc.edu/files/2017/03/oxfordhb-9780199335084-e-21.pdf

Meier, B. M., Kayser, G. L., Kestenbaum, J. G., Amjad, U. Q., Dalcanale, F., & Bartram, J. (2014). Translating the human right to water and sanitation into public policy reform. *Science and Engineering Ethics, 20*(4), 833-848.

Meierotto, L. (2015). Human rights in the context of environmental conservation on the US-Mexico border. *Journal of Human Rights, 14*(3), 401-418.

Mengistu, M. G., Simane, B., Eshete, G., & Workneh, T. S. (2015). A review on biogas technology and its contributions to sustainable rural livelihood in Ethiopia. *Renewable and Sustainable Energy Reviews, 48*(5), 306-316.

Meron, T. (2000). The humanization of humanitarian law. *American Journal of International Law*, 239-278.

Meyer, W. H. (1998). *Human rights and international political economy in third world nations: Multinational corporations, foreign aid, and repression.* Westport, CT: Praeger.

Miller, G. J. (2005). The political evolution of principal-agent models. *Annual Review of Political Science, 8*(1), 203-225.

Minear, L. (1999). The theory and practice of neutrality: Some thoughts on the tensions. *International Review of the Red Cross*(833).

Misiedjan, D., & Gupta, J. (2014). Indigenous communities: Analyzing their right to water under different international legal regimes. *Utrecht Law Review, 10*(2), 77-90.

Misra, A. K. (2014). Climate change and challenges of water and food security. *International Journal of Sustainable Built Environment* (0). doi:http://dx.doi.org/10.1016/j.ijsbe.2014.04.006

Mooney, A. (2012). Human rights: Law, language and the bare human being. *Language & Communication, 32*(3), 169-181. doi:http://dx.doi.org/10.1016/j.langcom.2011.12.001

Moore, S. A. (2001). Facility hostility: Sex discrimination and women's restrooms in the workplace. *Georgia Law Review, 36*(2), 599-634.

Moore, W. H., & Shellman, S. M. (2006). Refugee or internally displaced person?: To where should one flee? *Comparative Political Studies, 39*(5), 599-622.

Moravcsik, A. (2000). The origins of human rights regimes: Democratic delegation in postwar Europe. *International Organization, 54*(2), 217-252.

Morley, I. (2007). City chaos, contagion, Chadwick, and social justice. *The Yale Journal of Biology and Medicine, 80*(2), 61-72.

Motevallian, S., & Tabesh, M. (2011). *A framework for sustainability assessment of urban water systems using a participatory approach.* Paper presented at the Proceedings of 4th International Perspective on Water Resources and The Environment Conference, Singapore, National University of Singapore

Mottershaw, E., & Murray, R. (2012). National responses to human rights judgments: The need for government co-ordination and implementation. *European human rights law review*(6), 639-653.

Moyn, S. (2010). *The last utopia: Human rights in history.* Cambridge, Mass.: Belknap Press of Harvard University Press.

Mukherjee, A. (2005). *Engaging communities in public-private partnerships in the delivery of basic services to the poor: Inter-country models and perspectives.* Bangkok: United Nations Economic and Social Commission for Asia and the Pacific.

Mulenga, M., Manase, G., & Fawcett, B. (2004). *Building links: For improved sanitation in poor urban settlements- Recommendations from research in Southern Africa.* Southampton: Institute of Irrigation and Development Studies.

Müller, A., & Seidensticker, F. L. (2007). *The role of national human rights institutions and the United Nations treaty body process: Handbook.* Berlin: German Institute for Human Rights.

Müllerson, R. A. (1997). *Human rights diplomacy.* London; New York: Routledge.

Munamati, M., Nhapi, I., & Misi, S. (2016). Exploring the determinants of sanitation success in Sub-Saharan Africa. *Water Research, 103*, 435-443. doi:https://doi.org/10.1016/j.watres.2016.07.030

Murphy, J. B. (Ed.) (1999) Cambridge dictionary of philosophy. (2nd ed.). Cambridge, UK: Cambridge University Press.

Murthy, S. L. (2012). Land security and the challenges of realizing the human right to water and sanitation in the slums of Mumbai, India. *Health and Human Rights, 14*(2), 61-73.

Murthy, S. L. (2013). The human right (s) to water and sanitation: History, meaning and the controversy over privatization. *Berkeley Journal of International Law (BJIL), 31*(1).

Musgrave, R. A. (1957). Principles of budget determination. In Joint Economic Committee (Ed.), *Federal expenditure policy for economic growth and stability* (pp. 108–115). Washington, DC: Government Printing Office.

Musgrave, R. A. (1959). *The theory of public finance.* New York: McGraw-Hill Book Company.

Mutua, M. (Ed.) (2013). *Human rights NGOs in East Africa: Political and normative tensions.* Philadelphia: University of Pennsylvania Press.

Naisbett, J., & Aburdene, P. (1990). Megatrend 2000: Ten new directions for the 1990's. New York: William and Morrow

Nakagiri, A., Niwagaba, C. B., Nyenje, P. M., Kulabako, R. N., Tumuhairwe, J. B., & Kansiime, F. (2016). Are pit latrines in urban areas of Sub-Saharan Africa performing? A review of usage, filling, insects and odour nuisances. *BMC Public Health, 16*(1), 120-136.

Nakhaei, M., Khankeh, H. R., Masoumi, G. R., Hosseini, M. A., Parsa-Yekta, Z., Kurland, L., & Castren, M. (2015). Impact of disaster on women in Iran and implication for emergency nurses volunteering to provide urgent humanitarian aid relief: A qualitative study. *Australasian Emergency Nursing Journal, 18*, 165–172.

National Emergency Management Agency. (2013). National disaster management framework. Abuja: NEMA.

National Population Commission. (2010). *Population distribution by sex & types of household (state & local government area): Table HH1 – 2006 population and housing census priority table volume III.* Abuja, Nigeria: National Population Commission.

National Population Commission (NPC) [Nigeria] & ICF International. (2014). *Nigeria demographic and health survey 2013.* Abuja, Nigeria; Rockville, Maryland, USA: NPC and ICF International.

National Research Council. (2006). *Review of the draft research and restoration plan for Arctic-Yukon-Kuskokwim (Western Alaska) Salmon.* Washington, DC: The National Academies Press.

Nelson, M. J. (2010). Persistent legal pluralism and the challenge of universal human rights. *Journal of Human Rights Practice, 2*(3), 401-407. https://doi.org/10.1093/jhuman/huq012.

Neumayer, E. (2005). Do international human rights treaties improve respect for human rights? *Journal of Conflict Resolution, 49*(6), 925-953.

Nicole, W. (2015). The WASH approach: Fighting waterborne diseases in emergency situations. *Environmental Health Perspectives, 123*(1), A6-A15. doi:10.1289/ehp.123-A6

Nielsen, R. A., & Simmons, B. A. (2015). Rewards for ratification: Payoffs for participating in the international human rights regime? *International Studies Quarterly, 59*(2), 197-208.

Nieto-Navia, R. (2003). International peremptory norms (jus cogens) and international humanitarian law: *Man's inhumanity to man- Essays on international law in honour of Antonio Cassese* (pp. 595-640). The Hague Kluwer Law International.

Nirupama, N. (2013). Disaster risk management. In P. T. Bobrowsky (Ed.), *Encyclopedia of Natural Hazards* (pp. 164-170). Dordrecht, Heidelberg, New York, London: Springer.

Nkhata, M. J., & Mwenifumbo, A. W. (2010). The more things seem to change, the more they stay the same: The strained matrix for the protection and enforcement of human rights in Africa. *East African Journal of Peace and Human Rights, 16*(2), 139-215.

Nwabueze, R. N. (2010). Securing widows' sepulchral rights through the Nigerian constitution. *Harvard Human Rights Journal, 23,* 141–55.

Nwachukwu, N. C., Orji, F. A., & Ugbogu, O. C. (2013). Health care waste management: Public health benefits, and the need for effective environmental regulatory surveillance in Federal Republic of Nigeria. *Current Topics in Public Health.*

Nwakanma, P. C., & Nnamdi, K. C. (2013). Income status and homeownership: Micro-econometric evidence on Nigerian households. *West African Journal of Industrial and Academic Research, 8*(1), 182-191.

Nygaard, I., & Linder, M. (1997). Thirst at work: An occupational hazard? *International Urogynecology Journal, 8*(6), 340-343. doi:10.1007/bf02765593

Nzeadibe, T. C., & Anyadike, R. N. (2012). Social participation in city governance and urban livelihoods: Constraints to the informal recycling economy in Aba, Nigeria. *City, Culture and Society, 3*(4), 313-325.

Obani, P. (2015). The human rights to water and sanitation in courts worldwide: A selection of national, regional, and international case law, WaterLex, 2014, ISBN: 978-2-940526-00-0. *International Environmental Agreements: Politics, Law and Economics, 15*(2), 237-239.

Obani, P. (2017). Inclusiveness in humanitarian action access to water, sanitation & hygiene in focus. *Current Opinion in Environmental Sustainability, 24,* 24-29.

Obani, P., & Gupta, J. (2014a). The human right to water and sanitation: Reflections on making the system effective. In A. Bhaduri, J. Bogardi, J. Leentvaar, & S. Marx (Eds.), *The global water system in the anthropocene: Challenges for science and governance* (pp. 385-399). Cham: Springer Verlag.

Obani, P., & Gupta, J. (2014b). Legal pluralism in the area of human rights: Water and sanitation. *Current Opinion in Environmental Sustainability, 11*, 63-70.

Obani, P., & Gupta, J. (2015). The evolution of the right to water and sanitation: Differentiating the implications. *Review of European, Comparative & International Environmental Law, 24*(1), 27-39. doi:10.1111/reel.12095

Obani, P., & Gupta, J. (2016a). Human right to sanitation in the legal and non-legal literature: The need for greater synergy. *Wiley Interdisciplinary Reviews: Water, 3*(5), 678-691.

Obani, P., & Gupta, J. (2016b). Human security and access to water, sanitation, and hygiene: Exploring the drivers and nexus. In C. Pahl-Wostl, A. Bhaduri, & J. Gupta (Eds.), *Handbook on water security* (pp. 201-214). Glos; Massachusetts: Edward Elgar Publishing

Obani, P. C., & Gupta, J. (2016c). The impact of economic recession on climate change: Eight trends. *Climate and Development, 8*(3), 211-223. doi:10.1080/17565529.2015.1034226.

Odhiambo, F. O. (2010). *Water and sanitation monitoring platforms: Lessons learned from implementation*. Loughborough: WEDC, Loughborough University.

Offenheiser, R. C., & Holcombe, S. H. (2003). Challenges and opportunities in implementing a rights-based approach to development: An Oxfam America perspective. *Nonprofit and Voluntary Sector Quarterly, 32*(2), 268-301.

Office of the United Nations High Commissioner for Human Rights. (2012). Principles and Guidelines for a Human Rights Approach to Poverty Reduction Strategies HR/PUB/06/12.

Office of the Senior Special Assistant to the President on SDGs [OSSAP-SDGs] & National Bureau of Statistics [NBS]. (2017). *Nigeria: Sustainable Development Goals (SDGs) indicators baseline report 2016*. Abuja: OSSAP-SDGs; NBS.

O'Flaherty, M., Kędzia, Z., Müller, A., & Ulrich, G. (Eds.). (2011). *Introduction: Human rights diplomacy- Contemporary perspectives*. Leiden; Boston: Martinus Nijhoff Publishers.

Ogbonna, D. N., Chindah, A., & Ubani, N. (2012). Waste management options for health care wastes in Nigeria: A case study of Port Harcourt hospitals. *Journal of Public health and Epidemiology, 4*(6), 156-169.

Opschoor, J. H. (2006). Water and merit goods. *International Environmental Agreements: Politics, Law and Economics, 6*(4), 423-428.

Oredola, T. (2016). "Nigeria losses N500 billion on poor sanitation yearly, says group". *The Guardian,* 12 May 2016. Retrieved from https://guardian.ng/features/nigeria-loses-n500b-on-poor-sanitation-yearly-says-group/

Organisation for Economic Co-Operation and Development [OECD]. Recommendation on the Implementation of the Polluter Pays Principle 1974, C(74)223.

Organisation for Economic Co-Operation and Development [OECD]. (2017). *Development aidf rises again in 2016*. Paris: OECD. Retrieved from https://www.oecd.org/dac/financing-sustainable-development/development-finance-data/ODA-2016-detailed-summary.pdf

O'Rourke, E. (1992). The international drinking water supply and sanitation decade: Dogmatic means to a debatable end. *Water Science and Technology, 26*(7-8), 1929-1939.

Ostrom, E. (2008). The challenge of common-pool resources. *Environment: Science and Policy for Sustainable Development, 50*(4), 8-21.

Otaki, Y., Otaki, M., & Sakura, O. (2007). Water systems and urban sanitation: A historical comparison of Tokyo and Singapore. *Journal of Water and Health, 5*(2), 259-265.

Oyebade, W. (2016, January 10). 839 persons await trial for environmental offences in Lagos. *Guardian.* Retrieved from https://guardian.ng/news/839-persons-await-trial-for-environmental-offences-in-lagos/

Parahoo, K. (2006). *Nursing research: Principles, process and issues* (2nd ed.). Basingstoke: Palgrave Macmillan.

Parekh, S. (2007). Resisting "dull and torpid" assent: Returning to the debate over the foundations of human rights. *Human Rights Quarterly, 29*(3), 754-778.

Parkinson, J., Tayler, K., Colin, J., & Nema, A. (2008). *A guide to decisionmaking: Technology options for urban sanitation in India.* New Delhi: Water and Sanitation Program-South Asia, World Bank.

Paterson, C., Mara, D., & Curtis, T. (2007). Pro-poor sanitation technologies. *Geoforum, 38*(5), 901-907.

Pat-Mbano, E., & Ezirim, O. N. (2015). Capacity building strategy for sustainable environmental sanitation in Imo State, Nigeria. *Journal of Emerging Trends in Economics and Management Sciences, 6*(8), 395-402.

Piattoni, S. (2010). *The theory of multi-level governance: Conceptual, empirical, and normative challenges.* Oxford, U.K.; New York: Oxford University Press.

Pillay, N. (2009). Statement by Navanethem Pillay, United Nations High Commissioner for Human Rights, 'Human rights diplomacy: An oxymoron?'. Retrieved from OCHR http://www.ohchr.org/EN/NewsEvents/Pages/DisplayNews.aspx?NewsID=9569&LangID=e

Plummer, J. (2002). *Developing inclusive public-private partnerships: The role of small-scale independent providers in the delivery of water and sanitation services.* Paper presented at the DFID/World Bank Workshop on World Development Report 2004: Making Services Work for Poor People, Eynsham Hall, Oxfordshire. Retrieved from http://econ.worldbank.org/ wdr/wdr2004/

Popoola, A. O. (2010). Fundamental objectives and directives principles of state policy: Executive responsibility and the justiciability dilemma. In E. Azinge & B. Owasanoye (Eds.), *Justiciability and constitutionalism: An economic analysis of law.* Lagos: Nigerian Institute of Advanced Legal Studies.

Pories, L. (2016). Income-enabling, not consumptive: Association of household socio-economic conditions with safe water and sanitation. *Aquatic Procedia, 6*, 74-86. doi:http://dx.doi.org/10.1016/j.aqpro.2016.06.009

Potter, A., Klutse, A., Snehalatha, M., Batchelor, C., Uandela, A., Naafs, A., . . . Moriarty, P. (2011). *WASHCost: Assessing sanitation service levels* (2nd ed.). The Hague: IRC.

Poulos, C., & Whittington, D. (2000). Individuals' rates of time preference in developing countries: Results of a multi-country study. *Environmental Science & Technology, 43*(8), 1445-1455.

Prakash, A., & Ballabh, V. (2005). A win-some lose-all game. In D. Roth, R. Boelens, & M. Zwarteveen (Eds.), *Liquid relations: Contested water rights and legal complexity* (pp. 172-194). New Brunswick, New Jersey, London: Rutgers University Press.

Prüss-Ustün, A., Bartram, J., Clasen, T., Colford, J. M., Cumming, O., Curtis, V., . . . Fewtrell, L. (2014). Burden of disease from inadequate water, sanitation and hygiene in low-and middle-income settings: A retrospective analysis of data from 145 countries. *Tropical Medicine & International Health, 19*(8), 894-905.

Punch, K. F. (2005). *Introduction to social research: Quantitative and qualitative approaches*. London: Sage Publications Limited.

Qasim, S., Qasim, M., Shrestha, R. P., Khan, A. N., Tun, K., & Ashraf, M. (2016). Community resilience to flood hazards in Khyber Pukhthunkhwa province of Pakistan. *International Journal of Disaster Risk Reduction, 18,* 100-106.

Qerimi, Q. (2012). *Development in international law: A policy-oriented inquiry*. Leiden; Boston: Martinus Nijhoff Publishers.

Quane, H. (2013). Legal pluralism and international human rights law: Inherently incompatible, mutually reinforcing or something in between? *Oxford Journal of Legal Studies, 33*(4), 675-702.

Rabbani, M. G., Huq, S., & Rahman, S. H. (2013). Impacts of climate change on water resources and human health: Empirical evidences from a coastal district (Satkhira) in Bangladesh. In V. I. Grover (Ed.), *Impact of climate change on water and health* (pp. 272-285). Boca Raton: CRC Press.

Rahji, M. A. Y., & Oloruntoba, E. (2009). Determinants of households' willingness-to-pay for private solid waste management services in Ibadan, Nigeria. *Waste Management and Research, 27*(10), 961-965.

Rajamani, L. (2003). From Stockholm to Johannesburg: The anatomy of dissonance in the international environmental dialogue. *Review of European, Comparative & International Environmental Law, 12*(1), 23-32.

Rajan, R., Kak, K., & Vigil Public Opinion Forum. (2006). NGOs, activists & foreign funds: Anti-nation industry. Retrieved from http://catalog.hathitrust.org/api/volumes/oclc/71801443.html

Ramcharan, B. G. (1991). *The international law and practice of early-warning and preventive diplomacy: The emerging global watch*. Dordrecht; Boston: Martinus Nijhoff Publishers.

Ramesh, A., Blanchet, K., Ensink, J. H. J., & Roberts B. (2015). Evidence on the effectiveness of water, sanitation, and hygiene (WASH) interventions on health outcomes in humanitarian crises: A systematic review. *PLoS ONE 10*(9): e0124688. doi:10.1371/journal.pone.0124688.

Ramsey, M. (1994). Public Health in France. In D. Porter (Ed.), *The history of public health and the modern state* (pp. 45-118). Amsterdam-Atlanta, G.A.: Rodopi B.V. .

Rauniyar, G., & Kanbur, R. (2010a). Inclusive development: Two papers on conceptualization, application, and the ADB perspective. *Manila: Asian Development Bank.*

Rauniyar, G., & Kanbur, R. (2010b). Inclusive growth and inclusive development: A review and synthesis of Asian Development Bank literature. *Journal of the Asia Pacific Economy, 15*(4), 455-469. doi:10.1080/13547860.2010.517680

Rawls, J. (1999). *The law of peoples: With "The Idea of Public Reason Revisitied"*. Cambridge, MA: Havard University Press.

Raworth, K. (2012). A safe and just space for humanity. *Can We Live within the Doughnut.*

Rebitzer, G., Hunkeler, D., & Jolliet, O. (2003). LCC—the economic pillar of sustainability: methodology and application to wastewater treatment. *Environmental progress, 22*(4), 241-249.

Reddy, V., & Moletsane, R. (2009). Gender and poverty reduction in its African Feminist practice: An introduction. *Agenda, 23*(81), 3-13. doi:10.1080/10130950.2009.9676249

Reed, B. (2005). Technical Note No. 9. Draft Revised 7.1.05 - Minimum water quantity needed for domestic use in emergencies. Geneva: World Health Organization.

Reed, B., & Reed, B. (2011). Technical Notes on Drinking-Water, Sanitation and Hygiene in Emergencies: How much water is needed in emergencies. Geneva: World Health Organization.

Rheinländer, T., Konradsen F., Keraita, B., Apoya, P., & Gyapong, M. (2015). Redefining shared sanitation. *Bulletin of the World Health Organization*, *93*, 509-510. doi: http://dx.doi.org/10.2471/BLT.14.144980.

Riedel, E. (2006). The IBSA procedure as a tool of human rights monitoring *Background Paper to the Expert Symposium, Measuring Developments in the Realisation of the Right to Food - the IBSA Procedure.*

Riedel, E., Giacca, G. & Golay, C. (2014). Economic, Social, and Cultural Rights in International Law. In Riedel, E., Giacca, G. & Golay, C. (Ed.), *Economic, Social, and Cultural Rights in International Law: Contemporary Issues and Challenges*. Oxford: Oxford University Press.

Rivera, J. (2002). Assessing a voluntary environmental initiative in the developing world: The Costa Rican Certification for Sustainable Tourism. *Policy Sciences, 35*(4), 333-360.

Roaf, V., Khalfan, A., & Langford, M. (2005). *Global Issue Papers No. 14: Monitoring Implementation of the Right to Water: A Framework for Developing Indicators*. Berlin: Heinrich Boell Foundation.

Roberts, A. E. (2001). Traditional and modern approaches to customary international law: a reconciliation. *The American Journal of International Law, 95*(4), 757-791.

Robins, S. (2014). The 2011 Toilet Wars in South Africa: Justice and Transition between the Exceptional and the Everyday after Apartheid. *Development and Change, 45*(3), 479-501.

Rockström, J., Steffen, W., Noone, K., Persson, Å., Chapin III, F. S., Lambin, E., . . . Schellnhuber, H. (2009). Planetary boundaries: exploring the safe operating space for humanity. *Ecology and Society, 14*(2), 32.

Rockström, J., Steffen, W., Noone, K., Persson, Å., Chapin, F. S., Lambin, E. F., . . . Schellnhuber, H. J. (2009). A safe operating space for humanity. *Nature, 461*(7263), 472-475.

Rogers, R. S., & Kitzinger, C. (1986). Human rights: Bedrock or mosaic. *Operant Subjectivity: the International Journal of Q Methodology, 9*, 123-130.

Roosli, R., & Collins, A. E. (2016). Key Lessons and Guidelines for Post-Disaster Permanent Housing Provision in Kelantan, Malaysia. *Procedia Engineering, 145,* 1209–1217.

Rose-Sender, K. S., & Goodwin, M. (2010). Linking corruption and human rights: An unwelcome addition to the development discourse. *Tilburg University Legal Studies Working Paper Series No. 012/2010*. Retrieved from https://papers.ssrn.com/sol3/papers.cfm?abstract_id=1623225

Roseveare, C. (2013). The rule of law and international development. London: DFID.

Rosga, A., & Satterthwaie, M. L. (2009). The trust in indicators: Measuring human rights. *27 Berkeley J. Int'l Law, 27*, 253-315.

Roth, K. (2011). A façade of action: The misuse of dialogue and cooperation with rights abusers. In Human Rights Watch (Ed.), *Human Rights Watch world report 2011: Events of 2010*. New York; London: Seven Stories.

Ruggie, J. G. (2011). Guiding principles on business and human rights: Implementing the United Nations "protect, respect and remedy" framework. United Nations Human Rights Council. 17th Session. March. New York; Geneva: Office of the High Commissioner for Human Rights.

Rusca, M., Alda-Vidal, C., Hordijk, M., & Kral, N. (2017). Bathing without water, and other stories of everyday hygiene practices and risk perception in urban low-income areas:

The case of Lilongwe, Malawi. *Environment and Urbanization*, 0956247817700291. doi:10.1177/0956247817700291

Russell, A. F. (2010). International organizations and human rights: Realizing, resisting or repackaging the right to water? *Journal of Human Rights, 9*(1), 1-23.

Sahely, H. R., Kennedy, C. A., & Adams, B. J. (2005). Developing sustainability criteria for urban infrastructure systems. *Canadian Journal of Civil Engineering, 32*(1), 72-85.

Sala, S., Ciuffo, B., & Nijkamp, P. (2015). A systemic framework for sustainability assessment. *Ecological Economics, 119*, 314-325.

Salami, A., Stampini, M., & Kamara, A. (Eds.). (2012). *Development aid and access to water and sanitation in Sub-Saharan Africa*. Tunis-Belvedere: African Development Bank.

Salman, M. A. S. (2014). The human right to water and sanitation: Is the obligation deliverable? *Water International, 39*(7), 969-982.

Salter, D. (2008). *Identifying constraints to increasing sanitation coverage: Sanitation demand and supply in Cambodia*. Phnom Penh: Water and Sanitation Program.

Samaddar, S., Okada, N., Choi, J., & Tatano, H. (2017). What constitutes successful participatory disaster risk management?: Insights from post-earthquake reconstruction work in rural Gujarat, India. *Natural Hazards, 85*(1), 111-138. doi:10.1007/s11069-016-2564-x

Samuelson, P. A. (1954). The pure theory of public expenditure. *The review of economics and statistics, 36*(4), 387-389.

Sandin, P. (1999). Dimensions of the precautionary principle. *Human and Ecological Risk Assessment, 5*, 889–907.

Sandin, P., Peterson, M., Hansson, S. O., Rudén, C., & Juthe, A. (2002). Five charges against the precautionary principle. *Journal of Risk Research, 5*(4), 287-299. doi:10.1080/13669870110073729

Sands, P. (2008). Environmental protection in the twenty-first century: Sustainable development and international law. In R. L. Revesz, P. Sands, & R. B. Stewart (Eds.), *Environmental law, the economy and sustainable development: The United States, the European Union and the international community* (pp. 369-374). Cambridge: Cambridge University Press.

Sanitation. (n.d.). Retrieved from http://www.who.int/topics/sanitation/en/

Sanitation guidelines. (n.d.). Retrieved from http://www.who.int/water_sanitation_health/sanitation-waste/sanitation/sanitation-guidelines/en/

Saqib, S. E., Ahmad, M. M., Panezai, S., & Rana, I. A. (2016). An empirical assessment of farmers' risk attitudes in flood-prone areas of Pakistan. *International Journal of Disaster Risk Reduction, 18*, 107-114. doi:http://dx.doi.org/10.1016/j.ijdrr.2016.06.007

Sassòli, M., Bouvier, A. A., & Quintin, A. (2011). *How does law protect in war?* Geneva: ICRC.

Satterthwaite, D. (2016). Missing the millennium development goal targets for water and sanitation in urban areas. *Environment and Urbanization, 28*(1), 99-118.

Satterthwaite, D. (Ed.) (2003). *Millennium development goals and local processes: Hitting the target or missing the point?*. London: International Institute for Environment and Development.

Satterthwaite, M. (2014). On rights-based partnerships to measure progress in water and sanitation. *Science and Engineering Ethics, 20*(4), 877-884.

Saunders, M., Lewis, P., & Thornhill, A. (2009). *Research methods for business students*. New York: Prentice Hall.

Schachter, O. (1983). Human dignity as a normative concept. *American Journal of International Law, 77*(4), 848-854.

Scheinin, M. (2003). Economic and social rights as legal rights. In A. Eide, C. Krause, & A. Ronas (Eds.), *Economic, social and cultural rights: A textbook* (pp. 29-54). Dordrecht: Martinus Nijhoff Publishers.

Schindler, D. (1991). Transformations in the law of neutrality since 1945. In F. Kalshoven & G. J. Tanja (Eds.), *Humanitarian law of armed conflict: Challenges ahead - Essays in honour of Frits Kalshoven*. Dordrecht Martinus Nijhoff.

Schmitter, P. (2004). Neo-functionalism. In A. Wiener, Diez, T. (Ed.), *European integration theory* (pp. 45-74). Oxford: Oxford University Press.

Schmitz, H. P., & Sikkink, K. (2002). International human rights. In W. Carlsnaes, T. Risse, & B. A. Simmons (Eds.), *Handbook of international relations* (pp. 517-537). London: Sage.

Schouten, M. A. C., & Mathenge, R. W. (2010). Communal sanitation alternatives for slums: A case study of Kibera, Kenya. *Physics and Chemistry of the Earth, Parts A/B/C, 35*(13–14), 815-822. doi:http://dx.doi.org/10.1016/j.pce.2010.07.002

Schrijver, N. (2008). *The evolution of sustainable development in international law: Inception, meaning and status*. Leiden; Boston: Martinus Nijhoff.

Schroeder, D. (2012). Human rights and human dignity. *Ethical Theory and Moral Practice, 15*(3), 323-335. doi:10.1007/s10677-011-9326-3

Schuller, M., & Levey, T. (2014). Kabrit ki gen twòp mèt: Understanding gaps in WASH services in Haiti's IDP camps. *Disasters, 38*, S1-S24. doi:10.1111/disa.12053

Schuster, C. d. V. (2005). *Designing public toilets*. Savigliano: Gribaudo.

Schwartz, K., & Schouten, M. (2007). Water as a political good: Revisiting the relationship between politics and service provision. *Water Policy, 9*(2), 119-129.

Scott, P., Cotton, A., & Khan, M. S. (2013). Tenure security and household investment decisions for urban sanitation: The case of Dakar, Senegal. *Habitat International, 40*, 58-64.

Semba, R. D., de Pee, S., Kraemer, K., Sun, K., Thorne-Lyman, A., Moench-Pfanner, R., . . . Bloem, M. W. (2009). Purchase of drinking water is associated with increased child morbidity and mortality among urban slum-dwelling families in Indonesia. *International Journal of Hygiene and Environmental Health, 212*(4), 387-397.

Sen, A. (2004). Elements of a theory of human rights. *Philosophy & Public Affairs, 32*(4), 315-356.

Sen, A. (2011). *The idea of justice*. Cambridge, Massachusetts: The Belknap Press of Harvard University Press.

Sengupta, A. (2004). The human right to development. *Oxford Development Studies, 32*(2), 179-203.

Seraj, K. F. B. (2008). *Willingness to pay for improved sanitation services and its implication on demand responsive approach of BRAC water, sanitation and hygiene programme - Working Paper No. 1*. Dhaka: BRAC.

Shahabuddeen, M. (2011). Judicial creativity and joint criminal enterprise. In S. Darcy & J. Powderly (Eds.), *Judicial creativity at the international criminal tribunals* (pp. 184-203). Oxford: Oxford University Press.

Sharifi, A., & Yamagata, Y. (2016). On the suitability of assessment tools for guiding communities towards disaster resilience. *International Journal of Disaster Risk Reduction, 18*, 115–124.

Shelton, D. (2014). *Advanced introduction to international human rights law*. Cheltenham: Edward Elgar.

Shelton, D. (2003). Introduction: Law, non-law and the problem of 'soft law'. In D. Shelton (Ed.), *Commitment and compliance: The role of non-binding norms in the international legal system* (pp. 1-20). Oxford: Oxford University Press.

Sheng, H.-X., Ricci, P. F., & Fang, Q. (2015). Legally binding precautionary and prevention principles: Aspects of epistemic uncertain causation. *Environmental Science & Policy, 54*, 185-198. doi:https://doi.org/10.1016/j.envsci.2015.06.016

Simha, P., & Ganesapillai, M. (2016). Ecological sanitation and nutrient recovery from human urine: How far have we come? - A review. *Sustainable Environment Research*. doi:http://dx.doi.org/10.1016/j.serj.2016.12.001

Simiyu, S. (2016). Determinants of usage of communal sanitation facilities in informal settlements of Kisumu, Kenya. *Environment and Urbanization, 28*(1), 241-258.

Sinden, A. (2005). In defense of absolutes: Combating the politics of power in environmental law. *Iowa Law Review, 90*, 1405-1512.

Sinden, A. (2009). Power and responsibility: Why human rights should address corporate environmental wrongs. In D. J. Mcbarnet, A. Voiculescu, & T. Campbell (Eds.), *The new corporate accountability: Corporate social responsibility and the law.* Cambridge: Cambridge University Press.

Skogly, S. (2012). The requirement of using the 'maximum of available resources' for human rights realisation: A question of quality as well as quantity? *Human Rights Law Review, 12*, 393–420. doi:10.1093/hrlr/ngs022.

Slim, H. (1998). Sharing a universal ethic: The principle of humanity in war. *The International Journal of Human Rights, 2*(4), 28-48.

Smets, H. (2008). *Le prix abordable de l'eau potable de réseau dans la pratique des états.* Paris: Academie de l'Eau.

Smith, R. K. M. (2010). *Texts and materials on international law.* Oxon; New York Routledge.

Smith, R., & van den Anker, C. (2005). *The essentials of human rights: Everything you need to know about human rights.* London: Hodder Arnold.

Solo, T. M., Perez, E., & Joyce, S. (1993). *Constraints in providing water and sanitation services to the urban poor.* Washington, D.C.: Water and Sanitation for Heralth Project.

Sosa, M., & Zwarteveen, M. (2016). Questioning the effectiveness of planned conflict resolution strategies in water disputes between rural communities and mining companies in Peru. *Water International, 41*(3), 483-500. doi: 10.1080/02508060.2016.1141463

Sovacool, B. K. (2013). Expanding renewable energy access with pro-poor public private partnerships in the developing world. *Energy Strategy Reviews, 1*(3), 181-192. doi:https://doi.org/10.1016/j.esr.2012.11.003

Spears, D., Ghosh, A., Cumming, O., & Chaturvedi, V. (2013). Open defecation and childhood stunting in India: An ecological analysis of new data from 112 districts. *PLoS ONE, 8*(9), e73784.

Sphere Project. (2011). *Humanitarian charter and minimum standards in humanitarian response.* Geneva; Rugby: Sphere Project.

Srinivasan, R. (2015). *Lack of toilets and violence against Indian women: empirical evidence and policy implications.* Retrieved from https://sites.utexas.edu/raji-srinivasan/files/2014/07/NonFamilyViolenceAgainstIndianWomen.pdf

Stake, R. E. (1995). *The art of case study research.* Thousand Oaks: Sage Publications.

Steffen, W., Richardson, K., Rockström, J., Cornell, S. E., Fetzer, I., Bennett, E. M., . . . Sörlin, S. (2015). Planetary boundaries: Guiding human development on a changing planet. *Science, 347*, 736–746.

Steinberg, J. (1978). *Locke, Rosseau, and the idea of consent: An inquiry into the liberal-democratic theory of political obligation*. Westport, Conn.: Greenwood Press.

Stemler, S. (2001). An overview of content analysis. *Practical Assessment, Research & Evaluation, 7*(17). Retrieved from http://pareonline.net/getvn.asp?v=7&n=17

Stenner, P. (2010). Subjective dimensions of human rights: What do ordinary people understand by "human rights"? *The International Journal of Human Rights, 15*(8), 1215-1233.

Stern, N. (2011). *The economics of climate change: The Stern review*. Cambridge: Cambridge University Press.

Stewart, D. W., & Shamdasani, P. N. (1998). Focus group research: Exploration and discovery. In L. Bickman, & Rog, D. J. (Ed.s), *Handbook of applied social research methods*. Carlifonia, London, New Delhi: SAGE.

Strauss, A., & Corbin, J. (2007). *Basics of qualitative research: Techniques and procedures for developing grounded theory*: Sage Publications, Incorporated.

Sunstein, C. R. (2003). Beyond the precautionary principle. *University of Pennsylvania Law Review, 151*(3), 1003-1058.

Susser, M. (1993). Health as a human right: an epidemiologist's perspective on the public health. *American Journal of Public Health, 83*(3), 418-426.

Swyngedouw, E., Kaïka, M., & Castro, E. (2002). Urban water: A political-ecology perspective. *Built Environment, 28*(2), 124-137.

Tamanaha, B.Z. (2011). The rule of law and legal pluralism in development. *Hague J Rule Law 3*, 1-17.

Tanguy, J., & Terry, F. (1999). Humanitarian responsibility and committed action: Response to "principles, politics, and humanitarian action". *Ethics & International Affairs, 13*(1), 29-34.

The UN Refugee Agency. (2016). Access to justice project report: June-December 2016. The UN Refugee Agency. Retrieved from https://data2.unhcr.org/fr/documents/download/52663

Theobald, R. (1999). So what really is the problem about corruption?. *Third World Quarterly, 20*(3), 491-502.

Thielbörger, P. (2009). The human rights to water versus investor rights: double-dilemma or pseudo-conflict? In Dupuy, P. M., Francioni, F., & Petersmann, E. U. (Ed.s), *Human rights in international investment law and arbitration* (pp. 487-510). Oxford: Oxford University Press.

Thielbörger, P. (2014). The right(s) to water: The multi-level governance of a unique human right. Heidelberg; New York; Dordrecht; London: Springer. Retrieved from http://dx.doi.org/10.1007/978-3-642-33908-0

Thomalla, F., Downing, T., Spanger-Siegfried, E., Han, G., & Rockström, J. (2006). Reducing hazard vulnerability: Towards a common approach between disaster risk reduction and climate adaptation. *Disasters, 30*(1), 39-48. doi:10.1111/j.1467-9523.2006.00305.x

Thye, Y. P., Templeton, M. R., & Ali, M. (2011). A critical review of technologies for pit latrine emptying in developing countries. *Critical Reviews in Environmental Science and Technology, 41*(20), 1793-1819.

Tignino, M. (2011). The right to water and sanitation in post-conflict peacebuilding. *Water International, 36*(2), 242-249. doi:10.1080/02508060.2011.561523

Tilley, E., Zurbrügg, C., & Lüthi, C. (2010). A flowstream approach for sustainable sanitation systems. In B. van Vliet, G. Spaargaren, & P. Oosterveer (Eds.), *Social Perspectives on the Sanitation Challenge* (pp. 69-86). Netherlands: Springer.

Tilley, E., Ulrich, L., Lüthi, C., Reymond, P., & Zurbrügg, C. (2008). *Compendium of sanitation systems and technologies.* (2nd rev. ed.). Dübendorf, Switzerland: Swiss Federal Institute of Aquatic Science and Technology (Eawag).

Tillman, A.-M., Svingby, M., & Lundström, H. (1998). Life cycle assessment of municipal waste water systems. *The international journal of life cycle assessment, 3*(3), 145-157.

Tong, J. (2004). Questionable accountability: MSF and Sphere in 2003. *Disasters, 28*(2), 176-189.

Toubkiss, J. (2006). *Costing MDG Target 10 on water supply and sanitation: comparative analysis obstacles and recommendations.* Québec: World Water Council.

Traeger, C. P. (2011). Sustainability, limited substitutability, and non-constant social discount rates. *Journal of Environmental Economics and Management, 62*(2), 215-228. doi:http://dx.doi.org/10.1016/j.jeem.2011.02.001

Trémolet, S., & Rama, M. (2012). *Tracking national financial flows into sanitation, hygiene and drinking-water: GLAAS working paper.* Geneva: WHO.

Tsui, E. (2009). Analysis of normative developments in humanitarian resolutions since the adoption of 46/(1991). Retrieved from http://www.saarc-sadkn.org/downloads/analysis.pdf

Twining, W. (2009). Normative and legal pluralism: A global perspective. *Duke Journal of Comparative & International Law, 20*, 473-517.

Udas, P. B., Roth, D., & Zwarteveen, M. (2014). Informal privatisation of community taps: Issues of access and equity. *The International Journal of Justice and Sustainability, 19*(9), 1024-1041. http://dx.doi.org/10.1080/13549839.2014.885936

UN Chronicle. (1996). After Beijing: Emphasis on poverty eradication. *United Nations Chronicle, 33*(2), 74.

UN Committee on Economic Social and Cultural Rights (CESCR). (2002). *General Comment No. 15:The right to water (arts. 11 and 12 of the Covenant), 20 January 2003, E/C.12/2002/11.* UN Committee on Economic, Social and Cultural Rights (CESCR).

UN Committee on Economic, Social and Cultural Rights (CESCR). (2009). General comment No. 20: Non-discrimination in economic, social and cultural rights (art. 2, para. 2, of the International Covenant on Economic, Social and Cultural Rights), 2 July 2009, E/C.12/GC/20. Retrieved from http://www.refworld.org/docid/4a60961f2.html

UN General Assembly. (2010). *Resolution adopted by the General Assembly [without reference to a Main Committee (A/64/L.63/Rev.1 and Add.1)] 64/292. The human right to water and sanitation.* UN General Assembly (UNGA).

UN General Assembly. (2015). Resoultion adopted by the General Assembly on 17 Decemebr 2015 70/169: *The human rights to safe drinking water and sanitation.* United Nations A/RES/70/169.

UN General Assembly. (2010). *Resolution adotped by the General Assembly on 28 July 2010 64/292: The human right to water and sanitation.* United Nations A/RES/64/292.

UN-Habitat. (2016). *Urbanization and development: Emerging futures - World cities report 2016.* Nairobi, Kenya: United Nations Human Settlements Programme (UN-Habitat).

UN Human Rights Council. (2010). *Resolution adopted by the Human Rights Council 15/9: Human rights and access to safe drinking water and sanitation.* United Nations A/HRC/RES/15/9.

UNCED. (1992). *Report of the United Nations Conference on Environment and Development.* New York: UN.

United Kingdom. (2012). Statement on the human right to sanitation. Retrieved from https://www.gov.uk/government/uploads/system/uploads/attachment_data/file/36541/human-right-sanitation270612.pdf

United Nations. (1968a). Proclamation of Teheran, Final Act of the International Conference on Human Rights, Teheran, 22 April to 13 May 1968, U.N. Doc. A/CONF. 32/41. Retrieved from http://legal.un.org/avl/pdf/ha/fatchr/Final_Act_of_TehranConf.pdf

United Nations. (1968b). Resolution adopted by the General Assembly. 2450 (XXIII). U.N. Doc. A/10034.

United Nations. (1970). Human rights and scientific and technological developments. U.N. Doc. E/CN.4/1028/Add.6.

United Nations. (1975). Declaration on the use of scientific and technological progress in the interest of peace and for the benefit of mankind. General Assembly. Resolution 3384.

United Nations. (1992). *United Nations Conference on Environment & Development, Rio de Janerio, Brazil, 3 to 14 June 1992. Agenda 21*. New York: United Nations.

United Nations. (1993). Vienna Declaration and Program of Action.11, U.N. Doc. A/CONF.157/23 (July 12, 1993).

United Nations. (2009). *2009 UNISDR terminology on disaster risk reduction*. Geneva: United Nations International Strategy for Disaster Reduction (UNISDR).

United Nations. (2015a). *Global assessment report on disaster risk reduction 2015: Making development sustainable - The future of disaster risk management*. Geneva: United Nations Office for Disaster Risk Reduction (UNISDR).

United Nations. (2015b). *Sendai framework for disaster risk reduction 2015-2030*. Geneva: United Nations Office for Disaster Risk Reduction (UNISDR).

United Nations. (2017). *The Sustainable Development Goals Report: 2017*. New York: United Nations.

United Nations Children's Fund, & United Nations High Commissioner for Refugees. (2014). *Checklist for emergency proposals. WiE – 2.2 HO 1pP –Proposal Checklist.*

United Nations Children's Fund. (2012). *Assessment of community-led total sanitation (CLTS) in Nigeria: Final report*. Abuja: United Nations Children's Fund (UNICEF).

United Nations Children's Fund. (2015). *2014 Annual results report: Humanitarian action*. New York: United Nations Children's Fund.

United Nations Children's Fund. (2016). *UNICEF annual report 2015*. New York: United Nations Children's Fund (UNICEF).

United Nations Children's Fund. (2017). *UNICEF annual results report 2016: Humanitarian action*. New York: United Nations Children's Fund.United Nations Department of Economic Social Affairs [UNDESA], Population Division. (2015). *World population prospects: The 2015 revision - Key findings and advance tables*. New York: United Nations.

United Nations Development Program. (2016). Human development report 2016: Human development for everyone. New York: United Nations Development Programme.

United Nations Development Program. (2011). *Public private partnerships for service delivery (PPPSD)*. Johannesburg: UNDP Capacity Development Group.

United Nations Development Programme. (2006). *Human development report 2006: Beyond scarcity - Power, poverty and the global water crisis*. New York: Palgrave Macmillan.

United Nations Economic Commission for Europe, & World Health Organization Regional Office for Europe. (2012). *No one left behind: Good practices to ensure equitable access to water and sanitation in the Pan-European region*. Retrieved from http://www.unece.org/fileadmin/DAM/env/water/publications/PWH_No_one_left_behind/No_one_left_behind_E.pdf

United Nations High Commissioner for Human Rights. (2011). Report of the High Commissioner for Human Rights. UN Doc. E/2011/90.

United Nations High Commissioner for Human Rights. (2008). Report on indicators for promoting and monitoring the implementation of human rights. UN Doc. HRI/MC/2008/3.

United Nations Office for the Coordination of Humanitarian Affairs [OCHA]. (2012). OCHA on Message: Humanitarian Principles Geneva: OCHA. Retrieved from https://docs.unocha.org/sites/dms/documents/oom_humprinciple_english.pdf

United Nations Office for the Coordination of Humanitarian Affairs [OCHA]. (2016). Sirt Flash Appeal: September - December 2016. Retrieved from https://www.humanitarianresponse.info/system/files/documents/files/sirt_flash_appcal_16sept_en.pdf

United Nations Office for the Coordination of Humanitarian Affairs [OCHA]. (2017). Dominica Flash Appeal: Hurricane Maria - September - December 2017. Retrieved from
https://www.unocha.org/sites/unocha/files/Dominica_FlashAppeal_EN_20170929.pdf

United Nations, United Nations Economic Commission for Europe, & World Health Organization Regional Office for Europe. (2013). *The equitable access score-card: Supporting policy processes to achieve the human right to water and sanitation - Protocol on Water and Health to the Convention on the Protection and Use of Transboundary Wealth to the Convention on the Protection and Use of Transboundary Watercourses and International Lakes*. Geneva: United Nations.

UN-Water Decade Programme on Advocacy and Communication, & Water Supply and Sanitation Collaborative Council. (2010). The human right to water and sanitation: Media brief. Zaragoza: UNW-DPAC.

UN-Water. (2014). *Investing in water and sanitation: Increasing access, reducing inequalities - UN-Water global analysis and assessment of sanitation and drinking-water (GLAAS) 2014*. Geneva: World Health Organization.

United States General Accounting Office [US GAO], Program Evaluation and Methodology Division. (1996). *Content analysis: A methodology for structuring and analyzing written material*. U.S.: UU GAO.

USAID Egypt, Water Policy and Regulatory Reform Program. (2013). *Clean water for Egypt: Egypt water policy and regulatory reform - Final report*. Retrieved from http://pdf.usaid.gov/pdf_docs/PA00JPRX.pdf

Uwaegbulam, C. (2016). Privatisation not solution to water crisis in Lagos. Retrieved from http://guardian.ng/property/privatisation-not-solution-to-water-crisis-in-lagos/

van Boven, T. C. (1982). Distinguishing criteria of human rights. In K. Vasak & P. Alston (Eds.), *The international dimensions of human rights , Volume 1 - Volume 2* (pp. 43-59). Paris: UNESCO; Greenwood Press.

van de Guchte, C., & Vandeweerd, V. (2004). Targeting sanitation. *Our Planet, 14*(4), 19-21.

Van Ewijk, E., Baud, I., Bontenbal, M., Hordijk, M., van Lindert, P., Nijenhuis, G., & van Westen, G. (2014). Capacity development or new learning spaces through municipal international cooperation: Policy mobility at work? *Urban Studies, 52*(4), 756-774. doi:10.1177/0042098014528057

Van Minh, H., & Nguyen-Viet, H. (2011). Economic aspects of sanitation in developing countries. *Environmental health insights, 5*, 63-70.

Van Minh, H., Nguyen-Viet, H., Thanh, N. H., & Yang, J.-C. (2013). Assessing willingness to pay for improved sanitation in rural Vietnam. *Environmental Health and Preventive Medicine, 18*(4), 275-284. doi:10.1007/s12199-012-0317-3 van Stapele, N. (2013). Fighting over a public toilet: Masculinities, class and violence in a Nairobi ghetto. In G. Frerks, A. Ypeij, & R. König (Eds.), *Gender and conflict: Embodiments, discourses and symbolic practices*. Surrey: Ashgate.

Varma, M. K., Satterthwaite, M. L., Klasing, A. M., Shoranick, T., Jean, J., Barry, D., . . . Lyon, E. (2008). Wòch nan soley: The denial of the right to water in Haiti. *Health and Human Rights Journal*, 67-89.

Vasak, K. (1982). Human rights: As a legal reality. In K. Vasak (Ed.), *The International Dimensions of Human Rights* (pp. 1 - 47). Paris: UNESCO.

Vedung, E. (1998). Policy instruments: Typologies and theories. In M. Bemelmans-Videc, R. List, & E. Vedung (Eds.), *Carrots, sticks and sermons: Policy instruments and their evaluation* (pp. 21-58). London: Transaction Publishers.

Veeramany, A., Unwin, S. D., Coles, G. A., Dagle, J. E., Millard, D. W., Yao, J., . . . Gourisetti, S. N. G. (2016). Framework for modeling high-impact, low-frequency power grid events to support risk-informed decisions. *International Journal of Disaster Risk Reduction, 18*, 125-137. doi:http://dx.doi.org/10.1016/j.ijdrr.2016.06.008

Ver Eecke, W. (1999). Public goods: An ideal concept. *The Journal of Socio-Economics, 28*(2), 139-156.

Victor, M. E., & Ernest, W. S. (2007). *Municipal and rural sanitation* (7th ed.). New York: McGraw Hill Publishers.

Vietnam Red Cross. (2011). *UNJP/VIE/037/UNJ: "Strengthening capacities to enhance coordinated and integrated disaster risk reduction actions and adaptation to climate change in agriculture in the northern mountain regions of Viet Nam": Training manual for disaster risk management systems at the community level - CBDRM.* Hanoi: FAO.

Vinnerås, B. (2007). Comparison of composting, storage and urea treatment for sanitising of faecal matter and manure. *Bioresour Technol, 98*(17), 3317-3321.

Viola de Azevedo Cunha, M., Gomes de Andrade, N. N., Lixinski, L., & Féteira, L. T. (2013). *New technologies and human rights: Challenges to regulation*. Farnham: Ashgate.

Voigt, C. (2009). *Sustainable development as a principle of international law: Resolving conflicts between climate measures and WTO law*. Boston: Martinus Nijhoff Publishers.

von Sperling, M. (2007). *Biological wastewater treatment series: Volume two - Basic principles of wastewater treatment*. London: IWA Publishing.

Vuorinen, H. S. (2007). The emergence of the idea of water-borne diseases. In P. S. Juuti, T. S. Katko, & H. S. Vuorinen (Eds.), *Environmental history of water*. London: IWA Publishing.

Wampler, B. (2017). *The spread of participatory budgeting across Latin America: Efforts to improve social well-being*. Boise State University. Retrieved from http://www.tfd.org.tw/opencms/files/download/conf_170218_en-5.pdf

Wang, Y. B., Wu, P. T., Zhao, X. N., & Engel, B. A. (2014). Virtual water flows of grain within China and its impact on water resource and grain security in 2010. *Ecol Eng, 69*, 255–264. Doi:10.1016/j.ecoleng.2014.03.057.69.

WASHCost. (2012). Providing a basic level of water and sanitation services that last: Cost benchmarks - *WASHCost global infosheet*. The Hague, The Netherlands: IRC.

Water Engineering Development Centre. (2013). *Updated WHO/WEDC technical notes on WASH in emergencies*. Geneva: World Health Organization.

Water Lex. (2014). *The human rights to water and sanitation in courts worldwide: A selection of national, regional, and international case law*. Geneva: Water Lex.

WaterAid. (2015). *Essential element: Why International aid for water, sanitation and hygiene is still a critical source of finance for many countries*. WaterAid. Retrieved from www.wateraid.org/ppa

Waterkeyn, J., & Cairncross, S. (2005). Creating demand for sanitation and hygiene through community health clubs: A cost-effective intervention in two districts in Zimbabwe. *Social Science & Medicine, 61*(9), 1958-1970. doi:http://dx.doi.org/10.1016/j.socscimed.2005.04.012

Watt, K., & Weinstein, P. (2013). Casualties following natural hazards. In P. T. Bobrowsky (Ed.), *Encyclopedia of natural hazards* (pp. 59-64). Dordrecht, Heidelberg, New York, London: Springer.

Wekesa, S. M. (2013). Right to clean and safe water under the Kenyan Constitution 2010. *ESR Review : Economic and Social Rights in South Africa, 14*(1), 3-6.

Weiss, T. G. (1999). Principles, politics and humanitarian action. *Ethics & International Affairs, 13*, 1-22.

Weil, P. (1983). Towards relative normativity in international law? *The American Journal of International Law, 77*(3), 413-442.

Weschler, J. (Ed.) (2011). *Human rights diplomacy of the united nations security council*. Leiden: Nijhoff.

White, L., & Perelman, J. (2011). *Stones of hope: How African activists reclaim human rights to challenge global poverty*. Stanford, California: Stanford University Press.

Wicken, J., Verhagen, J., Sijbesma, C., Da Silva, C., & Ryan, P. (2008). *Beyond construction: Use by all - A collection of case studies from sanitaion and hygiene promotion practitioners in South Asia*: WaterAid, IRC.

Winara, A., Hutton, G., Oktarinda, Purnomo, E., Hadiwardoyo, K., Merdykasari, I., . . . Albrecht, M. (2011). *Economic assessment of sanitation interventions in Indonesia*. Jakarta: The World Bank, Water and Sanitation Program - East Asia & the Pacific Regional Office.

Winkler, I. T. (2012). *The human right to water: Significance, legal status and implications for water allocation*. Oxford: Hart Publishing.

Winkler, I. T. (2016). The human right to sanitation. *University of Pennsylvania Journal of International Law, 37*(4), 1331-1406.

Winkler, I. T., Satterthwaite, M. L., & Albuquerque, C. d. (2014). Treasuring what we measure and measuring what we treasure: Post-2015 monitoring for the promotion of equality in the water, sanitation, and hygiene sector. *Wisconsin International Law Journal, 32*(3), 547-594.

Winpenny, J., Trémolet, S., Cardone, R., Kolker, J., & Mountsford, L. (2016). *Aid flows to the water sector: Overview and recommendations*. Washington, DC: World Bank.

Wirseen, C., Munch, E. V., Patel, D., Wheaton, A., & Jachnow, A. (2009). Safer sanitation in slums and emergency settings with peepoo bags. *Appropriate Technology, 36*(4), 38-41.

Wiseberg, L. S. I. E. (1996). Introductory Essay. In E. Lawson (Ed.), *Encyclopedia of human rights* (pp. xix-xxxviii). Washington, D.C.: Taylor & Francis.

Woodside, A. G. (2010). Bridging the chasm between survey and case study research: Research methods for achieving generalization, accuracy, and complexity. *Industrial Marketing Management, 39*(1), 64-75. doi:https://doi.org/10.1016/j.indmarman.2009.03.017

World Bank. (1999). *Willing to pay, but unwilling to charge: Do "willingness-to-pay" studies make a difference? - Water and Sanitation Program field note*. Washington, D.C.: World Bank Group.

World Bank Group. (2014). *Towards sustainable peace, poverty eradication, and shared prosperity: Colombia policy notes*. Washington, DC: World Bank. Retrieved from http://EconPapers.repec.org/RePEc:wbk:wboper:21037

World Commission on Environment and Development. (1987). *Our common future*. Oxford; New York: Oxford University Press.

World Economic Forum. (2016). *The global risks report 2016*. Geneva: World Economic Forum.

World Health Organization, Female genital mutilation: Factsheet (2017). Retrieved from http://www.who.int/mediacentre/factsheets/fs241/en/

World Health Organization. Environmental health in healthcare and other settings: Preventing environmental health-related disease in health care and other settings. Geneva: Department of Public Health and Environment, WHO. Retrieved from http://www.who.int/phe/events/wha_66/flyer_wsh_health_care.pdf

World Health Organization, & UN-Water. (2015). *TrackFin Initiative: Tracking financing to sanitation, hygiene and drinking-water at the national level - Guidance document summary for decision makers*. Geneva: WHO.

World Health Organization. (2016). Sanitation safety planning: Manual for safe use and disposal of wastewater, greywater and excreta. Geneva: World Health Organization. Retrieved from http://www.who.int/topics/sanitation/en/

World Health Organization/UNICEF Joint Monitoring Programme for Water Supply and Sanitation. (2015). JMP green paper: Global monitoring of water, sanitation and hygiene post-2015 - Draft, Updated Oct-2015.

World Health Organization, & UNICEF. (2015). *Progress on sanitation and drinking water: 2015 update and MDG assessment*. Geneva: World Health Organization.

Wronka, J. (2008). *Human rights and social justice*. California: Sage Publications.

Wrońska, I. (2011). *Białystok law books 1: Fundamental rights protection in the Council of Europe - The role of the European Court of Human Rights (Vol. 1)*. Bialystock: Wydawnictwo Temida 2, 2011.

Yatmo, Y. A., & Atmodiwirjo, P. (2012). Communal toilet as a collective spatial system in high density urban Kampung. *Procedia - Social and Behavioral Sciences, 36*(0), 677-687. doi:http://dx.doi.org/10.1016/j.sbspro.2012.03.074

Yin, R. K. (2009). *Case study research: Design and methods*. Los Angeles, Calif.: Sage Publications.

Yin, R. K. (2014). *Case study research: Design and methods*. London: Sage Publication.

Young, O. R. (2002). *The institutional dimensions of environmental change: Fit, interplay, and scale*. Cambridge, Mass.: MIT Press.

Young, O. R., Agrawal, A., King, L. A., Sand, P. H., Underdal, A., & Wasson, M. (2005). *Institutional Dimensions of Global Environmental Change (IDGEC) Science Plan*. Bonn: IHDP.

Yunus, M. (2003). *Banker to the poor: Micro-lending and the battle against world poverty*. New York: PublicAffairs.

Zachos, L. G., Swann, C. T., Altinakar, M. S., McGrath, M. Z., & Thomas, D. (2016). Flood vulnerability indices and emergency management planning in the Yazoo Basin, Mississippi. *International Journal of Disaster Risk Reduction, 18*, 89-99. doi:http://dx.doi.org/10.1016/j.ijdrr.2016.03.012

Zahar, A. (2014). Mediated versus cumulative environmental damage and the International Law Association's legal principles on climate change. *Climate Law, 4*(3-4), 217-233. doi:doi:https://doi.org/10.1163/18786561-00404002

Zakaria, F., Garcia, H. A., Hooijmans, C. M., & Brdjanovic, D. (2015). Decision support system for the provision of emergency sanitation. *Science of the Total Environment, 512-513*, 645-658.

Zimmer, A., Winkler, I. T., & de Albuquerque, C. (2014). Governing wastewater, curbing pollution, and improving water quality for the realization of human rights. *Waterlines, 33*(4), 337-356.

Zommer, M. T. (2014). Operationalizing international humanitarian law: A decision-making process model for assessing state practice. *International Law Research, 3*, 150-158.

Zorn, C. R., & Shamseldin, A. Y. (2015). Post-disaster infrastructure restoration: A comparison of events for future planning. *International Journal of Disaster Risk Reduction, 13*, 158-166.

Table of cases

A.G. Bendel State v A. G. Federation & 22 Others (1982) 3 NCLR 1 SC

A.G. Cross River State v Esin (1991) 6 NWLR Pt. 197 at 365 CA

Abasi v State (1992) 8 NWLR Pt 260 at 383

Advisory Opinion, Legal Consequences for States for the Continued Presence of South Africa in Namibia (South West Africa) notwithstanding Security Council Resolution 276 (1970), ICJ, 21 June 1971 at 50

Afghanistan, Goatherd Saved from Attack, Case No. 257

Agbara v Amara (1995) 7 NWLR Pt. 410 at 712

Awolowo v Federal Minister of Internal Affairs (1962) LLR 177

Barcelona Traction Light and Power Company Limited Case (1970)

Becke v Smith (1836) 150 ER 724 at 726

Beja and Others v Premier of the Western Cape and Others [2011] High Court (Western Cape) 21332/10, [2011] ZAWCHC 97 (South Africa)

Beja and Others v Premier of the Western Cape and Others [2011] High Court (Western Cape) 21332/10, [2011] ZAWCHC 97

Belgian Court of Arbitration, A.s.b.l. Syndicat national des propriétaires et autre, Case No.9/1996, 8 February 1996, in: Moniteur belge, 1996(02)00035, section I, par III, 2

Board of Custom & Excise v Barau (1982) 10 SC 48

Canada - Halalt First Nation v British Columbia (Environment) [2011] Supreme Court (British Columbia) S098232, 2011 BCSC 945

Case T-13/99 Pfizer Animal Health SA v Council [2002] ECR 11-3305; Case T-70/99 Alpharma Inc. v Council [2002] ECR II-3495

City of Cape Town v Strümpher [2012] Supreme Court of Appeal 104/2011, [2012] ZASCA 54 (South Africa)

Commune de Wemmel, Cour d'arbitrage, Arrêt N°36/1998, 1 Avril 1998, Moniteur belge, 24 April 1998

Compagnie de dervices dénvironment v. Association des consommateurs de la Fontauliére Case No. 9800223, 5 March 1998

Comunidades Indigenas del Canton de Sisimitepet y Pushtan del Municipio de Nahuizalco c/Presidencia de la Republica de El Salvador y Otros [2008] Tribunal Latinoamericano del Agua

Coutts & Co. v IRC (1953) 2 WLR 364 at 368

Ejoh v IGP (1963) All NLR 248

Faith Okafor v. Attorney General of Lagos State [2016] LPELR-41066(CA)

Fasakin v Fasakin (1994) 4 NWLR Pt. 304 at 597 SC

FCSC v Laoye (1989) 2 NWLR Pt 106 at 652 SC

Federation for Sustainable Environment and Others v Minister of Water Affairs and Others [2012] High Court (North Gauteng, Pretoria) 35672/12, [2012] ZAGPPHC 128 (South Africa)

Federation for Sustainable Environment and Others v Minister of Water Affairs and Others [2012] High Court (North Gauteng, Pretoria) 35672/12, [2012] ZAGPPHC 128

France - Federation Departementale des Syndicats d'Exploitants Agricoles du Finistere [2012] Conseil constitutionnel 2012-270 QPC

Fundacion Chadileuvu c/ Estado Nacional Argentino y Provincia de Mendoza [2012] Tribunal Latinoamericano del Agua

Gabc̆íkovo-Nagymaros Project (Hungary v. Slovakia), [1997] ICJ Rep 7–84

Georgia/Russia, Human Rights Watch's Report on the Conflict in South Ossetia, Case No. 290 [Paras 28-30, 41-47]

Grupo de Formacion e Intervencion para el Desarrollo (Gufides) y Plataforma Interinstitutcional Celendina (PIC) c/ Estado Peruano y Minera Yanacocha SRL [2012] Tribunal Latinoamericano del Agua

Hernan Galeano Diaz c/ Empresas Publicas de Medellin ESP,y Marco Gomez Otero y Otros c/Hidropacifico SA ESP y Otros [2010] Corte Constitucional T-616/10

I.R.C. v Pemsel (1891) AC 531 at 542

Ibrahim v JSC (1998) 14 NWLR Pt. 584 at 1 SC

In the Arbitration Regarding the Iron Rhine ('Ijzeren Rijn') Railway (Belgium v. Netherlands), Award of 2005, Permanent Court of Arbitration, UNRIAA, vol. XXVII, 1941, pp. 35–125

Inter-American Court, Advisory Opinion No. 4, para. 57

Ishola v Ajiboye (1994) 6 NWLR Pt. 352 at 506 SC

Israel, Operation Cast Lead Case No. 124 [Part I, paras 120-126; Part II, paras 230-232]

Israel, Separation Wall/Security Fence in the Occupied Palestinian Territory, Case No. 123 ICJ [Part B., paras 36-85]

Jacobs v Belgium (2004) The Human Rights Committee Communication No. 943/2000

Lawal-Osula v Lawal-Osula (1993) 2 NWLR Pt. 274 at 158 CA

Lindiwe Mazibuko and Others v. City of Johannesburg and Others Case No. CCT 39/09, Judgement of 8 October 2009 (Constitutional Court of South Africa)

Mandla Bushula v Ukhahlamba District Municipality [2012] High Court (Eastern Cape Division) 2200/09, [2012] ZAECGHC 1

Marckx v. Belgium (1979) European Court Application No. 6833/74

Municipal Council, Ratlam v. Shri Vardhichand and Others, (1981) 1 S.C.R. 97 (India)

Nabham v Nabham (1967) All NLR 51

Nasr v Bouari (1969) 1 All LNR 35

Nicaragua v. United States (1986) Case No. 153, ICJ

NIPC Ltd v Bank of West Africa (1962) 1 All NLR 551

Okeke v A.G. Anambra State (1992) 1 NWLR Pt. 215 at 164

Oloyo v Alegbe (1983) 2 SCNLR 35 at 37

Ondo State v Folayan (1994) 7 NWLR Pt. 354 at 1 SC

Opeola v Opadiran (1994) 5 NWLR Pt. 344 at 368 SC

Osadebay v A.G. Bendel State (1991) 1 NWLR Pt 169 at 525 SC

Osawe v Registrar of Trade Unions (1985) 1 NWLR Pt. 4 at 755 SC 1

Prosecutor v. Zoran Kupreskic, Mirjan Kupreskic, Vlatko Kupreskic, Drago Josipovic, Dragan Papic, Vladimir Santic, Judgment, IT-95-16-T, ICTY Trial Chamber, 14 Jan. 2000, at 540 (International Criminal Tribunal for the Former Yugoslavia)

Prosecutor v. Issa Hassan Sesay, Morris Kallon, Augustine Gbao, Judgment, Case No. SCSL-04-15-T, SCSL Trial Chamber, 2 Mar. 2009, at 295 (Statute of the Special Court for Sierra Leone)

Pulp Mills on the River Uruguay (Argentina v. Uruguay), Judgment, ICJ Reports 2010, pp. 14–107

The Social and Economic Rights Action Center [SERAC] and the Center for Economic and Social Rights [CESCR] v. Nigeria (2001) [2001] African Commission on Human and Peoples' Rights Communication 155/96

Socio-Economic Rights and Accountability Projects (SERAP) v Nigeria [2010] Judgment No. ECW/CCJ/JUD/07/10

Slovenia - Ruling no Up-156/98 Ustavno Sodišče, Ruling No Up-156/98 [1999] Constitutional Court Official Gazette RS, no 17/99; OdlUS VIII, 118

Tillmans & Co. v S.S. Knustford (1908) 2 QB 385

Trail Smelter (United States v. Canada) Arbitration, [1938/1941] 13 RIAA 1905

Tribunal de Grande Instance (District Court) of Meaux, xxx.v.xxx., 28 February 2001, in: Droit Monde, No. 37-28, 2004 at 77

Usuarios y Consumidores en Defensa de sus Derechos Asociacion Civil c/ Aguas del Gran Bueno Aires SA [2002] Juez de paz (Moreno, Buenos Aires) 44.453 (Argentina)

UTC Nig Ltd v Pamotei (1989) 2 NWLR Pt. 103 at 244 SC

Utih v Onoyuvwe (1991) 1 NWLR Pt. 166 SC

Venezuela - Condominio del Conjunto Residencial Parque Choroní II c/ Compañía Anónima Hidrológica del Centro (Hidrocentro) Corte Primera de lo Contencioso Administrativo (2005)

Yabugbe v COP (1992) 4 NWLR Pt. 234 at 152

Table of constitutions and laws

Angola, Water Act, 21 June 2002 (Unofficial translation), articles 10(2), 22, 23

Australia, Australian Utilities Act 2000 No. 65, 2002, last amended by A2010-54 of 16 December 2010, sections 83, 84, 85 & 86

Belgium, Décret relatif au Livre II du Code de l'Environnement constituant le Code de l'Eau, Moniteur belge, 23 July 2004, Article 1.1 (Walloon region of Belgium) (unofficial translation).

Benin, Law No. 87-015, Public Hygiene Code, article 20

Benin, Public Hygiene Code, Law No. 87-015 of 21 September 1987 (Unofficial translation), article 20

Benin, Public Hygiene Code, Law No. 87-015 of 21 September 1987 (Unofficial translation), article 93

Benin, Water Code, Law No. 87-016 of 21 September 1987 (Unofficial translation), article 54

Brazil, Consumers Defence Code, Law 8078 of 11 September 1990, as last amended by Law 12.039 of 2009 (Unofficial translation), article 42

Brazil, Law on Basic Sanitation, 2007, article 40(3)

Brazil, Law on Basic Sanitation, 2007, articles 26, 27 & 53; Colombia, Law 142 establishing the regime for public household services of 11 July 1994 (Unofficial translation), article 9

Burkina Faso, Decree No. 2005-191/PRES/PM/MAHRH 4 April 2005 regarding priority uses and authority of government to control and allocate water in case of water shortage (Unofficial translation), Articles 2 and 3

Burundi, Government Decree No. 1/41 of 26 November 1992 on the establishment and organisation of the public water domain (Unofficial translation), articles 1 & 14

Cambodia, Law on Water Resources Management of the Kingdom of Cambodia 2007, article 11

Cape Verde, Water Code, Law No.41/II/84, as amended by Decree No. 5/99 (Unofficial translation), article 61

Cape Verde, Water Code, Law No.41/II/84, as amended by Decree No. 5/99 (Unofficial translation), articles 58(a) & 59

Central African Republic, Water Code 2006; LAW No 06.001 of 12 April 2006 (Unofficial translation), article 44

Chad, Water Code, Law 016/PR of 18 August 1999 (Unofficial translation), articles 149 & 150

Chile, Law 18778 Establishing Services for the Payment of Drinking Water Consumption and Sanitation Services 1989/1994, article 10

Congo DR, Constitution of the Democratic Republic of the Congo, 2006, Article 48

Constitution of the Republic of South Africa (Act No. 108 of 1996), Articles 27.1(b) and 27.2.

Ecuador, Constitution of the Republic of the Ecuador 2008, Article 12

England and Wales, the Education (School Premises) Regulations (1999 No. 2)

England and Wales, Workplace (Health, Safety and Welfare) Regulations (1992 No. 3004)

Ethiopia, Ethiopian Water Resources Management Proclamation, Proclamation No.197/2000, article 7(1)

European Convention on Human Rights 1950, Article 2

European Union Directive 2008/1/C Official Journal of the EU, 2008

European Commission White Paper on Strategy for a New Chemicals Policy, COM (2001)

European Union Directive 2008/1/C Official Journal of the EU, 2008.

Guatemala, New Health Code, Decreto 90-97 (Unofficial translation), article 89

Guinea Bissau Water Code, Law No. 5-A/92 (Unofficial translation), articles 7, 29, 29(1)(2)(5)

Guinea Bissau, Water Code, Law No. L/94/005/CTRN of 14 February 1994 (Unofficial translation), article 20

Guyana, Water and Sewerage Act, 2002, article 25(a)(d)

Kenya, Constitution of Kenya, 2010, Articles 43.1(b) and 43.1(d)

Kenya, National School Health Policy 2009.

Kenya, The Water Act 2002, No 8 of 2002, article 47(c)

Lesotho, Water Act, 2008, articles 5(2) & 13(2)(a)

Madagascar, Decree No. 2003-941 concerning monitoring of water resources, control of water destined for human consumption and priorities of access to water resources (Unofficial translation), articles 1 & 2

Mali, Decree No. 01-395/P-RM of 06 September 2001 determining the modalities for the management of wastewater and silt (Unofficial translation), article 8

Mauritania, Water Code, Law No. 2005-030 (Unofficial translation), article 37

Mauritania, Water Code, Law No. 2005-030 (Unofficial translation), article 5(1)(2)

Mexico, Constitution of the United Mexican States (as amended), Article 4

Morocco, Constitution of the Kingdom of Morocco, 2011, Article 31

Namibia, Water Resources Management Act, Act No. 24 of 2004, 26

Nicaragua, General Law on Drinking Water and Sanitation Services, article 40

Nicaragua, Political Constitution of the Republic of Nicaragua, 1986 (as amended), Article 105.

Nigeria, Benue State Water Supply and Sanitation Policy, First Edition, January 2008, p. 12

Nigeria, Criminal Code Act 1916, Section 245

Nigeria, Constitution of the Federal Republic of Nigeria 1999, Section 33

Nigeria, Environmental Management (Miscellaneous Provisions) Law (Ogun State) 2005 section 34(1)(f)

Nigeria, Sanitation and Pollution Management Law (Edo State) sections 4-8, 13, 14, 21-25, 27-31, 45

Nigeria, Waste Management Agency Bill (Rivers State) 2013 section 45, 48, 51, 54

Nigeria, Waste Management Agency Bill (Rivers State) 2013, section 53(5)(6)(7)(8)(9) and section 54

Nigeria, Water Resources Act, 1993, sections 2(iii), 18(i)

Nigeria, Water Sector Road Map, 2011

Peru, Water Resources Act, June 2009 (Unofficial translation), Articles 35, 36

Republic of Maldives, Functional Translation of the Constitution of the Republic of Maldives, 2008, Articles 23(a) and 23(f)

Republic of Niger, Constitution of the Republic of Niger, 2010, Article 12

Republic of Niger, Water Law, Law No. 93-014 of 2 March 1993 (Unofficial translation), Article 9

Senegal, Water Code, Law No. 81-13, 1981 (Unofficial translation), articles 75&76

Somalia, Provisional Constitution of the Federal Republic of Somalia, 2012, Article 27.1

South Africa, Constitution of the Republic of South Africa (Act No. 108 of 1996)

South Africa, National Water Act, Act 36 of 1998, section 22, Schedule 1: Permissible Use of Water [Sections 4(1) and 22(1)(a)(i) and Item 2 of Schedule 3]

South Africa, the Water Services Act, Act 108 of 1997, Section 3, Chapter 1, section 4(3)

South Africa, Water Services Act, Act 108 of 1997, amended 2004, sections 62 and 63

Swaziland, Water Act, Act No. 7 of 2003, Interpretation

Tanzania, The Water Utilization (Control and Regulations) Act, 1974, as revised 1993, section 10

The Netherlands, Drinking Water Law, of 19 July 2009, (Unofficial Translation) article 9

Tunisia, Constitution of the Republic of Tunisia, 2014, Article 39

Uganda, Constitution of the Republic of Uganda (as at 15 February 2006), Articles XIV(b) and XXI.

Uganda, The Water Statute, Statute No. 9 of 1995, sections 7(1)(2)

United Kingdom, generally Environmental Protection Act 1990, Pt 2A

Uruguay, Law No. 18.840 of 2011, articles 6 and 7

Uruguay, Uruguayan Constitution, 1967 (as amended), Article 47

Zambia, Water Act, 1949, sections 2 & 8; Zimbabwe, Water Act 1998, Preliminary

Zimbabwe, Constitution of Zimbabwe, Amendment (No. 20) Act, 2013, Section 77(a)

Table of subsidiary legislations

Bangladesh, National Policy for Safe Water Supply and Sanitation 1998

Belgium, Décret de la Communauté Flamande concernant diverses mesures dáccompagnement du budget 1997, 20 December 1996, Moniteur belge, 31 December 1996, 3rd edition (unofficial translation), Articles 3.1 and 34

European Commission White Paper on Strategy for a New Chemicals Policy, COM (2001)

Ghana, Public Utilities (Complaints Procedure) Regulation 1999

Kenya, The National Water Services Strategy (NWSS), 2007-2015, paragraph 3.3

Nigeria, 4th Draft Strategy, Scaling-up Sanitation Nigeria (2007)

Nigeria, National Water Supply and Sanitation Policy 2000, Para. 20(c) (vii)

Panama, Executive Decree 393 of 2005, article 12

Rwanda, Policy and Strategy for Water Supply and Sanitation Services 2010, section 4.6.3.

South Africa, Department of Provincial and Local Government, National framework for municipal indigent policies, 2006

South Africa, Indigent Policy of Mbombela´, section 14.4

South Africa, Regulations relating to compulsory national standards and measures to conserve water 2001

United States of America, State of Georgia, Rules and Regulations for Assisted Living Communities

Table of treaties

Annex A.   Resolution adopted by the UN General Assembly on 28 July 2010

UNITED NATIONS

A/RES/64/292

 **General Assembly**

Distr.: General

3 August 2010

Sixty-fourth session

Agenda item 48

Resolution adopted by the General Assembly on 28 July 2010

[*without reference to a Main Committee (A/64/L.63/Rev.1 and Add.1)*]

64/292. The human right to water and sanitation

The General Assembly,

Recalling its resolutions 54/175 of 17 December 1999 on the right to development, 55/196 of 20 December 2000, by which it proclaimed 2003 the International Year of Freshwater, 58/217 of 23 December 2003, by which it proclaimed the International Decade for Action, "Water for Life", 2005–2015, 59/228 of 22 December 2004, 61/192 of 20 December 2006, by which it proclaimed 2008 the International Year of Sanitation, and 64/198 of 21 December 2009 regarding the midterm comprehensive review of the implementation of the International Decade for Action, "Water for Life"; Agenda 21 of June 1992;[296] the Habitat Agenda of 1996;[297] the Mar del Plata Action Plan of 1977 adopted by the United Nations Water Conference;[298] and the Rio Declaration on Environment and Development of June 1992,[299]

Recalling also the Universal Declaration of Human Rights,[300] the International Covenant on Economic, Social and Cultural Rights,[301] the International Covenant on Civil and Political Rights,[6] the International

---

[296] Report of the United Nations Conference on Environment and Development, Rio de Janeiro, 3–14 June 1992, vol. I, Resolutions Adopted by the Conference (United Nations publication, Sales No. E.93.I.8 and corrigendum), resolution 1, annex II.

[297] Report of the United Nations Conference on Human Settlements (Habitat II), Istanbul, 3–14 June 1996 (United Nations publication, Sales No. E.97.IV.6), chap. I, resolution 1, annex II.

[298] Report of the United Nations Water Conference, Mar del Plata, 14–25 March 1977 (United Nations publication, Sales No. E.77.II.A.12), chap. I.

[299] Report of the United Nations Conference on Environment and Development, Rio de Janeiro, 3–14 June 1992, vol. I, Resolutions Adopted by the Conference (United Nations publication, Sales No. E.93.I.8 and corrigendum), resolution 1, annex I.

[300] Resolution 217 A (III).

[301] See resolution 2200 A (XXI), annex.

Convention on the Elimination of All Forms of Racial Discrimination,[302] the Convention on the Elimination of All Forms of Discrimination against Women,[303] the Convention on the Rights of the Child,[304] the Convention on the Rights of Persons with Disabilities[305] and the Geneva Convention relative to the Protection of Civilian Persons in Time of War, of 12 August 1949,[306]

Recalling further all previous resolutions of the Human Rights Council on human rights and access to safe drinking water and sanitation, including Council resolutions 7/22 of 28 March 2008[307] and 12/8 of 1 October 2009,[308] related to the human right to safe and clean drinking water and sanitation, general comment No. 15 (2002) of the Committee on Economic, Social and Cultural Rights, on the right to water (articles 11 and 12 of the International Covenant on Economic, Social and Cultural Rights)[309] and the report of the United Nations High Commissioner for Human Rights on the scope and content of the relevant human rights obligations related to equitable access to safe drinking water and sanitation under international human rights instruments,[310] as well as the report of the independent expert on the issue of human rights obligations related to access to safe drinking water and sanitation,[311]

Deeply concerned that approximately 884 million people lack access to safe drinking water and that more than 2.6 billion do not have access to basic sanitation, and alarmed that approximately 1.5 million children under 5 years of age die and 443 million school days are lost each year as a result of water- and sanitation-related diseases,

Acknowledging the importance of equitable access to safe and clean drinking water and sanitation as an integral component of the realization of all human rights,

Reaffirming the responsibility of States for the promotion and protection of all human rights, which are universal, indivisible, interdependent and interrelated, and must be treated globally, in a fair and equal manner, on the same footing and with the same emphasis,

Bearing in mind the commitment made by the international community to fully achieve the Millennium Development Goals, and stressing, in that context, the resolve of Heads of State and Government, as expressed in the United Nations Millennium Declaration,[312] to halve, by 2015, the proportion of people who are unable to reach or afford safe drinking water and, as agreed in the Plan of Implementation of the World Summit on Sustainable Development ("Johannesburg Plan of Implementation"),[313] to halve the proportion of people without access to basic sanitation,

---

[302] United Nations, *Treaty Series*, vol. 660, No. 9464.

[303] Ibid. vol. 1249, No. 20378.

[304] Ibid. vol. 1577, No. 27531.

[305] Resolution 61/106, annex I.

[306] United Nations, *Treaty Series*, vol. 75, No. 973.

[307] See Official Records of the General Assembly, Sixty-third Session, Supplement No. 53 (A/63/53), chap. II.

[308] See A/HRC/12/50 and Corr.1, part one, chap. I.

[309] See Official Records of the Economic and Social Council, 2003, Supplement No. 2 (E/2003/22), annex IV.

[310] A/HRC/6/3.

[311] A/HRC/12/24.

[312] See resolution 55/2.

[313] See Report of the World Summit on Sustainable Development, Johannesburg, South Africa, 26 August– 4 September 2002 (United Nations publication, Sales No. E.03.II.A.1 and corrigendum), chap. I, resolution 2, annex.

1.   *Recognizes* the right to safe and clean drinking water and sanitation as a human right that is essential for the full enjoyment of life and all human rights;

2.   *Calls upon* States and international organizations to provide financial resources, capacity-building and technology transfer, through international assistance and cooperation, in particular to developing countries, in order to scale up efforts to provide safe, clean, accessible and affordable drinking water and sanitation for all;

3.   *Welcomes* the decision by the Human Rights Council to request that the independent expert on human rights obligations related to access to safe drinking water and sanitation submit an annual report to the General Assembly,[13] and encourages her to continue working on all aspects of her mandate and, in consultation with all relevant United Nations agencies, funds and programmes, to include in her report to the Assembly, at its sixty-sixth session, the principal challenges related to the realization of the human right to safe and clean drinking water and sanitation and their impact on the achievement of the Millennium Development Goals.

108th plenary meeting
28 July 2010

Annex B.   Resolution adopted by the UN Human Rights Council on 6 October 2010

United Nations

A/HRC/RES/15/9

General Assembly

6 October 2010

Distr.: General

Original: English

Human Rights Council
Fifteenth session
Agenda item 3
Promotion and protection of all
human rights, civil political,
economic, social and cultural rights,
including the right to development

Resolution adopted by the Human Rights Council*
15/9
Human rights and access to safe drinking water and sanitation

The Human Rights Council,

Reaffirming all previous resolutions of the Council on human rights and access to safe drinking water and sanitation, in particular resolution 7/22 of 28 March 2008 and resolution 12/8 of 1 October 2009,

Recalling the Universal Declaration of Human Rights, the International Covenant on Economic, Social and Cultural Rights, the International Covenant on Civil and Political Rights, the International Convention on the Elimination of All Forms of Racial Discrimination, the Convention on the Elimination of All Forms of Discrimination against Women, the Convention on the Rights of the Child and the Convention on the Rights of Persons with Disabilities,

Recalling also relevant provisions of declarations and programmes with regard to access to safe drinking water and sanitation adopted by major United Nations conferences and summits, and by the General Assembly at its special sessions and during follow-up meetings, inter alia, the Mar del Plata Action Plan on Water and Development and Administration, adopted at the United Nations Water Conference in March 1977, Agenda 21 and the Rio Declaration on Environment and Development, adopted at the United Nations Conference on Environment and Development in June 1992, and the Habitat Agenda, adopted at the second United Nations Conference on Human Settlements in 1996, Assembly resolutions 54/175 of 17 December 1999 on the right to development, and 58/271 of 23 December 2003 proclaiming the International Decade for Action, "Water for Life" (2005-2015),

---

* The resolutions and decisions adopted by the Human Rights Council will be contained in the report of the Council on its fifteenth session (A/HRC/15/60), chap. I.

GE.10-16633

A/HRC/RES/15/9

Noting with interest regional commitments and initiatives promoting the further realization of human rights obligations related to access to safe drinking water and sanitation, including the Protocol on Water and Health, adopted by the Economic Commission for Europe in 1999, the European Charter on Water Resources, adopted by the Council of Europe in 2001, the Abuja Declaration, adopted at the first Africa-South America summit in 2006, the message from Beppu, adopted at the first Asian-Pacific Water Summit in 2007, the Delhi Declaration, adopted at the third South Asian Conference on Sanitation in 2008, and the Sharm el-Sheikh Final Document, adopted at the Fifteenth Summit Conference of Heads of State and Government of the Movement of Non-Aligned Countries in 2009,

Bearing in mind the commitments made by the international community to achieve fully the Millennium Development Goals, and stressing, in that context, the resolve of Heads of State and Government, as expressed in the United Nations Millennium Declaration, to halve, by 2015, the proportion of people unable to reach or afford safe drinking water, and to halve the proportion of people without access to basic sanitation, as agreed in the Plan of Implementation of the World Summit on Sustainable Development ("Johannesburg Plan of Implementation"),

Deeply concerned that approximately 884 million people lack access to improved water sources as defined by the World Health Organization and the United Nations Children's Fund in their 2010 Joint Monitoring Programme report, and that over 2.6 billion people do not have access to basic sanitation, and alarmed that approximately 1.5 million children under 5 years of age die and 443 million school days are lost every year as a result of water and sanitation-related diseases,

Reaffirming the fact that international human rights law instruments, including the International Covenant on Economic, Social and Cultural Rights, the Convention on the Elimination of All Forms of Discrimination against Women, the Convention on the Rights of the Child and the Convention on the Rights of Persons with Disabilities entail obligations for States parties in relation to access to safe drinking water and sanitation,

Recalling resolution 8/7 of 18 June 2008, in which the Council established the mandate of the Special Representative of the Secretary-General on the issue of human rights and transnational corporations and other business enterprises,

1.*Welcomes* the work of the independent expert on the issue of human rights obligations related to access to safe drinking water and sanitation, including the progress in collecting good practices for her compendium,[314] and the comprehensive, transparent and inclusive consultations conducted with relevant and interested actors from all regions for her thematic reports, as well as the undertaking of country missions;

---

[314] A/HRC/15/31/Add.1.

2

2.*Recalls* General Assembly resolution 64/292 of 28 July 2010, in which the Assembly recognized the right to safe and clean drinking water and sanitation as a human right that is essential for the full enjoyment of life and all human rights;

3.*Affirms* that the human right to safe drinking water and sanitation is derived from the right to an adequate standard of living and inextricably related to the right to the highest attainable standard of physical and mental health, as well as the right to life and human dignity;

4.*Calls upon* the independent expert to continue to pursue her work regarding all aspects of her mandate, including to clarify further the content of human rights obligations, including non-discrimination obligations in relation to safe drinking water and

—————————— **A/HRC/RES/15/9**

sanitation, in coordination with States, United Nations bodies and agencies, and relevant stakeholders;

5.*Acknowledges with appreciation* the second annual report of the independent expert[315] and takes note with interest of her recommendations and clarifications with regard to both the human rights obligations of States and the human rights responsibilities of nonState service providers in the delivery of water and sanitation services;

6.*Reaffirms* that States have the primary responsibility to ensure the full realization of all human rights, and that the delegation of the delivery of safe drinking water and/or sanitation services to a third party does not exempt the State from its human rights obligations;

7.*Recognizes* that States, in accordance with their laws, regulations and public policies, may opt to involve non-State actors in the provision of safe drinking water and sanitation services and, regardless of the form of provision, should ensure transparency, non-discrimination and accountability;

8.*Calls upon* States:

(a)   To develop appropriate tools and mechanisms, which may encompass legislation, comprehensive plans and strategies for the sector, including financial ones, to achieve progressively the full realization of human rights obligations related to access to safe drinking water and sanitation, including in currently unserved and underserved areas;

(b)   To ensure full transparency of the planning and implementation process in the provision of safe drinking water and sanitation and the active, free and meaningful participation of the concerned local communities and relevant stakeholders therein;

(c)   To pay particular attention to persons belonging to vulnerable and marginalized groups, including by respecting the principles of non-discrimination and gender equality;

(d)   To integrate human rights into impact assessments throughout the process of ensuring service provision, as appropriate;

(e)   To adopt and implement effective regulatory frameworks for all service providers in line with the human rights obligations of States, and to allow public regulatory institutions of sufficient capacity to monitor and enforce those regulations;

———————————

[315] A/HRC/15/31.

3

(f)   To ensure effective remedies for human rights violations by putting in place accessible accountability mechanisms at the appropriate level;

9.   *Recalls* that States should ensure that non-State service providers:

(a)   Fulfil their human rights responsibilities throughout their work processes, including by engaging proactively with the State and stakeholders to detect potential human rights abuses and find solutions to address them;

(b)   Contribute to the provision of a regular supply of safe, acceptable, accessible and affordable drinking water and sanitation services of good quality and sufficient quantity;

(c)   Integrate human rights into impact assessments as appropriate, in order to identify and help address human rights challenges;

A/HRC/RES/15/9

(*d*) Develop effective organizational-level grievance mechanisms for users, and refrain from obstructing access to State-based accountability mechanisms;

10.   S*tresses* the important role of the international cooperation and technical assistance provided by States, specialized agencies of the United Nations system, international and development partners as well as by donor agencies, in particular in the timely achievement of the relevant Millennium Development Goals, and urges development partners to adopt a human rights-based approach when designing and implementing development programmes in support of national initiatives and action plans related to the enjoyment of access to safe drinking water and sanitation;

11.   *Requests* the independent expert to continue to report, on an annual basis, to the Council and to submit an annual report to the General Assembly;

12.   *Requests* the United Nations High Commissioner for Human Rights to continue to ensure that the independent expert receives the resources necessary to enable her to discharge her mandate fully;

13.   *Decides* to continue its consideration of this matter under the same agenda item and in accordance with its programme of work.

31st meeting
30 September 2010
[Adopted without a vote.]

4

Annex C. Resolution adopted by the UN General Assembly on 17 December 2015

UNITED NATIONS

A/RES/70/169

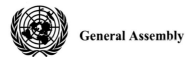

General Assembly

Distr.: General

22 February 2016

Seventieth session

Agenda item 72 (*b*)

Resolution adopted by the General Assembly on 17 December 2015

[on the report of the Third Committee (*A/70/489/Add.2*)]

70/169.   The human rights to safe drinking water and sanitation

The General Assembly,

Recalling its resolutions 64/292 of 28 July 2010, in which it recognized the right to safe and clean drinking water and sanitation as a human right that is essential for the full enjoyment of life and all human rights, and 68/157 of 18 December 2013, entitled "The human right to safe drinking water and sanitation",

Reaffirming the previous resolutions of the Human Rights Council regarding the human right to safe drinking water and sanitation, inter alia, Council resolutions 24/18 of 27 September 2013[316] and 27/7 of 25 September 2014,[317]

Recalling the Universal Declaration of Human Rights,[318] the International Covenant on Economic, Social and Cultural Rights,[319] the International Covenant on Civil and Political Rights,[319] the International Convention on the Elimination of All Forms of Racial Discrimination,[320] the Convention on the Elimination of All Forms of Discrimination against Women,[321] the Convention on the Rights of the Child[322] and the Convention on the Rights of Persons with Disabilities,[323]

---

[316] See Official Records of the General Assembly, Sixty-eighth Session, Supplement No. 53A (A/68/53/Add.1), chap. III.

[317] Ibid. *Sixty-ninth Session, Supplement No. 53A* and corrigenda (A/69/53/Add.1 and Corr.1 and 2), chap. IV, sect. A.

[318] Resolution 217 A (III).

[319] See resolution 2200 A (XXI), annex.

[320] United Nations, *Treaty Series*, vol. 660, No. 9464.

[321] Ibid. vol. 1249, No. 20378.

[322] Ibid. vol. 1577, No. 27531.

[323] Ibid. vol. 2515, No. 44910.

Welcoming the adoption of the 2030 Agenda for Sustainable Development,[324] including the reaffirmation of commitments regarding the human right to safe drinking water and sanitation therein,

Recalling the Rio Declaration on Environment and Development of June 1992[325] and its resolution 66/288 of 27 July 2012, entitled "The future we want", and emphasizing the critical importance of water and sanitation within the three dimensions of sustainable development,

Reaffirming its resolutions 58/217 of 23 December 2003, by which it proclaimed the period from 2005 to 2015 the International Decade for Action, "Water for Life", 61/192 of 20 December 2006, by which it established 2008 as the International Year for Sanitation, and 65/153 of 20 December 2010, by which it called upon Member States to support "Sustainable sanitation: the five-year drive to 2015", and recalling its resolution 65/154 of 20 December 2010, by which it declared 2013 the International Year of Water Cooperation,

Recalling the designation of 19 November as World Toilet Day, in the context of Sanitation for All, pursuant to General Assembly resolution 67/291 of 24 July 2013, in which the Assembly encouraged all Member States, as well as the organizations of the United Nations system and international organizations and other stakeholders, to approach the sanitation issue in a much broader context and to encompass all its aspects, including hygiene promotion, the provision of basic sanitation services, sewerage and wastewater treatment and reuse in the context of integrated water management,

Taking note of the relevant commitments and initiatives promoting the human right to safe drinking water and sanitation, including the Panama Declaration, adopted at the third Latin American Sanitation Conference, in 2013, the Kathmandu Declaration, adopted at the fifth South Asian Conference on Sanitation, in 2013, the 2015 Dushanbe Declaration of the High-level International Conference on the Implementation of the International Decade for Action "Water for Life", 2005–2015, the commitments made on the human right to safe drinking water and sanitation at the high-level meeting of the Sanitation and Water for All partnership in 2014 and the Ngor Declaration on Sanitation and Hygiene, adopted at the fourth AfricaSan Conference, in 2015,

Recalling general comment No. 15 (2002) of the Committee on Economic, Social and Cultural Rights on the right to water (articles 11 and 12 of the International Covenant on Economic, Social and Cultural Rights)[326] and the statement on the right to sanitation of the Committee of 19 November 2010,[327] as well as the reports of the Special Rapporteur of the Human Rights Council on the human right to safe drinking water and sanitation,

Welcoming the work of the World Health Organization and the United Nations Children's Fund in the 2015 update published by their Joint Monitoring Programme for Water Supply and Sanitation,[328]

---

[324] Resolution 70/1.

[325] Report of the United Nations Conference on Environment and Development, Rio de Janeiro, 3–14 June 1992, vol. I, Resolutions Adopted by the Conference (United Nations publication, Sales No. E.93.I.8 and corrigendum), resolution 1, annex I.

[326] Official Records of the Economic and Social Council, 2003, Supplement No. 2 (E/2003/22), annex IV.

[327] Ibid. *2011, Supplement No. 2* (E/2011/22), annex VI.

[328] World Health Organization/United Nations Children's Fund, *Progress on Sanitation and Drinking Water*, Geneva, 2015.

Welcoming also the fact that, according to Joint Monitoring Programme reports of the World Health Organization and the United Nations Children's Fund, the target on safe drinking water of the Millennium Development Goals was formally met, while being deeply concerned, however, that, according to the 2015 Joint Monitoring Programme update, 663 million people still lack access to improved drinking water sources and that 8 out of 10 people still without improved drinking water sources live in rural areas,

Deeply concerned that the world missed the sanitation component of Millennium Development Goal 7 by almost 700 million people and that more than 2.4 billion people still do not have access to improved sanitation facilities, including more than 946 million people who still practise open defecation, which is one of the clearest manifestations of poverty and extreme poverty,

Deeply concerned also that women and girls often face particular barriers in accessing water and sanitation and that they shoulder the main burden of collecting household water in many parts of the world, restricting their time for other activities, such as education and leisure, or for women earning a livelihood,

Deeply concerned further that the lack of access to adequate water and sanitation services, including for menstrual hygiene management, especially in schools, contributes to reinforcing the widespread stigma associated with menstruation, negatively affecting gender equality and women's and girls' enjoyment of human rights, including the right to education,

Deeply concerned that women and girls are particularly at risk and exposed to attacks, sexual and gender-based violence, harassment and other threats to their safety while collecting household water and when accessing sanitation facilities outside of their homes or practising open defecation,

Deeply alarmed that, every year, almost 700,000 children under 5 years of age die as a result of water and sanitation-related diseases, and underscoring that progress on reducing child mortality, morbidity and stunting is linked to children's and women's access to safe drinking water and sanitation,

Deeply concerned that official figures do not fully capture the dimensions of drinking water availability, safety, affordability of services and safe management of excreta and wastewater, as well as of inequality and discrimination in the access to safe drinking water and sanitation and therefore underestimate the numbers of those without access to safe and affordable drinking water and safely managed and affordable sanitation, and highlighting in this context the need to adequately monitor the safety of drinking water and sanitation in order to obtain data that capture those dimensions,

Deeply concerned also that inexistent or inadequate sanitation facilities as well as serious deficiencies in water management and wastewater treatment can negatively affect water provision and sustainable access to safe drinking water, and recognizing that, in progressively realizing the human rights to safe drinking water and sanitation as well as other human rights, States should increasingly pursue integrated approaches and strengthen their water resource management, including by improving their wastewater treatment and by preventing and reducing surface and groundwater pollution,

Affirming the importance of regional and international technical cooperation, where appropriate, as a means to promote the progressive realization of the human rights to safe drinking water and sanitation, without any prejudice to questions of international water law, including international watercourse law,

Reaffirming the responsibility of States to ensure the promotion and protection of all human rights, which are universal, indivisible, interdependent and interrelated, and must be treated globally, in a fair and equal manner, on the same footing and with the same emphasis,

Recalling the understanding by the Committee on Economic, Social and Cultural Rights and the Special Rapporteur on the human right to safe drinking water and sanitation that the rights to safe drinking water and sanitation are closely related, but have distinct features which warrant their separate treatment in order to address specific challenges in their implementation and that sanitation too often remains neglected if not addressed as a separate right, while being a component of the right to an adequate standard of living,

Recalling also that the human rights to safe drinking water and sanitation are derived from the right to an adequate standard of living and are inextricably related to the right to the highest attainable standard of physical and mental health, as well as to the right to life and human dignity,

Acknowledging the importance of equal access to safe drinking water and sanitation as an integral component of the realization of all human rights,

1. *Affirms* that the human rights to safe drinking water and sanitation as components of the right to an adequate standard of living are essential for the full enjoyment of the right to life and all human rights;

2. *Recognizes* that the human right to safe drinking water entitles everyone, without discrimination, to have access to sufficient, safe, acceptable, physically accessible and affordable water for personal and domestic use, and that the human right to sanitation entitles everyone, without discrimination, to have physical and affordable access to sanitation, in all spheres of life, that is safe, hygienic, secure, socially and culturally acceptable and that provides privacy and ensures dignity, while reaffirming that both rights are components of the right to an adequate standard of living;

3. *Welcomes* Goal 6 of the 2030 Agenda for Sustainable Development,[324] on ensuring the availability and sustainable management of water and sanitation for all, which includes important dimensions related to the human rights to safe drinking water and sanitation;

4. *Also welcomes* the work of the Special Rapporteur of the Human Rights Council on the human right to safe drinking water and sanitation, and takes note with appreciation, in particular, of his first reports on affordability of water and sanitation services[329] and on the analysis of the different types of water and sanitation services from the perspective of the human right to safe drinking water and sanitation;[330]

5. *Calls upon* States:

(*a*) To ensure the progressive realization of the human rights to safe drinking water and sanitation for all in a non-discriminatory manner while

---

[329] A/HRC/30/39.
[330] A/70/203.

eliminating inequalities in access, including for individuals belonging to groups at risk and to marginalized groups, on the grounds of race, gender, age, disability, ethnicity, culture, religion and national or social origin or on any other grounds, with a view to progressively eliminating inequalities based on factors such as rural-urban disparities, residence in a slum, income levels and other relevant considerations;

(*b*)   To give due consideration to the commitments regarding the human rights to safe drinking water and sanitation when implementing the 2030 Agenda for Sustainable Development, including through the full implementation of Goal 6;

(*c*)   To continuously monitor and regularly analyse the status of the realization of the human rights to safe drinking water and sanitation;

(*d*)   To identify patterns of failure to respect, protect or fulfil the human rights to safe drinking water and sanitation for all persons without discrimination and to address their structural causes in policymaking and budgeting within a broader framework, while undertaking holistic planning aimed at achieving sustainable universal access, including in instances where the private sector, donors and non-governmental organizations are involved in service provision;

(*e*)   To promote both women's leadership and their full, effective and equal participation in decision-making on water and sanitation management and to ensure that a gender-based approach is adopted in relation to water and sanitation programmes, including measures, inter alia, to reduce the time spent by women and girls in collecting household water, in order to address the negative impact of inadequate water and sanitation services on the access of girls to education and to protect women and girls from being physically threatened or assaulted, including from sexual violence, while collecting household water and when accessing sanitation facilities outside of their home or practising open defecation;

(*f*)   To progressively eliminate open defecation by adopting policies to increase access to sanitation, including for individuals belonging to vulnerable and marginalized groups;

(*g*)   To approach the sanitation issue in a much broader context, taking into account the need to pursue integrated approaches;

(*h*)   To consult and coordinate with local communities and other stakeholders, including civil society and the private sector, on adequate solutions to ensure sustainable access to safe drinking water and sanitation;

(*i*)   To provide for effective accountability mechanisms for all water and sanitation service providers to ensure that they respect human rights and do not cause human rights violations or abuses;

6.   *Calls upon* non-State actors, including business enterprises, both transnational and others, to comply with their responsibility to respect human rights, including the human rights to safe drinking water and sanitation, including by cooperating with State investigations into allegations of abuses of the human rights to safe drinking water and sanitation, and by progressively engaging with States to detect and remedy abuses of the human rights to safe drinking water and sanitation;

7.   *Invites* regional and international organizations to complement efforts by States to progressively realize the human rights to safe drinking water and sanitation;

8. *Calls upon* Member States to enhance global partnerships for sustainable development as a means to achieve and sustain the Goal and the targets of the 2030 Agenda for Sustainable Development, and highlights the need to develop adequate follow-up and review of progress on the 2030 Agenda, including on ensuring availability and sustainable management of water and sanitation for all;

9. *Reaffirms* that States have the primary responsibility to ensure the full realization of all human rights and to endeavour to take steps, individually and through international assistance and cooperation, especially economic and technical cooperation, to the maximum of their available resources, with a view to progressively achieving the full realization of the rights to safe drinking water and sanitation by all appropriate means, including, in particular, the adoption of legislative measures;

10. *Stresses* the important role of the international cooperation and technical assistance provided by States, specialized agencies of the United Nations system and international and development partners, as well as by donor agencies, and urges development partners to adopt a human rights-based approach when designing and implementing development programmes in support of national initiatives and plans of action related to the rights to safe drinking water and sanitation;

11. *Decides* to continue its consideration of the question at its seventy-second session.

80th plenary meeting
17 December 2015

Annex D.   Thesis log frame

MAIN RESEARCH QUESTION:

How can the human right to sanitation be interpreted and implemented to promote inclusive development?

	GAPS IN SCIENTIFIC KNOWLEDGE		SUB-RESEARCH QUESTIONS
1	Scholarly literature is largely more focused on the human right to water than on the human right to sanitation (HRS), leading to poor development of the normative content of the HRS.	1	What are the drivers of poor sanitation services and how are these currently being addressed in sanitation governance frameworks?
2	There are contestations over the meaning of the human right to sanitation.	2	How has the human right to sanitation evolved across different levels of governance, from international to local; how do the human right to sanitation principles address the drivers?
3	The human right to sanitation literature does not integrate perspectives from other fields on the drivers of poor sanitation services and the impact of the normative architecture of the human right to sanitation on inclusive development.	3	Which humanitarian law and any other non-human rights instruments, including principles and indicators, for sanitation governance promote the progressive realisation of the human right to sanitation, through addressing the drivers of poor sanitation services?
4	There is a paucity of indicators for assessing the progressive realisation of the human right to sanitation.	4	How does legal pluralism operate in sanitation governance, with the implementation of the human right to sanitation, alongside (other human rights regimes and) non-human rights instruments and principles?
5	There is incoherence between the legal and non-legal literature on the human right to sanitation.	5	How can the human right to sanitation institution be redesigned to advance ID outcomes across multiple levels of governance?

CHAPTERS	OBJECTIVES	ANSWERS RESEARCH QUESTION(S)	CLOSES KNOWLEDGE GAP(S)
1 **The Human Right to Sanitation and Inclusive Development Under an Uncertain Future**	To introduce the practical and theoretical problems; knowledge gaps; research questions; focus and limits and the structure of the thesis.	-	-
2 **Research Methodology and Theoretical Framework**	To explain the research methodology used and situate the research within the contemporary theoretical debates on relevant themes.	-	-
3 **Contextualizing the Sanitation Problem**	To expound on the meaning and economic classification of sanitation, the drivers of poor domestic sanitation services, and the main technologies for domestic sanitation services and the implications for the design of sanitation governance frameworks.	Sub-research Question (1)	Knowledge Gap (2)
4 **Human Rights Principles**	To assess how human rights principles for addressing key development challenges affect inclusive development.	Sub-research Question 2 (at the international level, as a background to analysing the human right to sanitation principles for addressing the drivers of poor sanitation services)	Knowledge Gap (3)
5 **Human Right to Sanitation Principles**	To assess how the human right to sanitation framework addresses the drivers of poor sanitation services to promote inclusive development.	Sub-research Questions (2) and (4)	Knowledge Gaps (1) – (5)

CHAPTERS	OBJECTIVES	ANSWERS RESEARCH QUESTION(S)	CLOSES KNOWLEDGE GAP(S)
6 **Human Right to Sanitation in Humanitarian Situations**	To assess how the current framework of humanitarian assistance supports the realisation of the human right to sanitation, addresses the drivers of poor sanitation services and contributes to inclusive development.	Sub-research Questions (3) and (4)	Knowledge Gaps (3) and (4)
7 **Non-human Rights Principles for Sanitation Governance**	To assess how non-human rights frameworks for sanitation governance support the realisation of the human right to sanitation, address the drivers of poor sanitation services and contribute to inclusive development.	Sub-research Questions (3) and (4)	Knowledge Gaps (3) and (4)
8 **Architecture for Sanitation Governance in Nigeria**	To assess how the current framework of sanitation governance in Nigeria supports the realisation of the human right to sanitation, addresses the drivers of poor sanitation services and contributes to inclusive development, and propose redesign where necessary.	Sub-research Questions (1) – (5) (for Nigeria)	Knowledge Gaps (1) – (5) (for Nigeria)
9 **Human Right to Sanitation and the Inclusive Development Imperative**	To discuss what kind of normative architecture for groundwater governance supports sustainable and inclusive development and what is required to achieve this architecture across multiple geographic levels.	Sub-research Question (5); Main Research Question	Knowledge Gaps (1) – (5) (across multiple levels of governance, from the international to the local levels)

Annex E.   Keywords searched in scientific databases

'Causes AND poor sanitation services'
'Drivers AND poor sanitation services'
'Effects of lack of sanitation'
'Human right to sanitation'
'Human right to water and sanitation'
'Inclusive development'
'Institutional analyses
'Instruments AND human right to sanitation'
'Legal pluralism AND human right to sanitation'
'Legal pluralism'
'Multi-level governance'
'Principles AND human right to sanitation'
'Sanitation AND economic good'
'Sanitation AND formal settlements'
'Sanitation AND governance'
'Sanitation AND humanitarian situations'
'Sanitation AND informal settlements'
'Sanitation AND Nigeria'
'Sanitation AND sustainable development goals'
'Sanitation crisis'

Annex F.   Criteria for coding sanitation governance principles

Code	Criteria
Acceptability	Requires sanitation facilities to be appropriate for use and meet users' preferences; social acceptance
Accessibility	Provides for sanitation facilities to be located in areas that can be easily accessed
Access to information	Entitles rightsholders to demand information concerning their sanitation services from both public utilities and private service providers
Accountability	Requires States to adopt effective regulatory framework for the realisation of the human right to sanitation and to prevent violations by non-State actors; mechanisms for rightsholders to demand action from the State and non-State actors involved in sanitation governance; requires good governance or compliance with constitutional provisions
Affordability	Requires that the direct and/or indirect costs of sanitation facilities should not interfere with meeting other human needs
Autonomy of Service Providers	Protects the service providers from undue political interference
Availability	Requires an adequate number of sanitation facilities for users
Capacity building	Provides for sanitation and hygiene training, finances or other resources that improve sanitation access
Cooperation	Requires States to cooperate in the management of transboundary water resources
Cost sharing	Allocates the financial cost of sanitation facilities to different stakeholders; requires users to own private facilities
Demand responsiveness	Requires sanitation facilities to be designed to meet users' stated preferences
Dignity	Protects privacy of users, especially vulnerable users like women and girls
Disaster risk reduction	Focuses on reducing the adverse impacts and/or possibility of disasters
Environmental Impact of Sanitation Facilities	Requires service providers and/or regulators to assess the negative impacts of sanitation facilities on water quality or the environment generally
Equality and Non-discrimination	Prohibits all forms of covert or overt discriminatory practices in sanitation governance processes and/or service delivery
Equity and Poverty Reduction	Requires the elimination of poverty and/or equal treatment of the poor; includes intra and inter-generational equity
Extra-territorial Obligation	Places obligations on States to respect, protect and fulfil the human right to sanitation outside their territories either directly or through the actions of non-State actors based on attribution ; requires international cooperation in the management of transboundary resources
Gender Responsiveness	Requires sanitation facilities to be designed to meet the special needs of women and girls especially

Code	Criteria
Human right to sanitation (express recognition)	Recognises the human right to access safe, affordable, accessible and acceptable sanitation facilities for are adequate for their personal needs
Human right to sanitation (implied recognition)	Recognises the need for sanitation as a requirement for the realisation of the human right to water or other related economic, social or cultural rights
Human right to water (express recognition)	Recognise the human right to access water for drinking, including other domestic consumption needs
Human right to water (implied recognition)	Recognises the human right to water, as a requirement for the realisation of the human right to water or other related economic, social or cultural rights
Integration	Requires considerations for environmental sustainability or ecosystems integrity in sanitation governance; monitors the environmental impact of sanitation facilities
Participation	Promotes the participation of stakeholders, especially users, in sanitation governance
Participation of Communities, Private Sector and Civil Society	Promotes the participation of communities, private sector and civil society specifically, in sanitation governance
Policy making and regulatory role of government	Highlights the role of the State to make policies and laws for sanitation governance
Polluter pays	Requires polluters to internalise the costs of pollution
Pollution prevention	Requires measures to prevent pollution
Precautionary principle	Requires action to prevent irreversible harm even without any conclusive scientific evidence and/or irrespective of the economic cost of taking action
Prioritization of human needs	Elevates basic human needs like drinking and sanitation to be satisfied before other uses of groundwater
Safety	Demands measures to protect users from harm caused by sanitation facilities
Subsidiarity	Requires sanitation governance and decision making at the lowest-appropriate level of governance
Sustainability	Requires sanitation facilities to the environmentally sustainable and/or meet the present needs
Social good	Protects water resources and the environment from contamination in order to meet human needs sustainably and derive the optimum long term benefits for the society; public trust doctrine
Transparency and Access to Information	Provides the public with access to information regarding the quality, status and other aspects of sanitation services which affects their enjoyment of their rights
Water as an economic good	Recognises water as an economic and/or promotes full cost recovery

Annex G.   List of interviewees

Code	Professional background/experience	Country of residence
Interviewee 1	Inter-governmental WASH agency	Nigeria
Interviewee 2	National regulatory agency	Nigeria
Interviewee 3	National regulatory agency	Nigeria
Interviewee 4	National regulatory agency	Nigeria
Interviewee 5	National regulatory agency	Nigeria
Interviewee 6	National regulatory agency – focused on humanitarian situations	Nigeria
Interviewee 7	Provincial sanitation regulatory agency	Nigeria
Interviewee 8	Provincial sanitation regulatory agency	Nigeria
Interviewee 9	Provincial WASH regulatory agency	Nigeria
Interviewee 10	Media activist – focused on informal settlements	Nigeria
Interviewee 11	Urban planner and environmental journalist	Nigeria
Interviewee 12	Journalist and development expert	Nigeria
Interviewee 13	Urban planner and WASH journalist	Nigeria
Interviewee 14	Civil geo-environmental consult & WASH service provider	Nigeria
Interviewee 15	Private sector sanitation service provider	Nigeria
Interviewee 16	International NGO – focused on informal settlements	Nigeria
Interviewee 17	National NGO – focused on informal settlements	Nigeria
Interviewee 18	National NGO – focused on humanitarian situations	Nigeria
Interviewee 19	National NGO – focused on informal settlements	Nigeria
Interviewee 20	International WASH NGO – focused on humanitarian situations	USA
Interviewee 21	International WASH NGO – focused on humanitarian situations	Pakistan
Interviewee 22	International WASH NGO – focused on humanitarian situations	South Africa
Interviewee 23	Law and international relations	West Indies
Interviewee 24	Sustainability science & governance	USA
Interviewee 25	Sustainability science & governance	Sweden
Interviewee 26	Sustainability science & governance	UK
Interviewee 27	Geographer; environmental & human rights law	UK
Interviewee 28	Water supply and sanitation governance	UK
Interviewee 29	Sustainability science/governance	UK
Interviewee 30	Law, natural resources and international security	Germany
Interviewee 31	Sustainability science & governance	UK
Interviewee 32	International development policy & management	Belgium
Interviewee 33	Sustainability science & governance	Canada
Interviewee 34	Sustainability science & governance	UK
Interviewee 35	Ecological and political philosophy	UK
Interviewee 36	Climate change law& policy	UK
Interviewee 37	Sustainability science & governance	Netherlands
Interviewee 38	Human geography and political ecology	Germany
Interviewee 39	Sustainability science & governance	USA
Interviewee 40	Sustainability science & governance	Netherlands
Interviewee 41	Sustainability science & governance	Japan
Interviewee 42	Sustainability science & governance	Netherlands
Interviewee 43	Sustainability science & governance	UK
Interviewee 44	Water and Sanitation engineer	India
Interviewee 45	Economist – focus on WASH	Netherlands
Interviewee 46	Waste management	Ghana
Interviewee 47	Water supply and sanitation governance	Netherlands

Annex H.   Background   information   for   the   respondent households

Variable	Frequency (n = 254)	Percent
Head of the household		
Head	71	28.0
Related to head	183	72.0
Type of housing		
Flat in a block of flats	100	39.4
House on a separate land	58	22.8
Rooms let in a house	50	19.7
Improvised dwelling	11	4.3
Bungalow	11	4.3
Semi-detached house	8	3.1
Duplex	8	3.1
Traditional structure	8	3.1
Method of payment for the house		
House owner	145	57.1
Normal rent	89	35.0
Free	15	5.9
Subsidized rent	5	2.0
Duration lived in the house		
< 1 year	27	10.6
1 – 5 years	107	42.1
6 – 10 years	63	24.8
> 10 years	57	22.4
Population living in the house		
< 7	183	72.0
7 – 10	53	20.9
> 10	18	7.1

Annex I.   States with human right to sanitation and/or water legislations, grouped according to their continents

Africa

Country	UNGA Vote	HRS	HRW
Algeria	√	√	√
Angola	√	X	X
Benin	√	√	√
Botswana	Abstention	√	√
Burkina Faso	√	√	√
Burundi	√	X	X
Cameroon	Absent	√	√
Cape Verde	√	0	0
Central African Republic	√	√	√
Chad	Absent	√	√
Comoros	√	0	0
Congo	√	X	X
Côte d'Ivoire	√	√	√
Democratic Republic of the Congo	√	0	√
Djibouti	√	0	0
Egypt	√	0	0
Equatorial Guinea	√	0	0
Eritrea	√	0	√
Ethiopia	Abstention	√	√
Gabon	√	X	X
Gambia	Absent	√	√
Ghana	√	√	√
Guinea	Absent	X	X
Guinea-Bissau	Absent	√	√
Kenya	Abstention	√	√
Lesotho	Abstention	X	0
Liberia	√	0	0
Libya	√	0	0
Madagascar	√	√	√
Malawi	Absent	0	0
Mali	√	√	√
Mauritania	Absent	X	√
Mauritius	√	0	0
Morocco	√	√	√
Mozambique	Absent	X	X
Namibia	Absent	0	0
Niger	√	√	√
Nigeria	√	√	√
Rwanda	Absent	√	√
Sao Tome and Principe	Absent	0	0
Senegal	√	X	X
Seychelles	√	0	0
Sierra Leone	Absent	√	√
Somalia	√	0	0
South Africa	√	√	√

Country	UNGA Vote	HRS	HRW
Sudan	√	√	√
Swaziland	Absent	0	0
Togo	√	X	√
Tunisia	√	√	X
Uganda	Absent	√	√
United Republic of Tanzania	Abstention	√	√
Zambia	Abstention	0	0
Zimbabwe	√	√	√
√ - Yes X – Made representation to the GLAAS 2014 but does not have national human right to sanitation legislation 0 - No representation to the GLAAS 2014 and no additional records of human right to sanitation laws found			

Asia

Country	UNGA Vote	HRS	HRW
Afghanistan	√	X	X
Bahrain	√		
Bangladesh	√	√	√
Bhutan	√	√	√
Brunei Darussalam	√	0	0
Cambodia	√	X	X
China	√	0	0
Democratic People's Republic of Korea	√	0	0
India	√	X	X
Indonesia	√	√	√
Iran	√	√	√
Iraq	√	0	0
Israel	Abstention	0	0
Japan	Abstention	0	0
Jordan	√	X	X
Kazakhstan	Abstention	√	√
Kuwait	√	0	0
Kyrgyzstan	√	√	√
Lao People's Democratic Republic	√	√	√
Lebanon	√	√	√
Malaysia	√	0	0
Maldives	√	√	√
Mongolia	√	√	√
Myanmar	√	X	X
Nepal	√	√	√
Oman	√	√	√
Pakistan	√	X	X
Philippines	Absent	√	√
Qatar	√	0	0
Republic of Korea	Abstention	0	0
Russian Federation	√	0	0
Saudi Arabia	√	0	0
Singapore	√	0	0
Sri Lanka	√	√	√

Country	UNGA Vote	HRS	HRW
Syria	√	0	0
Tajikistan	√	0	√
Thailand	√	√	√
Timor-Leste	√	√	0
Turkey	Abstention	0	0
Turkmenistan	Absent	0	0
United Arab Emirates	√	0	0
Uzbekistan	Absent	0	0
Viet Nam	√	√	√
Yemen	√	X	√

√ - Yes

X – Made representation to the GLAAS 2014 but does not have national human right to sanitation legislation

0 - No representation to the GLAAS 2014 and no additional records of human right to sanitation laws found

Europe

Country	UNGA Vote	HRS	HRW
Albania	Absent	0	0
Andorra	√	0	0
Armenia	Abstention	0	0
Austria	Abstention	0	0
Azerbaijan	√	√	√
Belarus	√	√	√
Belgium	√	0	0
Bosnia and Herzegovina	Abstention	0	0
Bulgaria	Abstention	0	0
Croatia	Abstention	0	0
Cyprus	Abstention	0	0
Czech Republic	Abstention	0	0
Denmark	Abstention	0	0
Estonia	Abstention	0	0
Finland	√	0	0
France	√	0	√
Georgia	√	√	√
Germany	√	0	0
Greece	Abstention	0	0
Hungary	√	0	0
Iceland	Abstention	0	0
Ireland	Abstention	0	0
Italy	√	0	0
Latvia	Abstention	0	0
Liechtenstein	√	0	0
Lithuania	Abstention	√	√
Luxembourg	Abstention	0	0
Malta	Abstention	0	0
Monaco	√	0	0
Montenegro	√	0	0
Netherlands	Abstention	0	0
Norway	√	0	0

Country	UNGA Vote	HRS	HRW
Poland	Abstention	0	0
Portugal	√	0	0
Republic of Moldova	Abstention	0	0
Romania	Abstention	0	0
San Marino	√	0	0
Serbia	√	0	0
Slovakia	Abstention	0	0
Slovenia	√	0	0
Spain	√	0	0
Sweden	Abstention	0	0
Switzerland	√	0	0
The former Yugoslav Republic of Macedonia	√	0	√
Ukraine	Abstention	√	√
United Kingdom	Abstention	0	0

√ - Yes
X – Made representation to the GLAAS 2014 but does not have national human right to sanitation legislation
0 - No representation to the GLAAS 2014 and no additional records of human right to sanitation laws found

North America

Country	UNGA Vote	HRS	HRW
Antigua and Barbuda	√	0	0
Bahamas	√	0	0
Barbados	√	0	0
Belize	Absent	0	0
Canada	Abstention	0	0
Costa Rica	√	√	√
Cuba	√	√	√
Dominica	√	0	0
Dominican Republic	√	√	√
El Salvador	√	0	0
Grenada	√	0	0
Guatemala	√	0	0
Haiti	√	√	√
Honduras	√	√	√
Jamaica	√	0	0
Mexico	√	√	√
Nicaragua	√	0	√
Panama	√	X	X
Saint Kitts and Nevis	Absent	0	0
Saint Lucia	√	0	0
Saint Vincent and the Grenadines	√	0	0
Trinidad and Tobago	Abstention	0	0
United States	Abstention	0	0

√ - Yes
X – Made representation to the GLAAS 2014 but does not have national human right to sanitation legislation
0 - No representation to the GLAAS 2014 and no additional records of human right to sanitation laws found

Oceania

Country	UNGA Vote	HRS	HRW
Australia	Abstention	0	0
Fiji	Absent	√	√
Kiribati	Absent	0	0
Marshall Islands	Absent	0	0
Micronesia (Federated States of)	Absent	0	0
Nauru	Absent	0	0
New Zealand	Abstention	0	0
Palau	Absent	0	0
Papua New Guinea	Absent	0	0
Samoa	√	0	0
Solomon Islands	√	0	0
Tonga	Absent	0	0
Tuvalu	√	0	0
Vanuatu	√	X	X

√ - Yes
X – Made representation to the GLAAS 2014 but does not have national human right to sanitation legislation
0 - No representation to the GLAAS 2014 and no additional records of human right to sanitation laws found

South America

Country	UNGA Vote	HRS	HRW
Argentina	√	√	√
Bolivia	√	√	√
Brazil	√	√	√
Chile	√	X	X
Colombia	√	√	√
Ecuador	√	0	0
Guyana	Abstention	0	0
Paraguay	√	X	√
Peru	√	X	X
Suriname	Absent	0	0
Uruguay	√	√	√
Venezuela	√	0	0

√ - Yes
X – Made representation to the GLAAS 2014 but does not have national human right to sanitation legislation
0 - No representation to the GLAAS 2014 and no additional records of human right to sanitation laws found

**Netherlands Research School for the
Socio-Economic and Natural Sciences of the Environment**

DIPLOMA

For specialised PhD training

The Netherlands Research School for the
Socio-Economic and Natural Sciences of the Environment
(SENSE) declares that

Pedi Chiemena Obani

born on 29 December 1985 in Port Harcourt, Nigeria

has successfully fulfilled all requirements of the
Educational Programme of SENSE.

Amsterdam, 16 May 2018

the Chairman of the SENSE board

Prof. dr. Huub Rijnaarts

the SENSE Director of Education

Dr. Ad van Dommelen

The SENSE Research School has been accredited by the Royal Netherlands Academy of Arts and Sciences (KNAW)

KONINKLIJKE NEDERLANDSE
AKADEMIE VAN WETENSCHAPPEN

The SENSE Research School declares that **Pedi Chiemena Obani** has successfully fulfilled all requirements of the Educational PhD Programme of SENSE with a work load of 36.9 EC, including the following activities:

SENSE PhD Courses

- o   SENSE writing week (2013)
- o   Environmental research in context (2013)
- o   Research in context activity: 'Organising the sense writing week' (2013)

Other PhD and Advanced MSc Courses

- o   IHE Delft PhD seminar, Delft, the Netherlands (2012)
- o   IHE Delft PhD seminar, Delft, the Netherlands (2014)
- o   University of Amsterdam PhD Day, Amsterdam, the Netherlands (2015)

Didactic Skills Training

- o   Supervising 11 final year LL.B. (Hons.) students with thesis (2013-2014)
- o   Lecturing in LL.B. (Hons.) course 'Legal Methodology' (2013-2014, 2017-2018), University of Benin
- o   Lecturing in LL.B. (Hons.) course ' Criminology' (2013-2014), University of Benin

Management Skills Training

- o   Vice-chair of UNESCO-IHE PhD board (2012-2014)
- o   SENSE strategy meeting (2015)

Oral Presentations

- o   *National development and security anchored on deregulation of exploitation of natural resources in Nigeria.* Nigeria association of law teachers conference, 11-16 June 2017, Awka, Nigeria
- o   Human rights and access in earth system governance: case of sanitation. Earth System Governance Conference, 01-03 July 2014, Norwich, United Kingdom
- o   *Legal pluralism in the area of human rights: water and sanitation.* European Association Of Development Research And Training Institutes, 23-26 June 2014, Bonn, Germany

SENSE Coordinator PhD Education

Dr. Peter Vermeulen